Computational Methods in Ordinary Differential Equations

Introductory Mathematics for Scientists and Engineers

A Wiley Series, edited by

D. S. JONES, M.B.E., F.R.S.E., F.R.S.

Ivory Professor of Mathematics in the University of Dundee

C. DIXON	Applied Mathematics of Science and Engineering
D. S. JONES and D. W. JORDAN	Introductory Analysis Vol. 1
D. S. JONES and D. W. JORDAN	Introductory Analysis Vol. 2
H. LIEBECK	Algebra
A. R. MITCHELL	Computational Methods in Partial Differential Equations
J. M. RUSHFORTH and J. Ll. MORRIS	Computers and Computing
R. C. SMITH and P. SMITH	Mechanics

Computational Methods in Ordinary Differential Equations

J. D. LAMBERT

Reader in Mathematics
University of Dundee, Scotland

JOHN WILEY & SONS CHICHESTER NEW YORK BRISBANE TORONTO

Library of Congress Catalog Card No. 72-5718

ISBN 0 471 51194 3

Reprinted September 1976
Reprinted April 1977
Reprinted April 1979
Reprinted June 1981
Reprinted with corrections January 1983

Printed in Great Britain by J. W. Arrowsmith Ltd, Bristol

To Heather

To Heather

Introductory Mathematics for Scientists and Engineers

Foreword to the Series

The past few years have seen a steady increase in the courses of mathematics and computing provided for students undertaking higher education. This is partly due to high speed digital computation being more readily available and partly because so many disciplines now find mathematics an essential element of their curriculum. Many of the students, e.g. those of physics, chemistry, engineering, biology and economics, will be concerned with mathematics and computing mainly as tools, but tools with which they must acquire proficiency. On the other hand, courses for mathematicians must take cognizance of the existence of electronic computers. All these students may therefore study similar material though possibly at different stages of their careers—some, perhaps, will encounter it shortly after commencing their training while others may not come to grips with it until after graduating. This series is designed to cater for these differing requirements; some of the books are appropriate to the basic mathematical training of students in many disciplines while others, dealing with more specialized topics, are intended both for those for whom such topics are an essential ingredient of their course and for those who, although not specialists, find the need for a working knowledge of these areas. However, the presentation of all the books has been planned so as to demand the minimal mathematical equipment for the topics discussed. Instructors will therefore often be able to extract a shorter introductory course when a fuller treatment is not desired.

The authors have, in general, avoided the strict axiomatic approach which is favoured by some writers, but there has been no dilution of the standard of mathematical argument. Learning to follow and construct a logical sequence of ideas is one of the important attributes of courses in mathematics and computing.

While the authors' purpose has been to stress mathematical ideas which are central to applications and necessary for subsequent investigations, they have attempted, when appropriate, to convey some notion of the connection between a mathematical model and the real world. They have also taken account of the fact that most students now have access to electronic digital computers.

The careful explanation of difficult points and the provision of large numbers of worked examples and exercises should ensure the popularity of the books in this series with students and teachers alike.

D. S. JONES
Department of Mathematics
University of Dundee

Preface

Computational methods for ordinary differential equations, although constituting one of the older established areas of numerical analysis, have been the subject of a great deal of research in recent years. It is hoped that this book will, in addition to its primary purpose of serving as a text for advanced undergraduates, provide postgraduate students and general uses of numerical analysis with a readable account of these developments. The only prerequisites required of the reader are a sound course in calculus and some acquaintance with complex numbers, matrices, and vectors.

There is no general agreement on how the phrase 'numerical analysis' should be interpreted. Some see 'analysis' as the key word, and wish to embed the subject entirely in rigorous modern analysis. To others, 'numerical' is the vital word, and the algorithm the only respectable product. In this book I have tried to take a middle course between these two extremes. On the one hand, few theorems are stated (and even fewer proved), while, on the other, no programmes will be found. The approach is rather analogous to that of the classical applied mathematician, whose genuine interest in real problems does not detract from his delight in watching mathematics at work. Thus, most of the exercises and worked examples are intended to cast light (and, in some cases, doubt) on our interpretation of why numerical methods perform in the way that they do. It is hoped that the reader will supplement such exercises by programming and running the methods, discussed in the text, when applied to specific differential equations or systems of equations, preferably arising from real problems.

Much of the material of this book is based on lecture courses given to advanced undergraduate and early postgraduate students in the Universities of St. Andrews, Aberdeen, and Dundee. Chapters 2, 3, and 4 develop the study of linear multistep methods and Runge–Kutta methods in some detail, and culminate in some of the most efficient forms of these methods currently available. These two classes form a convenient basis for the development of concepts, such as weak stability, which are widely applicable. Chapter 5 is concerned with hybrid methods—a class whose computational potentialities have probably not yet been fully exploited.

Chapter 6 deals with the highly efficient class of extrapolation methods, while chapter 7 is concerned with the somewhat neglected area of special methods for problems which exhibit special features other than stiffness. Up to this point, in the interests of ease of exposition, only the single first-order differential equation is considered. The first half of chapter 8 gives an account of precisely how much of the previous work goes over unaltered to the problem of a system of first-order equations, how much needs modification, and how much is invalid; the second half is devoted to a full account of the problem of stiffness, and examines certain connections with the theory of stability of finite difference methods for partial differential equations. The last chapter is concerned with a special class of second-order differential equations. The emphasis throughout is on initial value problems, since direct techniques for boundary value problems lead to a large and separate area of numerical analysis. An exception is made in the case of the shooting method for two-point boundary value problems, which is described in the appendix.

Any book on this topic necessarily owes a large debt to the well known book by P. Henrici, *Discrete Variable Methods in Ordinary Differential Equations*. This book, together with the celebrated papers of G. Dahlquist, has played a unique rôle in the developments of the last decade. Henrici's approach is somewhat more rigorous than that of the present book, but I have purposely adopted a notation consistent with Henrici's, in order that the reader may the more easily make the transition to Henrici's book.

Many people have influenced the development of this book. Foremost among these is Professor A. R. Mitchell, who, over many years in the successive rôles of teacher, supervisor, colleague—and always as friend—has greatly influenced my attitude to numerical analysis. It was my good fortune that, during the preparation of this book, a year-long seminar on numerical analysis was held in the University of Dundee, with the generous support of the Science Research Council. This book was much influenced by useful discussions with many of the distinguished numerical analysts who took part in that seminar. These included Professors G. Dahlquist, C. W. Gear, W. B. Gragg, M. R. Osborne, H. J. Stetter, and, in particular, Professor T. E. Hull, who suggested several important improvements in the manuscript. My thanks are also due to Professors J. C. Butcher, J. D. Lawson, Drs. A. R. Gourlay, J. J. H. Miller, and Mr. S. T. Sigurdsson for useful discussions. I am also grateful to several past and present research students in the University of Dundee, and in particular to Mr. K. M. Lovell, for help with computational examples. Finally, I am indebted to Professor D. S. Jones for his useful comments on an early draft, and to Mrs. M. L. Bennett for her accurate typing of part of the manuscript.

J. D. Lambert
Dundee, Scotland,
February, 1972.

Contents

1

Preliminaries

1.1 Notation

Throughout this book we shall denote scalars by y, ϕ, etc., and (column) vectors by $\mathbf{y}$, $\boldsymbol{\phi}$, etc. The components of the m-dimensional vector $\mathbf{y}$ will be denoted by ${}^i y$, $i = 1, 2, \ldots, m$; that is, we can write $\mathbf{y} = [{}^1 y, {}^2 y, \ldots, {}^m y]^T$, where T denotes the transpose.

We shall repeatedly use the notation

$$f(x) = O(\phi(x)) \quad \text{as} \quad x \to x_0,$$

which means that there exists a positive constant K such that $|f(x)| \leqslant K|\phi(x)|$ for x sufficiently close to x_0. Our most frequent use of this notation will be in the context $f(h) = O(h^p)$ as $h \to 0$, where h is the steplength associated with some numerical method. So frequently shall we write this, that it will be impracticable always to include the phrase 'as $h \to 0$', which is consequently to be taken as read. However, it is important to guard against the temptation mentally to debase the notation $f(h) = O(h^p)$ to mean '$f(h)$ is roughly the same size as h^p, *whatever the size of* h'. Apparent discrepancies between theoretical estimates and numerical results frequently stem from a failure to realize that the notation $f(h) = O(h^p)$ carries with it the implication 'as $h \to 0$'.

The closed interval $a \leqslant x \leqslant b$ will be denoted by $[a, b]$, and the open interval $a < x < b$ by (a, b). We shall occasionally use the notation $y(x) \in C^m[a, b]$ to indicate that $y(x)$ possesses m continuous derivatives for $x \in [a, b]$.

1.2 Prerequisites

Remarkably little is required by way of prerequisites for the study of computational methods for ordinary differential equations. It is assumed that the reader is familiar with the following topics; suggested references are given.

(i) Introductory analysis, including the geometry of the complex plane. (Jones and Jordan.[82])

(ii) Elementary numerical analysis, including the finite difference operators Δ, ∇, and elementary interpolation and quadrature formulae. (Morris.[135])

(iii) For the material of chapter 8, elementary properties of vector and matrix norms. (Mitchell,[134] chapter 1.)

1.3 Initial value problems for first-order ordinary differential equations

A first-order differential equation $y' = f(x, y)$ may possess an infinite number of solutions. For example, the function $y(x) = Ce^{\lambda x}$ is, for any value of the constant C, a solution of the differential equation $y' = \lambda y$, where λ is a given constant. We can pick out any particular solution by prescribing an *initial condition*, $y(a) = \eta$. For the above example, the particular solution satisfying this initial condition is easily found to be $y(x) = \eta e^{\lambda(x-a)}$. We say that the differential equation together with an initial condition constitutes an *initial value problem*,

$$y' = f(x, y), \qquad y(a) = \eta. \tag{1}$$

The following theorem, whose proof may be found in Henrici,[67] states conditions on $f(x, y)$ which guarantee the existence of a unique solution of the initial value problem (1).

Theorem 1.1 *Let $f(x, y)$ be defined and continuous for all points (x, y) in the region D defined by $a \leqslant x \leqslant b$, $-\infty < y < \infty$, a and b finite, and let there exist a constant L such that, for every x, y, y^* such that (x, y) and (x, y^*) are both in D,*

$$|f(x, y) - f(x, y^*)| \leqslant L|y - y^*|. \tag{2}$$

Then, if η is any given number, there exists a unique solution $y(x)$ of the initial value problem (1), where $y(x)$ is continuous and differentiable for all (x, y) in D.

The requirement (2) is known as a *Lipschitz condition*, and the constant L as a *Lipschitz constant*. This condition may be thought of as being intermediate between differentiability and continuity, in the sense that

$f(x, y)$ continuously differentiable w.r.t. y for all (x, y) in D

$\Rightarrow f(x, y)$ satisfies a Lipschitz condition w.r.t. y for all (x, y) in D

$\Rightarrow f(x, y)$ continuous w.r.t. y for all (x, y) in D.

In particular, if $f(x, y)$ possesses a continuous derivative with respect to y for all (x, y) in D, then, by the mean value theorem,

$$f(x, y) - f(x, y^*) = \frac{\partial f(x, \bar{y})}{\partial y}(y - y^*),$$

where $\bar{y}$ is a point in the interior of the interval whose end-points are y and y^*, and (x, y) and (x, y^*) are both in D. Clearly, (2) is then satisfied if we choose

$$L = \sup_{(x,y)\in D} \left| \frac{\partial f(x, y)}{\partial y} \right|. \tag{3}$$

1.4 Initial value problems for systems of first-order ordinary differential equations

In many applications, we are faced, not with a single differential equation, but with a system of m simultaneous first-order equations in m dependent variables ${}^1y, {}^2y, \dots, {}^my$. If each of these variables satisfies a given condition *at the same value a of x*, then we have an initial value problem for a first-order system, which we may write

$$\begin{aligned}
{}^1y' &= {}^1f(x, {}^1y, {}^2y, \dots, {}^my), \qquad & {}^1y(a) &= {}^1\eta, \\
{}^2y' &= {}^2f(x, {}^1y, {}^2y, \dots, {}^my), \qquad & {}^2y(a) &= {}^2\eta, \\
&\;\vdots & &\;\vdots \\
{}^my' &= {}^mf(x, {}^1y, {}^2y, \dots, {}^my), \qquad & {}^my(a) &= {}^m\eta.
\end{aligned} \tag{4}$$

(If the ${}^iy, i = 1, 2, \dots, m$, satisfy given conditions at *different* values $a, b, c, \dots$ of x, then we have a *multipoint boundary value problem*; if there are just two different values a and b of x, then we have a *two-point boundary value problem*.) Introducing the vector notation

$$\mathbf{y} = [{}^1y, {}^2y, \dots, {}^my]^T, \qquad \mathbf{f} = [{}^1f, {}^2f, \dots, {}^mf]^T = \mathbf{f}(x, \mathbf{y}),$$
$$\boldsymbol{\eta} = [{}^1\eta, {}^2\eta, \dots, {}^m\eta]^T,$$

we may write the initial value problem (4) in the form

$$\mathbf{y}' = \mathbf{f}(x, \mathbf{y}), \qquad \mathbf{y}(a) = \boldsymbol{\eta}. \tag{5}$$

Theorem 1.1 readily generalizes to give necessary conditions for the existence of a unique solution to (5); all that is required is that the region D now be defined by $a \leqslant x \leqslant b$, $-\infty < {}^iy < \infty$, $i = 1, 2, \dots, m$, and (2) be replaced by the condition

$$\|\mathbf{f}(x, \mathbf{y}) - \mathbf{f}(x, \mathbf{y}^*)\| \leqslant L\|\mathbf{y} - \mathbf{y}^*\|, \tag{6}$$

where $(x, \mathbf{y})$ and $(x, \mathbf{y}^*)$ are in D, and $\|\,.\,\|$ denotes a vector norm (Mitchell,[134] chapter 1). In the case when each of the ${}^if(x, {}^1y, {}^2y, \ldots, {}^my)$, $i = 1, 2, \ldots, m$, possesses a continuous derivative with respect to each of the ${}^jy$, $j = 1, 2, \ldots, m$, then we may choose, analogously to (3),

$$L = \sup_{(x,y)\in D} \left\|\frac{\partial \mathbf{f}}{\partial \mathbf{y}}\right\|, \tag{7}$$

where $\partial\mathbf{f}/\partial\mathbf{y}$ is the *Jacobian* of $\mathbf{f}$ with respect to $\mathbf{y}$—that is, the $m \times m$ matrix whose i–jth element is $\partial\, {}^if(x, {}^1y, {}^2y, \ldots, {}^my)/\partial\, {}^jy$, and $\|\,.\,\|$ denotes a matrix norm subordinate to the vector norm employed in (6) (see Mitchell,[134] chapter 1).

1.5 Reduction of higher order differential equations to first-order systems

Let us consider an initial value problem involving an ordinary differential equation of order m, which can be written in the form†

$$y^{(m)} = f(x, y^{(0)}, y^{(1)}, \ldots, y^{(m-1)}), \qquad y^{(t)}(a) = \eta_t, \qquad t = 0, 1, \ldots, m-1. \tag{8}$$

We define new variables ${}^iy$, $i = 1, 2, \ldots, m$, as follows:

$$\begin{aligned}
{}^1y &\equiv y\ (\equiv y^{(0)}),\\
{}^2y &\equiv {}^1y'\ (\equiv y^{(1)}),\\
{}^3y &\equiv {}^2y'\ (\equiv y^{(2)}),\\
&\ \vdots\\
{}^my &\equiv {}^{m-1}y'\ (\equiv y^{(m-1)}).
\end{aligned}$$

Then, on writing ${}^i\eta$ for η_{i-1}, $i = 1, 2, \ldots, m$, the initial value problem (8) may be written as an initial value problem for a first-order system, namely

$$\begin{aligned}
{}^1y' &= {}^2y, & {}^1y(a) &= {}^1\eta,\\
{}^2y' &= {}^3y, & {}^2y(a) &= {}^2\eta,\\
&\ \vdots & &\ \vdots\\
{}^{m-1}y' &= {}^my, & {}^{m-1}y(a) &= {}^{m-1}\eta,\\
{}^my' &= f(x, {}^1y, {}^2y, \ldots, {}^my), & {}^my(a) &= {}^m\eta,
\end{aligned}$$

† Superscripts in round brackets indicate the order of higher derivatives; that is, $y^{(0)}(x) \equiv y(x)$, $y^{(1)}(x) \equiv y'(x)$, $y^{(2)}(x) \equiv y''(x)$, etc.

or,

$$\mathbf{y}' = \mathbf{f}(x, \mathbf{y}), \quad \mathbf{y}(a) = \boldsymbol{\eta},$$

where

$$\mathbf{f} = [{}^2y, {}^3y, \ldots, {}^my, f(x, {}^1y, {}^2y, \ldots, {}^my)]^T.$$

With certain exceptions to be discussed in chapter 9, our normal procedure for dealing with an initial value problem of the form (8) will be to reduce it to an initial value problem for an equivalent first-order system. Note that the differential equation appearing in (8) is not the most general differential equation of order m; *implicit differential equations* of the form

$$F(x, y^{(0)}, y^{(1)}, \ldots, y^{(m)}) = 0$$

are also possible. Using the technique described above, an initial value problem involving such an equation may be written in the form

$$\mathbf{F}(x, \mathbf{y}, \mathbf{y}') = \mathbf{0}, \quad \mathbf{y}(a) = \boldsymbol{\eta}.$$

There exist very few numerical methods which tackle this form of initial value problem directly, and we shall always assume that we are given an initial value problem in one of the forms (1), (5), or (8).

1.6 First-order linear systems with constant coefficients

The first-order system $\mathbf{y}' = \mathbf{f}(x, \mathbf{y})$, where $\mathbf{y}$ and $\mathbf{f}$ are m-dimensional vectors, is said to be *linear* if $\mathbf{f}(x, \mathbf{y}) = A(x)\mathbf{y} + \boldsymbol{\phi}(x)$, where $A(x)$ is an $m \times m$ matrix and $\boldsymbol{\phi}(x)$ an m-dimensional vector; if, in addition, $A(x) = A$, a constant matrix, the system is said to be *linear with constant coefficients.* In chapter 8, we shall require the general solution of such a system,

$$\mathbf{y}' = A\mathbf{y} + \boldsymbol{\phi}(x). \tag{9}$$

Let $\hat{\mathbf{y}}(x)$ be the general solution of the corresponding homogeneous system

$$\mathbf{y}' = A\mathbf{y}. \tag{10}$$

If $\boldsymbol{\psi}(x)$ is any particular solution of (9), then $\mathbf{y}(x) = \hat{\mathbf{y}}(x) + \boldsymbol{\psi}(x)$ is the general solution of (9). A set of solutions $\mathbf{y}_t(x)$, $t = 1, 2, \ldots, M$, of (10) is said to be *linearly independent* if $\sum_{t=1}^{M} a_t\mathbf{y}_t(x) \equiv 0$ implies $a_t = 0$, $t = 1, 2, \ldots, M$. A set of m linearly independent solutions $\hat{\mathbf{y}}_t(x)$, $t = 1, 2, \ldots, m$, of (10) is said to form a *fundamental system* of (10), and the most general solution of (10) may be written as a linear combination of the members of the fundamental system. It is easily seen that $\hat{\mathbf{y}}_t(x) = e^{\lambda_t x}\mathbf{c}_t$, where $\mathbf{c}_t$ is an m-dimensional vector, is a solution of (10) if $\lambda_t \mathbf{c}_t = A\mathbf{c}_t$, that is, if λ_t is an eigenvalue of A and $\mathbf{c}_t$ is the corresponding eigenvector. It will be sufficient for our purposes to consider only the case where A possesses m *distinct*, possibly complex, eigenvalues λ_t, $t = 1, 2, \ldots, m$. The corresponding eigenvectors $\mathbf{c}_t$, $t = 1, 2, \ldots, m$, are then linearly independent (Mitchell,[134] chapter 1), and it follows that the solutions $\hat{\mathbf{y}}_t(x) = e^{\lambda_t x}\mathbf{c}_t$, $t = 1, 2, \ldots, m$, form a fundamental system of (10), whose most general solution is thus of the form $\sum_{t=1}^{m} k_t e^{\lambda_t x}\mathbf{c}_t$, where the $k_t, t = 1, 2, \ldots, m$ are arbitrary constants. The most general solution of (9) is then

$$\mathbf{y}(x) = \sum_{t=1}^{m} k_t e^{\lambda_t x}\mathbf{c}_t + \boldsymbol{\psi}(x). \tag{11}$$

We can now find the solution of the initial value problem

$$\mathbf{y}' = A\mathbf{y} + \boldsymbol{\phi}(x), \qquad \mathbf{y}(a) = \boldsymbol{\eta} \tag{12}$$

under the assumptions that A has m distinct eigenvalues, and that we know a particular solution $\boldsymbol{\psi}(x)$ of (9). By (11), the general solution of (9) satisfies the initial condition given in (12) if

$$\boldsymbol{\eta} - \boldsymbol{\psi}(a) = \sum_{t=1}^{m} k_t e^{\lambda_t a}\mathbf{c}_t. \tag{13}$$

Since the vectors $\mathbf{c}_t$, $t = 1, 2, \ldots, m$, form a basis of the m-dimensional vector space (Mitchell,[134] chapter 1), we may express $\boldsymbol{\eta} - \boldsymbol{\psi}(a)$ uniquely in the form

$$\boldsymbol{\eta} - \boldsymbol{\psi}(a) = \sum_{t=1}^{m} \kappa_t \mathbf{c}_t. \tag{14}$$

On comparing (13) with (14), we see that (11) is a solution of (12) if we choose $k_t = \kappa_t e^{-\lambda_t a}$. The solution of (12) is thus

$$\mathbf{y}(x) = \sum_{t=1}^{m} \kappa_t e^{\lambda_t (x-a)}\mathbf{c}_t + \boldsymbol{\psi}(x).$$

For a fuller treatment of first-order linear systems with constant coefficients, the reader is referred to Hurewicz.[78]

Example 1 *Solve the initial value problem* $\mathbf{y}' = A\mathbf{y}, \mathbf{y}(0) = [1, 0, -1]^T$, *where*

$$A = \begin{bmatrix} -21 & 19 & -20 \\ 19 & -21 & 20 \\ 40 & -40 & -40 \end{bmatrix}.$$

The eigenvalues of A are the roots of the equation $\det(A - \lambda I) = 0$, and are found to be $\lambda_1 = -2$, $\lambda_2 = -40 + 40i$, and $\lambda_3 = -40 - 40i$, $i^2 = -1$, and are distinct. The corresponding eigenvectors are the non-trivial solutions for $\mathbf{c}$ of the systems $(A - \lambda_t I)\mathbf{c} = 0$, $t = 1, 2, 3$, and are found to be $\mathbf{c}_1 = [1, 1, 0]^T$, $\mathbf{c}_2 = [1, -1, -2i]^T$ and $\mathbf{c}_3 = [1, -1, 2i]^T$. The general solution of $\mathbf{y}' = A\mathbf{y}$ is

$$\mathbf{y}(x) = k_1 e^{\lambda_1 x}\mathbf{c}_1 + k_2 e^{\lambda_2 x}\mathbf{c}_2 + k_3 e^{\lambda_3 x}\mathbf{c}_3.$$

For this problem, $\boldsymbol{\psi}(x)$ is identically zero, and the given initial vector $\boldsymbol{\eta} = [1, 0, -1]^T$ can be expressed as the following linear combination of $\mathbf{c}_1$, $\mathbf{c}_2$, and $\mathbf{c}_3$:

$$[1, 0, -1]^T = \tfrac{1}{2}[1, 1, 0]^T + \tfrac{1}{4}(1 - i)[1, -1, -2i]^T + \tfrac{1}{4}(1 + i)[1, -1, 2i]^T.$$

We thus choose $k_1 = \frac{1}{2}$, $k_2 = \frac{1}{4}(1 - i)$, and $k_3 = \frac{1}{4}(1 + i)$, giving the solution

$$\mathbf{y}(x) = \tfrac{1}{2}e^{-2x}\begin{bmatrix}1\\1\\0\end{bmatrix} + \tfrac{1}{4}(1 - i)e^{(-40+40i)x}\begin{bmatrix}1\\-1\\-2i\end{bmatrix} + \tfrac{1}{4}(1 + i)e^{(-40-40i)x}\begin{bmatrix}1\\-1\\2i\end{bmatrix}.$$

On writing $e^{i\theta}$ as $\cos\theta + i\sin\theta$, this solution may be expressed in the form

$$u(x) = \tfrac{1}{2}e^{-2x} + \tfrac{1}{2}e^{-40x}(\cos 40x + \sin 40x),$$
$$v(x) = \tfrac{1}{2}e^{-2x} - \tfrac{1}{2}e^{-40x}(\cos 40x + \sin 40x),$$
$$w(x) = -e^{-40x}(\cos 40x - \sin 40x),$$

where $\mathbf{y}(x) = [u(x), v(x), w(x)]^T$.

Exercises

1. Write the differential equation $y^{(3)} = ay^{(2)} + by^{(1)} + cy + \phi(x)$ in the form of a first-order system $\mathbf{y}' = A\mathbf{y} + \boldsymbol{\phi}(x)$. Show that the eigenvalues of A are the roots of the polynomial $\lambda^3 - a\lambda^2 - b\lambda - c$. Show also that if $\psi(x)$ is a particular solution of the given differential equation, then $\boldsymbol{\psi}(x) = [\psi(x), \psi'(x), \psi''(x)]^T$ is a particular solution of the equivalent first-order system.

2. The differential equation $y^{(3)} + y = x^2 + e^{-2x}$ has a particular solution $\psi(x) = x^2 - e^{-2x}/7$. Using the results of exercise 1, construct the equivalent first-order system and find its general solution.

1.7 Linear difference equations with constant coefficients

We shall occasionally need the general solution of the kth-*order linear difference equation with constant coefficients*,

$$\gamma_k y_{n+k} + \gamma_{k-1} y_{n+k-1} + \dots + \gamma_0 y_n = \phi_n, \; n = n_0, n_0 + 1, \dots, \tag{15}$$

where γ_j, $j = 0, 1, \dots, k$, are constants independent of n, and $\gamma_k \neq 0$, $\gamma_0 \neq 0$. A solution of such a difference equation will consist of a sequence $y_{n_0}, y_{n_0+1}, \dots$, which we shall indicate by $\{y_n\}$. Let $\{\hat{y}_n\}$ be the general solution of the corresponding homogeneous difference equation

$$\gamma_k y_{n+k} + \gamma_{k-1} y_{n+k-1} + \ldots + \gamma_0 y_n = 0, n = n_0, n_0 + 1, \ldots. \quad (16)$$

If $\{\psi_n\}$ is any particular solution of (15), then the general solution of (15) is $\{y_n\}$, where $y_n = \hat{y}_n + \psi_n$. In chapter 3, we shall be concerned with the case when $\phi_n (= \phi)$ is independent of n; if $\sum_{j=0}^{k} \gamma_j \neq 0$, we can then choose as particular solution

$$\psi_n = \phi \Big/ \sum_{j=0}^{k} \gamma_j. \quad (17)$$

A set of solutions $\{y_{n,t}\}$, $t = 1, 2, \ldots, K$, of (16) is said to be *linearly independent* if

$$a_1 y_{n,1} + a_2 y_{n,2} + \ldots + a_K y_{n,K} = 0 \text{ for } n = n_0, n_0 + 1, \ldots$$

implies $a_1 = a_2 = \ldots = a_K = 0$. A set of k linearly independent solutions $\{\hat{y}_{n,t}\}$, $t = 1, 2, \ldots, k$, solutions of (16), is said to form a *fundamental system* of (16), and the most general solution of (16) may be written in the form $\{\sum_{t=1}^{k} d_t y_{n,t}\}$. Let us try to find a solution of (16) of the form $y_{n,t} = r_t^n$. On substituting into (16), we find that this is indeed a solution if

$$\gamma_k r_t^k + \gamma_{k-1} r_t^{k-1} + \ldots + \gamma_0 = 0,$$

that is, if r_t is a root of the polynomial $P(r) \equiv \gamma_k r^k + \gamma_{k-1} r^{k-1} + \ldots + \gamma_0$. If $P(r)$ has k *distinct* roots r_t, $t = 1, 2, \ldots, k$, then it can be shown that the set of solutions $\{r_t^n\}$, $t = 1, 2, \ldots, k$, forms a fundamental system of (16). It follows that the most general solution of (15) is $\{y_n\}$, where

$$y_n = \sum_{t=1}^{k} d_t r_t^n + \psi_n,$$

the d_t, $t = 1, 2, \ldots, k$, being arbitrary constants. Suppose now that $r_1 (= r_2)$ is a double root of $P(r)$, and that all the other roots r_t, $t = 3, 4, \ldots, k$, are distinct. We would appear to have a set of only $k - 1$ solutions of the form $\{r_t^n\}$, and such a set cannot constitute a fundamental system. However, it is easily found by substitution that, if r_1 is a double root, then $\{n r_1^n\}$ is also a solution of (16). The set of solutions consisting of $\{r_1^n\}$, $\{n r_1^n\}$ and $\{r_t^n\}$, $t = 3, 4, \ldots, k$, forms a fundamental system of (16), and consequently the general solution of (15) is $\{y_n\}$, where

$$y_n = d_1 r_1^n + d_2 n r_1^n + \sum_{t=3}^{k} d_t r_t^n + \psi_n.$$

In general, if $P(r)$ has roots r_j, $j = 1, 2, \ldots, p$, and the root r_j has multiplicity μ_j, where $\sum_{j=1}^{p} \mu_j = k$, then the general solution of (15) is $\{y_n\}$, where

$$\begin{aligned} y_n = {} & [d_{1,1} + d_{1,2}n + d_{1,3}n(n-1) + \ldots \\ & \qquad + d_{1,\mu_1}n(n-1)\ldots(n-\mu_1+2)]r_1^n \\ & + [d_{2,1} + d_{2,2}n + d_{2,3}n(n-1) + \ldots \\ & \qquad + d_{2,\mu_2}n(n-1)\ldots(n-\mu_2+2)]r_2^n \\ & + \ldots \\ & + [d_{p,1} + d_{p,2}n + d_{p,3}n(n-1) + \ldots \\ & \qquad + d_{p,\mu_p}n(n-1)\ldots(n-\mu_p+2)]r_p^n + \psi_n, \end{aligned}$$

the k constants $d_{j,l}$, $l = 1, 2, \ldots, \mu_j$, $j = 1, 2, \ldots, p$, being arbitrary.

For a fuller discussion of linear difference equations, the reader is referred to Henrici,[67] page 210.

Example 2 Find the solution of the difference equation

$$y_{n+4} - 4y_{n+3} + 5y_{n+2} - 4y_{n+1} + 4y_n = 4 \tag{18}$$

which satisfies the initial conditions $y_0 = 5$, $y_1 = 0$, $y_2 = -4$, $y_3 = -12$.

By (17), a particular solution of this equation is $\psi_n = 4/(1 - 4 + 5 - 4 + 4) = 2$. The associated polynomial is

$$P(r) = r^4 - 4r^3 + 5r^2 - 4r + 4 = (r^2 + 1)(r - 2)^2,$$

whose roots are i, $-i$, 2, and 2, where $i^2 = -1$. Thus the general solution of (18) is $\{y_n\}$, where

$$y_n = d_1 i^n + d_2(-i)^n + d_3 2^n + d_4 n 2^n + 2.$$

The given initial conditions are satisfied if

$$\begin{aligned} d_1 + d_2 + d_3 \qquad\quad + 2 &= 5, \\ id_1 - id_2 + 2d_3 + 2d_4 + 2 &= 0, \\ -d_1 - d_2 + 4d_3 + 8d_4 + 2 &= -4, \\ -id_1 + id_2 + 8d_3 + 24d_4 + 2 &= -12. \end{aligned}$$

The solution of this set of equations is found to be $d_1 = 1 + i$, $d_2 = 1 - i$, $d_3 = 1$, $d_4 = -1$. The solution of (18) which satisfies the given initial conditions is thus $\{y_n\}$, where

$$y_n = (1 + i)i^n + (1 - i)(-i)^n + 2^n - n2^n + 2,$$

or, on writing $i = \exp(i\pi/2)$,

$$y_n = 2(\cos n\pi/2 - \sin n\pi/2) + (1 - n)2^n + 2.$$

Exercises

3. Compute y_4, y_5, y_6, and y_7 directly from the difference equation (18), using the given values for y_0, y_1, y_2, and y_3. Verify that the values so found coincide with those given by the general solution found in example 2.

4. The Fibonacci numbers $\phi_n, n = 0, 1, 2, \ldots$, are a sequence of integers, each of which equals the sum of the two preceding it, the first two being 0 and 1. Construct the first eleven Fibonacci numbers, and compute the ratios $\phi_{n+1}/\phi_n, n = 1, 2, \ldots, 9$. Find a closed expression for the nth Fibonacci number, and deduce that

$$\lim_{n \to \infty} \phi_{n+1}/\phi_n = (1 + \sqrt{5})/2.$$

5. Under what conditions on the roots of the polynomial $P(r)$ will the solution of (16) tend to zero as $n \to \infty$?

6. Find the general solution of the difference equation $y_{n+2} - 2\mu y_{n+1} + \mu y_n = 1$, $0 < \mu < 1$. Show that $y_n \to 1/(1 - \mu)$ as $n \to \infty$.

1.8 Iterative methods for non-linear equations

We shall frequently have to find an approximate solution of the equation

$$y = \phi(y), \tag{19}$$

where $\phi(y)$ is a non-linear function of y. We shall do this iteratively by constructing the sequence $\{y^{[s]}\}$ of iterates defined by

$$y^{[s+1]} = \phi(y^{[s]}), \; s = 0, 1, 2, \ldots, \; y^{[0]} \text{ arbitrary}. \tag{20}$$

The following theorem, whose proof may be found in Henrici[67] page 216, states conditions under which (19) will possess a unique solution to which the sequence defined by (20) will converge.

Theorem 1.2 Let $\phi(y)$ satisfy a Lipschitz condition

$$|\phi(y) - \phi(y^*)| \leqslant M|y - y^*|$$

for all y, y^*, where the Lipschitz constant M satisfies $0 \leqslant M < 1$. Then there exists a unique solution α of (19), and if $\{y^{[s]}\}$ is defined by (20), then $\lim_{s \to \infty} y^{[s]} = \alpha$.

A system of simultaneous non-linear equations of the form $\mathbf{y} = \boldsymbol{\phi}(\mathbf{y})$ can likewise be tackled iteratively by constructing the sequence $\{\mathbf{y}^{[s]}\}$ of vector iterates defined by $\mathbf{y}^{[s+1]} = \boldsymbol{\phi}(\mathbf{y}^{[s]}), s = 0, 1, 2, \ldots, \mathbf{y}^{[0]}$ arbitrary. Theorem 1.2 holds for the vector case if, throughout, absolute values of scalars are replaced by norms of corresponding vectors.

2

Linear multistep methods I: basic theory

2.1 The general linear multistep method

Consider the initial value problem for a single first-order differential equation

$$y' = f(x, y), \; y(a) = \eta. \tag{1}$$

We seek a solution in the range $a \leqslant x \leqslant b$, where a and b are finite, and we assume that f satisfies the conditions stated in theorem 1.1 which guarantee that the problem has a unique continuously differentiable solution, which we shall indicate by $y(x)$. Consider the sequence of points $\{x_n\}$ defined by $x_n = a + nh$, $n = 0, 1, 2, \ldots$. The parameter h, which will always be regarded as constant, except where otherwise indicated, is called the *steplength*. An essential property of the majority of computational methods for the solution of (1) is that of discretization; that is, we seek an approximate solution, not on the continuous interval $a \leqslant x \leqslant b$, but on the discrete point set $\{x_n | n = 0, 1, \ldots, (b - a)/h\}$. Let y_n be an approximation to the theoretical solution at x_n, that is, to $y(x_n)$, and let $f_n \equiv f(x_n, y_n)$. If a computational method for determining the sequence $\{y_n\}$ takes the form of a *linear* relationship between $y_{n+j}, f_{n+j}, j = 0, 1, \ldots, k$, we call it a *linear multistep method of stepnumber k*, or a *linear k-step method*. The general linear multistep method may thus be written

$$\sum_{j=0}^{k} \alpha_j y_{n+j} = h \sum_{j=0}^{k} \beta_j f_{n+j}, \tag{2}$$

where α_j and β_j are constants; we assume that $\alpha_k \neq 0$ and that not both α_0 and β_0 are zero. Since (2) can be multiplied on both sides by the same constant without altering the relationship, the coefficients α_j and β_j are arbitrary to the extent of a constant multiplier. We remove this arbitrariness by assuming throughout that $\alpha_k = 1$.

Thus the problem of determining the solution $y(x)$ of the, in general, non-linear initial value problem (1) is replaced by that of finding a sequence $\{y_n\}$ which satisfies the difference equation (2). Note that, since f_n $(= f(x_n, y_n))$ is, in general, a non-linear function of y_n, (2) is a *non-linear difference equation.* Such equations are no easier to handle theoretically than are non-linear differential equations, but they have the practical advantage of permitting us to compute the sequence $\{y_n\}$ numerically. In order to do this, we must first supply a set of *starting values*, $y_0, y_1, \ldots, y_{k-1}$. (In the case of a one-step method, only one such value, y_0, is needed, and we normally choose $y_0 = \eta$.) Methods for obtaining starting values will be discussed in section 3.2.

We say that the method (2) is *explicit* if $\beta_k = 0$, and *implicit* if $\beta_k \neq 0$. For an explicit method, equation (2) yields the current value y_{n+k} directly in terms of $y_{n+j}, f_{n+j}, j = 0, 1, \ldots, k-1$, which, at this stage of the computation, have already been calculated. An implicit method, however, will call for the solution, at each stage of the computation, of the equation

$$y_{n+k} = h\beta_k f(x_{n+k}, y_{n+k}) + g, \tag{3}$$

where g is a known function of the previously calculated values y_{n+j}, $f_{n+j}, j = 0, 1, \ldots, k-1$. When the original differential equation in (1) is linear, then (3) is also linear in y_{n+k}, and there is no problem in solving it. When f is non-linear, then by theorem 1.2, there exists a unique solution for y_{n+k}, which can be approached arbitrarily closely by the iteration

$$y_{n+k}^{[s+1]} = h\beta_k f(x_{n+k}, y_{n+k}^{[s]}) + g, \qquad y_{n+k}^{[0]} \text{ arbitrary,}\dagger$$

provided that $0 \leqslant M < 1$, where M is the Lipschitz constant with respect to y_{n+k} of the right-hand side of (3). If the Lipschitz constant of f with respect to y is L, then clearly we can take M to have the value $Lh|\beta_k|$. Thus a unique solution for y_{n+k} exists, and the above iteration converges to it, if

$$h < 1/L|\beta_k|. \tag{4}$$

(In cases where $L \gg 1$—see the sections on stiff equations in chapter 8—this requirement can impose a severe restriction on the size of the steplength.) Thus implicit methods in general entail a substantially greater computational effort than do explicit methods; on the other hand, for a given stepnumber k, implicit methods can be made more accurate than explicit ones and, moreover, enjoy more favourable stability properties.

† Iteration superscripts will always be enclosed in square brackets. Superscripts in round brackets indicate the order of higher total derivatives.

We now turn to the problem of determining the coefficients α_j, β_j appearing in (2). Any specific linear multistep method may be derived in a number of different ways; we shall consider a selection of different approaches which cast some light on the nature of the approximation involved.

2.2 Derivation through Taylor expansions

Consider the Taylor expansion for $y(x_n + h)$ about x_n:

$$y(x_n + h) = y(x_n) + hy^{(1)}(x_n) + \frac{h^2}{2!}y^{(2)}(x_n) + \ldots,$$

where

$$y^{(q)}(x_n) = \left.\frac{\mathrm{d}^q y}{\mathrm{d}x^q}\right|_{x=x_n}, \qquad q = 1, 2, \ldots.$$

If we truncate this expansion after two terms and substitute for $y^{(1)}(x)$ from the differential equation (1), we have

$$y(x_n + h) \doteqdot y(x_n) + hf(x_n, y(x_n)), \tag{5}$$

a relation which is in error by

$$\frac{h^2}{2!}y^{(2)}(x_n) + \frac{h^3}{3!}y^{(3)}(x_n) + \ldots. \tag{6}$$

Equation (5) expresses an *approximate* relation between *exact* values of the solution of (1). We can also interpret it as an *exact* relation between *approximate* values of the solution of (1) if we replace $y(x_n)$, $y(x_n + h)$ by y_n, y_{n+1} respectively, yielding

$$y_{n+1} = y_n + hf_n, \tag{7}$$

an explicit linear one-step method. It is, in fact, *Euler's rule*, the simplest of all linear multistep methods. The error associated with it is the expression (6) (multiplied by $+1$ or -1 according to the sense of the definition of error) and is called the *local truncation error* or *local discretization error*. We postpone until later a more precise definition of this error but observe that, for this method, the local truncation error is $O(h^2)$, and that it is identically zero if the solution of (1) is a polynomial of degree not exceeding one.

Consider now Taylor expansions for $y(x_n + h)$ and $y(x_n - h)$ about x_n:

$$y(x_n + h) = y(x_n) + hy^{(1)}(x_n) + \frac{h^2}{2!}y^{(2)}(x_n) + \frac{h^3}{3!}y^{(3)}(x_n) + \dots,$$

$$y(x_n - h) = y(x_n) - hy^{(1)}(x_n) + \frac{h^2}{2!}y^{(2)}(x_n) - \frac{h^3}{3!}y^{(3)}(x_n) + \dots.$$

Subtracting, we get

$$y(x_n + h) - y(x_n - h) = 2hy^{(1)}(x_n) + \frac{h^3}{3}y^{(3)}(x_n) + \dots.$$

Arguing as previously yields the associated linear multistep method

$$y_{n+1} - y_{n-1} = 2hf_n.$$

This can be brought into the standard form (2) by replacing n by $n + 1$ to give

$$y_{n+2} - y_n = 2hf_{n+1}. \tag{8}$$

This is the *Mid-point rule* and its local truncation error is

$$\pm\tfrac{1}{3}h^3y^{(3)}(x_n) + \dots.$$

Similar techniques can be used to derive any linear multistep method of given specification. Thus if we wish to find the most accurate one-step implicit method,

$$y_{n+1} + \alpha_0 y_n = h(\beta_1 f_{n+1} + \beta_0 f_n),$$

we write down the associated approximate relationship

$$y(x_n + h) + \alpha_0 y(x_n) \doteqdot h[\beta_1 y^{(1)}(x_n + h) + \beta_0 y^{(1)}(x_n)], \tag{9}$$

and choose α_0, β_1, β_0 so as to make the approximation as accurate as possible. The following expansions are used:

$$y(x_n + h) = y(x_n) + hy^{(1)}(x_n) + \frac{h^2}{2!}y^{(2)}(x_n) + \dots$$

$$y^{(1)}(x_n + h) = y^{(1)}(x_n) + hy^{(2)}(x_n) + \frac{h^2}{2!}y^{(3)}(x_n) + \dots.$$

Substituting in (9) and collecting the terms on the left-hand side gives

$$C_0 y(x_n) + C_1 hy^{(1)}(x_n) + C_2 h^2 y^{(2)}(x_n) + C_3 h^3 y^{(3)}(x_n) + \dots \doteqdot 0,$$

where

$$C_0 = 1 + \alpha_0, \qquad C_1 = 1 - \beta_1 - \beta_0,$$
$$C_2 = \tfrac{1}{2} - \beta_1, \qquad C_3 = \tfrac{1}{6} - \tfrac{1}{2}\beta_1.$$

Thus, in order to make the approximation in (9) as accurate as possible, we choose $\alpha_0 = -1$, $\beta_1 = \beta_0 = \frac{1}{2}$. C_3 then takes the value $-\frac{1}{12}$. The linear multistep method is now

$$y_{n+1} - y_n = \frac{h}{2}(f_{n+1} + f_n), \tag{10}$$

the *Trapezoidal rule*: its local truncation error is

$$\pm\tfrac{1}{12}h^3 y^{(3)}(x_n) + \dots.$$

This method of derivation leaves some unanswered questions. If we take the Taylor expansions about some new point (say about $x_n + h$ in the last example) do we get the same values for the coefficients α_j, β_j? Do we get the same values for the coefficients in the infinite series representing the local truncation error? How can we remove the ambiguity of sign in the truncation error? A further difficulty concerns the existence of higher derivatives of the solution $y(x)$ implied in the Taylor series: theorem 1.1 implies only that $y(x) \in C^1[a, b]$. In section 2.6 we shall describe a formalization of the above process which will resolve these problems.

Exercise

1. Find the most accurate implicit linear two-step method. Find also the first term in the local truncation error.

2.3 Derivation through numerical integration

Consider the identity

$$y(x_{n+2}) - y(x_n) \equiv \int_{x_n}^{x_{n+2}} y'(x)\,dx. \tag{11}$$

Using the differential equation (1), we can replace $y'(x)$ by $f(x, y)$ in the integrand. If our aim is to derive, say, a linear two-step method, then the only available data for the approximate evaluation of the integral will be the values f_n, f_{n+1}, f_{n+2}. Let $P(x)$ be the unique polynomial of degree two passing through the three points (x_n, f_n), (x_{n+1}, f_{n+1}), (x_{n+2}, f_{n+2}). By the Newton–Gregory forward interpolation formula,

$$P(x) = P(x_n + rh) = f_n + r\Delta f_n + \frac{r(r-1)}{2!}\Delta^2 f_n.$$

We now make the approximation

$$\int_{x_n}^{x_{n+2}} y'(x)\,dx \doteqdot \int_{x_n}^{x_{n+2}} P(x)\,dx = \int_0^2 [f_n + r\Delta f_n + \tfrac{1}{2}r(r-1)\Delta^2 f_n]h\,dr$$

$$= h(2f_n + 2\Delta f_n + \tfrac{1}{3}\Delta^2 f_n).$$

Expanding Δf_n and $\Delta^2 f_n$ in terms of f_n, f_{n+1}, f_{n+2} and substituting in (11) gives

$$y_{n+2} - y_n = \frac{h}{3}(f_{n+2} + 4f_{n+1} + f_n), \tag{12}$$

which is *Simpson's rule*, the most accurate implicit linear two-step method (see exercise 1). This derivation is, of course, very close to the derivation of a Newton–Cotes quadrature formula for the numerical evaluation of $\int_a^b f(x)\,dx$. Indeed, (12) is such a formula, and all Newton–Cotes formulae can be regarded as linear multistep methods. However, linear multistep methods which are not identical with Newton–Cotes quadrature formulae can be derived in a similar way. Thus if we replace (11) by the identity

$$y(x_{n+2}) - y(x_{n+1}) \equiv \int_{x_{n+1}}^{x_{n+2}} y'(x)\,dx,$$

and replace $y'(x)$ by $P(x)$, defined as above, we derive the method

$$y_{n+2} - y_{n+1} = \frac{h}{12}(5f_{n+2} + 8f_{n+1} - f_n), \tag{13}$$

the *two-step Adams–Moulton method.*

Clearly this technique can be used to derive only a subclass of linear multistep methods consisting of those methods for which

$$\alpha_k = +1, \quad \alpha_j = -1, \quad \alpha_i = 0, \quad i = 0, 1, \ldots, j-1, j+1, \ldots, k, \qquad j \neq k.$$

The importance of the technique is that it establishes a link between the concepts of polynomial interpolation and linear multistep methods. In the next method of derivation, we shall make use of a more fundamental connection that exists between these concepts, more fundamental in the sense that the interpolation will take place in the x–y plane rather than in the x–f plane.

2.4 Derivation through interpolation

We illustrate the method by deriving again the implicit two-step method (12). Let $y(x)$, the solution of (1), be approximated locally in the range $x_n \leqslant x \leqslant x_{n+2}$ by a polynomial $I(x)$. What conditions should be imposed on $I(x)$ in order that it be a good representation of $y(x)$? We ask that $I(x)$ should interpolate the points (x_{n+j}, y_{n+j}), $j = 0, 1, 2$, and, moreover, that the derivative of $I(x)$ should coincide with the prescribed derivative f_{n+j} for $j = 0, 1, 2$. This defines $I(x)$ as an *osculatory* or *Hermite* interpolant. The conditions imposed on $I(x)$ are thus

$$I(x_{n+j}) = y_{n+j}; \qquad I'(x_{n+j}) = f_{n+j}, \qquad j = 0, 1, 2. \tag{14}$$

There are six conditions in all; let $I(x)$ have five free parameters, that is, let it be a polynomial of degree four, namely

$$I(x) = ax^4 + bx^3 + cx^2 + dx + e.$$

Eliminating the five undetermined coefficients $a, b, \ldots, e$ between the six equations (14) yields the identity

$$y_{n+2} - y_n = \frac{h}{3}(f_{n+2} + 4f_{n+1} + f_n),$$

which is the linear multistep method (12). Derivation of (12) by the method of Taylor expansions shows that the local truncation error is

$$\pm\tfrac{1}{90}h^5 y^{(5)}(x_n) + \ldots$$

(see exercise 1). It follows that the method will be exact when applied to an initial value problem whose theoretical solution, $y(x)$, is a polynomial of degree not greater than four. In such a case, $I(x)$ is identical with $y(x)$.

Exercise

2. Use the above method to derive the Mid-point rule (8). Hint: require $I(x)$ to satisfy $I(x_{n+j}) = y_{n+j}, j = 0, 1, 2, I'(x_{n+1}) = f_{n+1}$.

The technique described above is not an efficient way of determining the coefficients in a linear multistep method. Moreover, it gives no information on the coefficients in the local truncation error associated with the method. Its interest lies in the fact that it indicates that a single application of a linear multistep method is equivalent to locally representing the solution by a polynomial. In situations where osculatory interpolation of $y(x)$ by a polynomial would be inappropriate, we can expect linear multistep methods to perform badly. This happens in the case of stiff equations (which we shall discuss in chapter 8) for which a typical solution is a rapidly decaying exponential. The error in interpolating such functions by a polynomial of high degree is very large, and correspondingly we find that linear multistep methods of high stepnumber cannot be used successfully for such problems.

If we seek to represent $y(x)$ by an interpolant not *locally* on $[x_n, x_{n+k}]$, but *globally* on $[x_0, x_n]$, then we can again establish a connection with linear multistep methods, if we choose the interpolant to be a *spline* function. Consider an interval $[a, b]$ of x divided into n subintervals by inserting the *knots* x_j, $j = 0, 1, \ldots, n$, where $a = x_0 < x_1 < x_2 < \ldots < x_n = b$. Then $S_m(x)$ is a *spline of degree m* if it is a polynomial of degree m in each of the subintervals $[x_j, x_{j+1}]$, $j = 0, 1, \ldots, n-1$, and if $S_m(x) \in C^{m-1}[a, b]$. One way of representing such a spline $S_m(x)$ is

$$S_m(x) = \sum_{j=0}^{m} a_j x^j + \sum_{j=1}^{n} c_j (x - x_j)_+^m,$$

where (15)

$$z_+ = \begin{cases} z, & \text{if } z \geqslant 0, \\ 0, & \text{if } z < 0, \end{cases}$$

from which it is clear that $S_m(x)$ and its first $(m - 1)$ derivatives are indeed continuous, but that the mth derivative suffers a discontinuity at each of the knots.

First consider the quadratic polynomial

$$I_j(x) = a_{j2}x^2 + a_{j1}x + a_{j0},$$

which we use to approximate $y(x)$ on the interval $[x_j, x_{j+1}]$. Then, proceeding as before, we impose the conditions

$$I_j(x_j) = y_j, \qquad I_j(x_{j+1}) = y_{j+1},$$
$$I'_j(x_j) = f_j, \qquad I'_j(x_{j+1}) = f_{j+1}.$$

Eliminating the three coefficients a_{j2}, a_{j1} and a_{j0} between these four equations yields the Trapezoidal rule

$$y_{j+1} - y_j = \frac{h}{2}(f_{j+1} + f_j).$$

Thus, repeatedly applying the Trapezoidal rule to the intervals $[x_j, x_{j+1}]$, $[x_{j+1}, x_{j+2}], \ldots$ is equivalent to representing $y(x)$ by $I_j(x)$ in $[x_j, x_{j+1}]$, by

$I_{j+1}(x)$ in $[x_{j+1}, x_{j+2}]$, etc. These interpolating polynomials are not independent of each other for, at a typical point x_{j+1}, we have

$$\begin{aligned} I_j(x_{j+1}) &= I_{j+1}(x_{j+1}) \,(= y_{j+1}), \\ I'_j(x_{j+1}) &= I'_{j+1}(x_{j+1}) \,(= f_{j+1}). \end{aligned} \tag{16}$$

Thus it follows that using the Trapezoidal rule repeatedly on the intervals $[x_0, x_1]$, $[x_1, x_2], \ldots, [x_{n-1}, x_n]$ is equivalent to approximating $y(x)$ *globally* on $[x_0, x_n]$ by a quadratic spline with knots at $x_0, x_1, \ldots, x_n$. In the interval $[x_j, x_{j+1}]$ the global approximant reduces to the quadratic polynomial $I_j(x)$ and the equations (16) supply the remaining conditions for the global approximant to be a quadratic spline.

When we consider the global approximant equivalent to repeated use of a linear multistep method of stepnumber greater than one, the situation is rather different. We have seen that a single application of Simpson's rule

$$y_{j+2} - y_j = \frac{h}{3}(f_{j+2} + 4f_{j+1} + f_j)$$

is equivalent to locally representing $y(x)$ on $[x_j, x_{j+2}]$ by a *quartic* polynomial $I_j(x)$.

Similarly, applying Simpson's rule at the next step,

$$y_{j+3} - y_{j+1} = \frac{h}{3}(f_{j+3} + 4f_{j+2} + f_{j+1})$$

is equivalent to representing $y(x)$ on $[x_{j+1}, x_{j+3}]$ by a quartic polynomial $I_{j+1}(x)$. Thus, on the overlapping interval $[x_{j+1}, x_{j+2}]$ there are two representations for $y(x)$, namely $I_j(x)$ and $I_{j+1}(x)$, and these are not necessarily identical, since they are related by only four equations

$$\begin{aligned} I_j(x_{j+1}) &= I_{j+1}(x_{j+1}) \,(= y_{j+1}), \\ I_j(x_{j+2}) &= I_{j+1}(x_{j+2}) \,(= y_{j+2}), \\ I'_j(x_{j+1}) &= I'_{j+1}(x_{j+1}) \,(= f_{j+1}), \\ I'_j(x_{j+2}) &= I'_{j+1}(x_{j+2}) \,(= f_{j+2}). \end{aligned}$$

These are not enough to allow us to deduce that the two quartics $I_j(x)$ and $I_{j+1}(x)$ are identical. Hence we cannot use our knowledge of a local representation of $y(x)$ equivalent to a single application of Simpson's rule to deduce the nature of a global representation of $y(x)$ equivalent to repeated applications of Simpson's rule. However, a single application of

Simpson's rule is also equivalent to a local representation of $y(x)$ by another osculatory interpolant, namely a *cubic* spline. Let $S_3^j(x)$ be a cubic spline with knots x_j, x_{j+1} and x_{j+2}. (The superscript j indicates the local nature of the representation.) From (15) we see that

$$\begin{aligned} S_3^j(x) &= a_{j3}x^3 + a_{j2}x^2 + a_{j1}x + a_{j0} \quad \text{for } x \in [x_j, x_{j+1}] \\ &= a_{j3}x^3 + a_{j2}x^2 + a_{j1}x + a_{j0} + c_{j+1}(x - x_{j+1})^3 \\ &\qquad\qquad\qquad\qquad\qquad\qquad \text{for } x \in [x_{j+1}, x_{j+2}]. \end{aligned}$$

There are now five undetermined coefficients and elimination of these between the six equations

$$S_3^j(x_{j+i}) = y_{j+i}, \qquad S_3^{j\prime}(x_{j+i}) = f_{j+i}, \qquad i = 0, 1, 2,$$

yields, once again, Simpson's rule

$$y_{j+2} - y_j = \frac{h}{3}(f_{j+2} + 4f_{j+1} + f_j).$$

Similarly, applying Simpson's rule at the next step

$$y_{j+3} - y_{j+1} = \frac{h}{3}(f_{j+3} + 4f_{j+2} + f_{j+1})$$

is equivalent to representing $y(x)$ on $[x_{j+1}, x_{j+3}]$ by the cubic spline $S_3^{j+1}(x)$ with knots x_{j+1}, x_{j+2}, and x_{j+3}. Again we have two representations for $y(x)$ on the overlapping interval $[x_{j+1}, x_{j+2}]$, namely $S_3^j(x)$ and $S_3^{j+1}(x)$. However, in $[x_{j+1}, x_{j+2}]$ both of these are *cubic* polynomials and the four conditions

$$\begin{aligned} S_3^j(x_{j+1}) &= S_3^{j+1}(x_{j+1}) \ (= y_{j+1}), \\ S_3^j(x_{j+2}) &= S_3^{j+1}(x_{j+2}) \ (= y_{j+2}), \\ S_3^{j\prime}(x_{j+1}) &= S_3^{j+1\prime}(x_{j+1}) \ (= f_{j+1}), \\ S_3^{j\prime}(x_{j+2}) &= S_3^{j+1\prime}(x_{j+2}) \ (= f_{j+2}) \end{aligned}$$

are sufficient to imply that $S_3^j(x)$ and $S_3^{j+1}(x)$ are identical in $[x_{j+1}, x_{j+2}]$. It follows that repeated application of Simpson's rule on the intervals $[x_0, x_2], [x_1, x_3], \ldots, [x_{n-2}, x_n]$ is equivalent to representing $y(x)$ *globally* on $[x_0, x_n]$ by a cubic spline. In a series of papers Loscalzo,[117] Loscalzo and Talbot,[119,120] and Loscalzo and Schoenberg[118] have developed a method for generating spline approximations to the solution of (1). A theorem, due to Schoenberg, uses properties of splines to establish a result formally linking spline approximants with certain linear multistep methods. The method proposed in these papers actually generates the global spline, not just its values at the knots x_j: thus a *continuous* global approximation to the solution $y(x)$ is produced. A FORTRAN subroutine package (SPLINDIF) for this process is available (Loscalzo[116]).

Exercises

3. (Alternative derivation of the connection between Simpson's rule and global approximation by a cubic spline.)

Consider the interpolant $I_j(x) = \lambda_j x^3 + a_{j2}x^2 + a_{j1}x + a_{j0}$. Impose the conditions $I_j(x_{j+i}) = y_{j+i}$, $I'_j(x_{j+i}) = f_{j+i}$, $i = 0, 1$, and eliminate a_{j2}, a_{j1}, a_{j0}, retaining λ_j as a parameter. The resulting one-step formula is *not* a linear multistep method, since it contains the parameter λ_j. Write this formula down again with j replaced by $j + 1$ and add the two formulae to get a two-step formula (*) which contains the parameter $\lambda_j + \lambda_{j+1}$. Show that the condition $I''_j(x_{j+1}) = I''_{j+1}(x_{j+1})$ holds only if

$$\lambda_j + \lambda_{j+1} = (f_{j+2} - 2f_{j+1} + f_j)/3h^2,$$

and that (*) now reduces to a linear multistep method, namely Simpson's rule.

4. An *extended spline*, $\tilde{S}_m(x)$, of degree m with knots x_j, $j = 0, 1, \ldots, n$ is defined to be a polynomial of degree m in each of the intervals $[x_j, x_{j+1}]$, $j = 0, 1, \ldots, n - 1$, and to satisfy $\tilde{S}_m(x) \in C^{m-p}[x_0, x_n]$, $1 < p \leqslant m$.

Simpson's rule is also a Newton–Cotes integration formula for the approximate evaluation of $\int_{x_n}^{x_{n+2}} f(x)\,dx$. The integral $\int_{x_0}^{x_n} f(x)\,dx$ may be evaluated by applying Simpson's rule in each of the intervals $[x_0, x_2]$, $[x_2, x_4], \ldots, [x_{n-2}, x_n]$, n even. (Note the essential difference from the application of Simpson's rule as a linear multistep method.) Show that this process is equivalent to integrating a global interpolant, in the x–f plane, of the points (x_i, f_i), $i = 0, 1, \ldots, n$, and that the interpolant is an extended spline. What is the degree of the extended spline? What value does p take? What are the knots?

2.5 Convergence

A basic property which we shall demand of an acceptable linear multistep method is that the solution $\{y_n\}$ generated by the method converges, in some sense, to the theoretical solution $y(x)$ as the steplength h tends to zero. In converting this intuitive concept into a precise definition, the following points must be kept in mind.

(i) It is inappropriate to consider n as remaining fixed while $h \to 0$. For example, consider a fixed point $x = x^*$, and let the initial choice of steplength h_0 be such that $x^* = a + 2h_0$. In the special case when the steplength is successively halved, the situation illustrated by figure 1 holds. If we use the notation $y_n(h)$ to denote the value y_n given by the linear multistep method (2) when the steplength is h, then we are interested *not* in the convergence of the

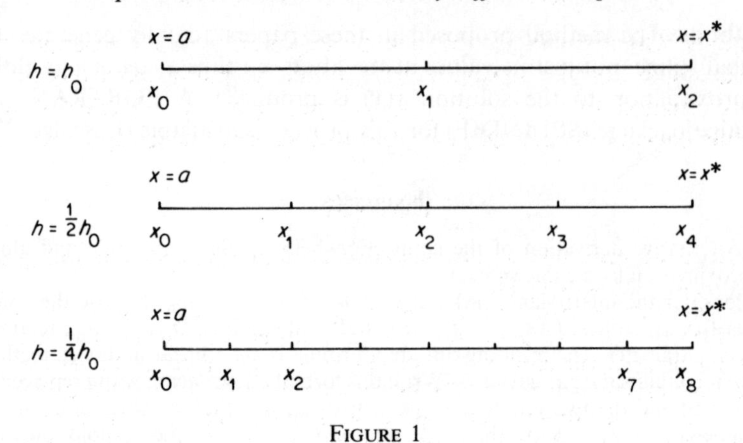

FIGURE 1

sequence $y_2(h_0), y_2(\frac{1}{2}h_0), y_2(\frac{1}{4}h_0), \ldots$, but in the convergence of the sequence $y_2(h_0)$, $y_4(\frac{1}{2}h_0)$, $y_8(\frac{1}{4}h_0), \ldots$, to $y(x^*)$. (In general, of course, we consider the case when h tends continuously to zero.) We are thus led to the idea of 'fixed station convergence' by which we mean convergence in the limit as $h \to 0$, $n \to \infty$, $nh = x - a$ remaining fixed. Such a limit will be written

$$\lim_{\substack{h \to 0 \\ nh = x - a}}$$

(ii) The definition must take account of the additional starting values $y_1, y_2, \ldots, y_{k-1}$, which must be supplied when $k \geqslant 2$.

(iii) If the term 'convergent' is to be applied to the *method* (2), then the convergence property must hold for *all* initial value problems (1) subject to the hypotheses of theorem 1.1.

Definition The linear multistep method (2) is said to be **convergent** *if, for all initial value problems (1) subject to the hypotheses of theorem 1.1, we have that*

$$\lim_{\substack{h \to 0 \\ nh = x - a}} y_n = y(x_n)$$

holds for all $x \in [a, b]$, and for all solutions $\{y_n\}$ of the difference equation (2) satisfying starting conditions $y_\mu = \eta_\mu(h)$ for which $\lim_{h \to 0} \eta_\mu(h) = \eta$, $\mu = 0, 1, 2, \ldots, k - 1$.

It should be observed that this definition is very generous in the conditions it imposes on the starting values y_μ, $\mu = 0, 1, \ldots, k - 1$. It does not demand that these be exact solutions of the initial value problem (1) at the appropriate values for x, but only that, regarded as functions of h, they all tend to the given initial value η as $h \to 0$. Note that it is not even demanded that $y_0 = \eta$, although this is almost invariably the choice we make for y_0 in practice.

2.6 Order and error constant

In this section we are in effect formalizing the Taylor series method, described in section 2.2, for the derivation of linear multistep methods.

With the linear multistep method (2), we associate the linear difference operator $\mathscr{L}$ defined by

$$\mathscr{L}[y(x); h] = \sum_{j=0}^{k} [\alpha_j y(x + jh) - h\beta_j y'(x + jh)], \tag{17}$$

where $y(x)$ is an arbitrary function, continuously differentiable on $[a, b]$. The reason for introducing this operator is that, by allowing it to operate on an *arbitrary* test function $y(x)$, which we may assume to have as many higher derivatives as we require, we can formally define the order of accuracy of the operator and of the associated linear multistep method, without invoking the solution of the initial value problem (1) which, as we have already observed, may possess only a first derivative. Expanding the test function $y(x + jh)$ and its derivative $y'(x + jh)$ as Taylor series about x, and collecting terms in (17) gives

$$\mathscr{L}[y(x); h] = C_0 y(x) + C_1 h y^{(1)}(x) + \ldots + C_q h^q y^{(q)}(x) + \ldots, \tag{18}$$

where the C_q are constants.

Definition *The difference operator (17) and the associated linear multistep method (2) are said to be of* **order** p *if, in (18),* $C_0 = C_1 = \ldots = C_p = 0$, $C_{p+1} \neq 0$.

A simple calculation yields the following formulae for the constants C_q in terms of the coefficients α_j, β_j:

$$\begin{aligned} C_0 &= \alpha_0 + \alpha_1 + \alpha_2 + \ldots + \alpha_k, \\ C_1 &= \alpha_1 + 2\alpha_2 + \ldots + k\alpha_k - (\beta_0 + \beta_1 + \beta_2 + \ldots + \beta_k), \\ C_q &= \frac{1}{q!}(\alpha_1 + 2^q\alpha_2 + \ldots + k^q\alpha_k) \\ &\quad - \frac{1}{(q-1)!}(\beta_1 + 2^{q-1}\beta_2 + \ldots + k^{q-1}\beta_k), \qquad q = 2, 3, \ldots. \end{aligned} \tag{19}$$

These formulae can be used to derive a linear multistep method of given structure and maximal order.

Example 1 Construct an implicit linear two-step method of maximal order, containing one free parameter, and find its order.

$k = 2$; $\alpha_2 = +1$, by hypothesis. Let $\alpha_0 = a$ be the free parameter. There remain four undetermined coefficients $\alpha_1, \beta_0, \beta_1$, and β_2, and we can thus set $C_0 = C_1 = C_2 = C_3 = 0$.

$$C_0 = a + \alpha_1 + 1 = 0,$$

$$C_1 = \alpha_1 + 2 - (\beta_0 + \beta_1 + \beta_2) = 0,$$

$$C_2 = \frac{1}{2!}(\alpha_1 + 4) - (\beta_1 + 2\beta_2) = 0,$$

$$C_3 = \frac{1}{3!}(\alpha_1 + 8) - \frac{1}{2!}(\beta_1 + 4\beta_2) = 0.$$

Solving this set of equations gives

$$\alpha_1 = -1 - a, \quad \beta_0 = -\tfrac{1}{12}(1 + 5a), \quad \beta_1 = \tfrac{2}{3}(1 - a), \quad \beta_2 = \tfrac{1}{12}(5 + a),$$

and the method is

$$y_{n+2} - (1 + a)y_{n+1} + ay_n = \frac{h}{12}[(5 + a)f_{n+2} + 8(1 - a)f_{n+1} - (1 + 5a)f_n]. \tag{20}$$

Moreover,

$$C_4 = \frac{1}{4!}(\alpha_1 + 16) - \frac{1}{3!}(\beta_1 + 8\beta_2) = \frac{-1}{4!}(1 + a),$$

$$C_5 = \frac{1}{5!}(\alpha_1 + 32) - \frac{1}{4!}(\beta_1 + 16\beta_2) = -\frac{1}{3.5!}(17 + 13a).$$

If $a \neq -1$, then $C_4 \neq 0$, and method (20) is of order 3.
If $a = -1$, then $C_4 = 0$, $C_5 \neq 0$, and method (20), which is now Simpson's rule, is of order 4.
Note that when $a = 0$, (20) is the two-step Adams–Moulton method (13), while if $a = -5$, it is an explicit method.

We turn now to the problem, mentioned earlier, of the origin of the Taylor expansions used to derive (18). Suppose we choose to expand $y(x + jh)$ and $y'(x + jh)$ about $x + th$, where t need not necessarily be an integer. In place of (18) we obtain

$$\mathscr{L}[y(x); h] = D_0 y(x + th) + D_1 h y^{(1)}(x + th) + \ldots + D_q h^q y^{(q)}(x + th) + \ldots. \tag{21}$$

The right-hand side can be written as an expression of the exact form of the right-hand side of (18) by employing the Taylor expansions

$$y^{(q)}(x + th) = y^{(q)}(x) + thy^{(q+1)}(x) + \ldots + \frac{t^s h^s}{s!} y^{(q+s)}(x) + \ldots, \quad q = 0, 1, 2, \ldots,$$

where $y^{(0)}(x) \equiv y(x)$. Making this substitution and equating term by term with (18) gives

$$
\begin{aligned}
C_0 &= D_0,\\
C_1 &= D_1 + tD_0,\\
C_2 &= D_2 + tD_1 + \frac{t^2}{2!}D_0,\\
&\vdots\\
C_p &= D_p + tD_{p-1} + \dots + \frac{t^p}{p!}D_0,\\
C_{p+1} &= D_{p+1} + tD_p + \dots + \frac{t^{p+1}}{(p+1)!}D_0,\\
C_{p+2} &= D_{p+2} + tD_{p+1} + \dots + \frac{t^{p+2}}{(p+2)!}D_0.
\end{aligned}
\tag{22}
$$

It follows that $C_0 = C_1 = \dots = C_p = 0$ if and only if $D_0 = D_1 = \dots = D_p = 0$. Hence we can define the order of $\mathscr{L}$ to be p if precisely the first $p + 1$ coefficients in the expansion (21) vanish, and this definition is independent of the choice for t. Moreover, if $C_0 = C_1 = \dots = C_p = 0$, then, from (22), $D_{p+1} = C_{p+1}$, $D_{p+2} = C_{p+2} - tC_{p+1}$. Hence the first non-vanishing coefficient in the expansion (21) is independent of the choice for t, but subsequent coefficients are functions of t.

The formulae giving the constants D_q, defined by (21), in terms of the coefficients α_j, β_j are

$$
\begin{aligned}
D_0 &= \alpha_0 + \alpha_1 + \alpha_2 + \dots + \alpha_k,\\
D_1 &= -t\alpha_0 + (1-t)\alpha_1 + (2-t)\alpha_2 + \dots + (k-t)\alpha_k\\
&\qquad -(\beta_0 + \beta_1 + \beta_2 + \dots + \beta_k),\\
D_q &= \frac{1}{q!}[(-t)^q\alpha_0 + (1-t)^q\alpha_1 + (2-t)^q\alpha_2 + \dots + (k-t)^q\alpha_k]\\
&\qquad -\frac{1}{(q-1)!}[(-t)^{q-1}\beta_0 + (1-t)^{q-1}\beta_1\\
&\qquad + (2-t)^{q-1}\beta_2 + \dots + (k-t)^{q-1}\beta_k], \qquad q = 2, 3, \dots.
\end{aligned}
\tag{23}
$$

It is seen that these formulae revert to (19) in the case $t = 0$. The labour in deriving linear multistep methods can sometimes be reduced by a judicious choice for t.

Example 2 Re-derive the method of example 1 by using Taylor expansion about $x + h$.

$k = 2$; $\alpha_2 = +1$, by hypothesis; $\alpha_0 = a$. Setting $t = 1$ in (23) gives

$$D_0 = a + \alpha_1 + 1,$$

$$D_1 = -a + 1 - (\beta_0 + \beta_1 + \beta_2),$$

$$D_2 = \frac{1}{2!}(a + 1) - (-\beta_0 + \beta_2),$$

$$D_3 = \frac{1}{3!}(-a + 1) - \frac{1}{2!}(\beta_0 + \beta_2),$$

$$D_4 = \frac{1}{4!}(a + 1) - \frac{1}{3!}(-\beta_0 + \beta_2),$$

$$D_5 = \frac{1}{5!}(-a + 1) - \frac{1}{4!}(\beta_0 + \beta_2).$$

Solving the set of equations $D_j = 0, j = 0, 1, 2, 3$, is slightly easier than solving the corresponding set in example 1, for now α_1, as well as β_1, appears in only one equation. It is readily established that the solution coincides with that found in example 1. However, it is now possible to draw conclusions about attainable order without first solving these equations. From inspection of the equations $D_2 = D_3 = 0$, it is clear that a solution for the set of equations $D_j = 0, j = 0, 1, 2, 3$, exists for all a, so that an order of at least 3 is always attainable. Moreover, by inspection it is clear that D_2 and D_4 cannot both vanish unless $a = -1$, so that when $a \neq -1$, the maximum attainable order is 3. When $a = -1$, $D_2 = 0$ implies $D_4 = 0$, so that order of at least 4 is attainable. That order 5 is not attainable when $a = -1$ follows from observing that D_3 and D_5 cannot both vanish unless $a = +1$.

Observe, also, that with the values for $\alpha_1, \beta_0, \beta_1$, and β_2 which satisfy the equations $D_j = 0, j = 0, 1, 2, 3$, we find that $D_4 = (-1/4!)(a + 1)$ and $D_5 = (2/3.5!) \times (a - 1)$. Thus in the case when the order p is 3, $D_4 = C_4$ but $D_5 \neq C_5$, whereas in the case $a = -1$, when the order p is 4, $D_4 = C_4 = 0$ and $D_5 = C_5$, demonstrating that the coefficient C_{p+1}, where p is the order, is independent of the choice for t.

From the above discussion it is clear that only the first of the non-vanishing coefficients in the expansion (18), namely C_{p+1}, has any absolute significance. We shall call C_{p+1} the *error constant*. We note that, had we not made the stipulation $\alpha_k = +1$, then by multiplying across (2) by an arbitrary constant—which would not alter the performance of the linear multistep method—it would have been possible arbitrarily to change the value of the constant C_{p+1} and thus rob it of its significance. Our fixing of α_k precludes this possibility, but some authors, who do not assume that $\alpha_k = +1$, define the error constant to be C_{p+1} multiplied by some normalizing factor; thus Henrici[67] takes the error constant to be $C_{p+1}/\sum_{j=0}^{k} \beta_j$.

Exercises

5. Find the order and error constant of *Quade's method*,

$$y_{n+4} - \frac{8}{19}(y_{n+3} - y_{n+1}) - y_n = \frac{6h}{19}(f_{n+4} + 4f_{n+3} + 4f_{n+1} + f_n).$$

What is the most efficient value for t to use in this exercise?

6. Show that the operator $\mathscr{L}$ defined by (17), and hence the associated linear multistep method (2), have order p if and only if

$$\mathscr{L}[x^r; h] \equiv 0, \qquad r = 0, 1, \ldots, p, \qquad \mathscr{L}[x^{p+1}; h] \not\equiv 0.$$

Show also that the error constant C_{p+1} is then given by

$$(p + 1)!h^{p+1}C_{p+1} = \mathscr{L}[x^{p+1}; h].$$

2.7 Local and global truncation error

The phrase 'local truncation error', used without definition in section 2.2, can be made precise as follows:

Definition *The* **local truncation error** *at* x_{n+k} *of the method (2) is defined to be the expression* $\mathscr{L}[y(x_n); h]$ *given by (17), when* $y(x)$ *is the theoretical solution of the initial value problem (1).*

We shall occasionally use the notation T_{n+k} to denote the local truncation error at x_{n+k}. Note that this definition holds for all initial value problems of the class (1), since (17) requires only that $y(x) \in C^1[a, b]$, and this is so when $y(x)$ is the solution of (1). On the other hand, expansion (18) will not hold, in general, when $y(x)$ is the solution of (1).

The truncation error T_{n+k} is *local* in the following sense. Consider the application of (2) to yield y_{n+k} under the simplifying assumption (henceforth called 'the localizing assumption') that *no previous truncation errors have been made*. In particular, assume that $y_{n+j} = y(x_{n+j})$, $j = 0, 1, \ldots, k - 1$. From (17)

$$\begin{aligned}\sum_{j=0}^{k} \alpha_j y(x_n + jh) &= h \sum_{j=0}^{k} \beta_j y'(x_n + jh) + \mathscr{L}[y(x_n); h] \\ &= h \sum_{j=0}^{k} \beta_j f(x_n + jh, y(x_n + jh)) + \mathscr{L}[y(x_n); h],\end{aligned}$$

since, in this context, $y(x)$ is taken to be the exact solution of (1). The value for y_{n+k} given by (2) satisfies

$$\sum_{j=0}^{k} \alpha_j y_{n+j} = h \sum_{j=0}^{k} \beta_j f(x_{n+j}, y_{n+j}).$$

Subtracting, and using the localizing assumption stated above, gives

$$y(x_{n+k}) - y_{n+k} = h\beta_k[f(x_{n+k}, y(x_{n+k})) - f(x_{n+k}, y_{n+k})] + \mathscr{L}[y(x_n); h].$$

By the mean value theorem,

$$f(x_{n+k}, y(x_{n+k})) - f(x_{n+k}, y_{n+k}) = [y(x_{n+k}) - y_{n+k}]\frac{\partial f(x_{n+k}, \eta_{n+k})}{\partial y}$$

where η_{n+k} is an interior point of the interval whose end-points are y_{n+k} and $y(x_{n+k})$. Hence

$$\left[1 - h\beta_k\frac{\partial f(x_{n+k}, \eta_{n+k})}{\partial y}\right][y(x_{n+k}) - y_{n+k}] = \mathscr{L}[y(x_n); h] = T_{n+k}. \quad (24)$$

Thus for an explicit method (that is, one for which $\beta_k = 0$), the local truncation error is the difference between the theoretical solution and the solution given by the linear multistep method under the localizing assumption stated in italics; for an implicit method, the local truncation error is (approximately) proportional to this difference. If we make the further assumption that the theoretical solution $y(x)$ has continuous derivatives of sufficiently high order, then, for both explicit and implicit methods, we deduce from (24) that

$$y(x_{n+k}) - y_{n+k} = C_{p+1}h^{p+1}y^{(p+1)}(x_n) + O(h^{p+2}), \quad (25)$$

where p is the order of the method. The term $C_{p+1}h^{p+1}y^{(p+1)}(x_n)$, frequently called the *principal local truncation error*, is known if the order p, and the error constant C_{p+1} of the method (2) are known. We prefer to indicate local accuracy by specifying order and error constant, since these entities, unlike principal local truncation error, are defined solely in terms of the coefficients of the method (2) and do not require us to make restrictive assumptions about the existence of higher derivatives of the theoretical solution $y(x)$ of the initial value problem.

It must be emphasized that results (24) and (25) hold only when the simplifying—and unrealistic—localizing assumption italicized above holds. If we make no such assumption, then the error $y(x_{n+k}) - y_{n+k} = e_{n+k}$ is the *global* or *accumulated truncation error*. This error involves all the truncation errors made at each application of the method, and depends in a complicated way on the coefficients of the method and on the initial value problem. It is *this* error which convergence demands shall tend to zero as $h \to 0$, $n \to \infty$, $nh = x_n - a$ remaining fixed.

In chapter 3 we shall discuss local truncation error further and consider the problem of establishing bounds for both local and global truncation errors.

Example 3 Verify, by direct calculation, the validity of (24) using (i) Euler's rule and (ii) the Trapezoidal rule, to solve the initial value problem $y' = Ay$, $y(0) = 1$.

The theoretical solution is $y(x) = \exp(Ax)$.
(i) For Euler's rule it is easily established that

$$\mathscr{L}[y(x);h] = \frac{1}{2!}h^2y^{(2)} + \frac{1}{3!}h^3y^{(3)} + \ldots + \frac{1}{q!}h^qy^{(q)} + \ldots.$$

Substituting $\exp(Ax)$ for $y(x)$ gives

$$\begin{aligned}\mathscr{L}[y(x_n):h] &= [\exp(nhA)]\left[\frac{1}{2!}h^2A^2 + \frac{1}{3!}h^3A^3 + \ldots + \frac{1}{q!}h^qA^q + \ldots\right]\\ &= [\exp(nhA)][\exp(hA) - 1 - hA]\end{aligned}$$

Also

$$y_{n+1} = y_n + hAy_n = (1 + hA)\exp(nhA),$$

by the localizing assumption that $y_n = y(x_n)$. Hence

$$\begin{aligned}y(x_{n+1}) - y_{n+1} &= \exp[(n+1)hA] - (1 + hA)\exp(nhA)\\ &= [\exp(nhA)][\exp(hA) - 1 - hA] = \mathscr{L}[y(x_n);h].\end{aligned}$$

Since $\beta_2 = 0$ for Euler's rule, the validity of (24) is verified.
(ii) For the Trapezoidal rule, we find

$$\begin{aligned}\mathscr{L}[y(x);h] = &\left(\frac{1}{3!} - \frac{1}{2.2!}\right)h^3y^{(3)} + \left(\frac{1}{4!} - \frac{1}{2.3!}\right)h^4y^{(4)} + \ldots\\ &+ \left(\frac{1}{q!} - \frac{1}{2.(q-1)!}\right)h^qy^{(q)} + \ldots.\end{aligned}$$

Substituting $\exp(Ax)$ for $y(x)$ gives

$$\begin{aligned}\mathscr{L}[y(x_n);h] &= [\exp(nhA)]\left[\left(\frac{1}{3!} - \frac{1}{2.2!}\right)h^3A^3 + \left(\frac{1}{4!} - \frac{1}{2.3!}\right)h^4A^4 + \ldots\right.\\ &\qquad\left. + \left(\frac{1}{q!} - \frac{1}{2(q-1)!}\right)h^qA^q + \ldots\right]\\ &= [\exp(nhA)]\left\{\exp(hA) - 1 - hA - \frac{h^2A^2}{2}\right.\\ &\qquad\left. - \tfrac{1}{2}hA[\exp(hA) - 1 - hA]\right\}\\ &= [\exp(nhA)][(1 - \tfrac{1}{2}hA)\exp(hA) - (1 + \tfrac{1}{2}hA)].\end{aligned}$$

Also

$$y_{n+1} - y_n = \tfrac{1}{2}hA(y_{n+1} + y_n)$$

or $(1 - \frac{1}{2}hA)y_{n+1} = (1 + \frac{1}{2}hA)\exp(nhA)$, by the localizing assumption that $y_n = y(x_n)$. Hence

$$\begin{aligned}(1 - \tfrac{1}{2}hA)[y(x_{n+1}) - y_{n+1}] &= (1 - \tfrac{1}{2}hA)\exp[(n+1)hA] - (1 + \tfrac{1}{2}hA)\exp(nhA) \\ &= [\exp(nhA)][(1 - \tfrac{1}{2}hA)\exp(hA) - (1 + \tfrac{1}{2}hA)] \\ &= \mathscr{L}[y(x_n); h],\end{aligned}$$

verifying (24).

2.8 Consistency and zero-stability

Definition The linear multistep method (2) *is said to be* **consistent** *if it has order* $p \geqslant 1$.

From (19) it follows that the method (2) is consistent if and only if

$$\sum_{j=0}^{k} \alpha_j = 0; \qquad \sum_{j=0}^{k} j\alpha_j = \sum_{j=0}^{k} \beta_j. \tag{26}$$

The following heuristic argument indicates the significance of the property of consistency, and, incidentally explains why the property is so named. Throughout this argument, we shall consider only limits in the sense used in the definition of convergence, namely limits as $h \to 0$, $n \to \infty$, $nh = x - a$ remaining fixed. Let us assume that, in this limit, $y_n \to y(x)$. Since k is fixed, we also have that $y_{n+j} \to y(x), j = 0, 1, \ldots, k$ or

$$y(x) = y_{n+j} + \theta_{j,n}(h), \qquad j = 0, 1, \ldots, k,$$

where $\lim \theta_{j,n}(h) = 0$, $j = 0, 1, \ldots, k$.

Hence

$$\sum_{j=0}^{k} \alpha_j y(x) = \sum_{j=0}^{k} \alpha_j y_{n+j} + \sum_{j=0}^{k} \alpha_j \theta_{j,n}(h),$$

or

$$y(x) \sum_{j=0}^{k} \alpha_j = h \sum_{j=0}^{k} \beta_j f_{n+j} + \sum_{j=0}^{k} \alpha_j \theta_{j,n}(h).$$

In the limit, both terms on the right-hand side vanish, whereas the left-hand side is unaffected by the limiting process and must, therefore, be zero. Since $y(x)$ is not in general zero, we conclude that $\sum_{j=0}^{k} \alpha_j = 0$, and this is the first of the conditions (26). It should be observed that we have not made any use of the differential equation so far. Indeed, the above argument holds if we merely assume that $\{y_n\}$ tends to *some* function $y(x)$.

The second condition of (26) ensures that the function $y(x)$ does in fact satisfy the differential equation. For, under the limiting process,

$$(y_{n+j} - y_n)/jh \to y'(x), \qquad j = 1, 2, \ldots, k,$$

or

$$y_{n+j} - y_n = jhy'(x) + jh\phi_{j,n}(h), \qquad j = 1, 2, \ldots, k,$$

where $\lim \phi_{j,n}(h) = 0$.

Hence

$$\sum_{j=0}^{k} \alpha_j y_{n+j} - \sum_{j=0}^{k} \alpha_j y_n = h \sum_{j=0}^{k} j\alpha_j y'(x) + h \sum_{j=0}^{k} j\alpha_j \phi_{j,n}(h),$$

or

$$h \sum_{j=0}^{k} \beta_j f_{n+j} - y_n \sum_{j=0}^{k} \alpha_j = hy'(x) \sum_{j=0}^{k} j\alpha_j + h \sum_{j=0}^{k} j\alpha_j \phi_{j,n}(h).$$

Since $\sum_{j=0}^{k} \alpha_j = 0$, we have, on dividing through by h,

$$\sum_{j=0}^{k} \beta_j f_{n+j} = y'(x) \sum_{j=0}^{k} j\alpha_j + \sum_{j=0}^{k} j\alpha_j \phi_{j,n}(h).$$

Under the limiting process, $f_{n+j} \to f(x, y(x))$, and hence, in the limit,

$$f(x, y(x)) \sum_{j=0}^{k} \beta_j = y'(x) \sum_{j=0}^{k} j\alpha_j.$$

Thus $y(x)$ satisfies the differential equation (1) if and only if $\sum_{j=0}^{k} j\alpha_j = \sum_{j=0}^{k} \beta_j$, the second of the conditions (26). The above argument has proved—albeit non-rigorously—that if the sequence $\{y_n\}$ converges to the solution of the initial value problem (1) then the conditions (26) must hold, that is, a convergent linear multistep method is necessarily consistent. We shall see presently that consistency alone is not sufficient for convergence.

We now introduce the *first* and *second characteristic polynomials* of the linear multistep method (2), defined as $\rho(\zeta)$ and $\sigma(\zeta)$ respectively, where

$$\begin{aligned} \rho(\zeta) &= \sum_{j=0}^{k} \alpha_j \zeta^j, \\ \sigma(\zeta) &= \sum_{j=0}^{k} \beta_j \zeta^j. \end{aligned} \tag{27}$$

It follows from (26) that the linear multistep method is consistent if and only if

$$\rho(1) = 0, \qquad \rho'(1) = \sigma(1). \tag{28}$$

Thus, for a consistent method, the first characteristic polynomial $\rho(\zeta)$ always has a root at $+1$. We shall call this root the *principal root* and always label it ζ_1. The remaining roots, ζ_s, $s = 2, 3, \ldots, k$, are *spurious roots* and arise only when the stepnumber of the method is greater than one, that is, when we choose to replace a first-order differential equation by a difference equation of order greater than one. It is not surprising to find that the location of these spurious roots must be carefully controlled if the method is to be convergent.

Consider the trivial initial value problem $y' = 0$, $y(0) = 0$, whose solution is $y(x) \equiv 0$. The linear multistep method (2) applied to this problem gives the difference equation $\sum_{j=0}^{k} \alpha_j y_{n+j} = 0$. Let us consider the case when all the roots ζ_s of $\rho(\zeta)$ are real and distinct. Then, by section 1.7, a solution of the difference equation for y_n which satisfies the requirement that the starting values $y_\mu \to y(0)$ as $h \to 0$, $\mu = 0, 1, \ldots, k-1$, is

$$y_n = h(d_1\zeta_1^n + d_2\zeta_2^n + \ldots + d_k\zeta_k^n),$$

where the d_s are arbitrary constants, and we recall that $\zeta_1 = +1$. If the method is to be convergent we must have that $y_n \to 0$ as $h \to 0$, $n \to \infty$, $nh = x$ remaining fixed. However, since

$$\lim_{\substack{h\to 0\\ nh=x}} h\zeta_s^n = x \lim_{n\to\infty} \zeta_s^n/n = 0 \quad \text{if and only if} \quad |\zeta_s| \leqslant 1,$$

it is clear that the method will not be convergent if any of the roots ζ_s has modulus greater than one. Also, consider the case when ζ_s is a real root of $\rho(\zeta)$ with multiplicity $m > 1$. Then, by section 1.7, the contribution of ζ_s to the solution for y_n is of the form $h[d_{s,1} + d_{s,2}n + d_{s,3}n(n-1) + \ldots + d_{s,m}n(n-1)\ldots(n-m+2)]\zeta_s^n$. Since for $q \geqslant 1$,

$$\lim_{\substack{h\to 0\\ nh=x}} hn^q\zeta_s^n = x \lim_{n\to\infty} n^{q-1}\zeta_s^n = 0 \quad \text{if and only if} \quad |\zeta_s| < 1,$$

it is clear that the method will not be convergent if $\rho(\zeta)$ has a multiple root of modulus greater than or equal to one. The argument extends to the case when the roots of $\rho(\zeta)$ are complex, and motivates our next definition. In passing, we note that since consistency controls only the position of the principal root and not of the spurious roots, we have demonstrated that a consistent method is not necessarily convergent.

Definition *The linear multistep method (2) is said to be* **zero-stable** *if no root of the first characteristic polynomial* $\rho(\zeta)$ *has modulus greater than one, and if every root with modulus one is simple.*

We emphasize the fact that the roots of $\rho(\zeta)$ are in general complex by remarking that for a zero-stable method all the roots of $\rho(\zeta)$ lie in or on the unit circle, those on the circle being simple.

A linear multistep method which is zero-stable is frequently called simply 'stable'.† However, the word 'stable' is used widely in the literature of the numerical solution of ordinary differential equations to mean several different things, and we shall indeed discuss other forms of stability later in this book. In an attempt to minimize confusion with the literature, we shall never use the word 'stable' without some descriptive epithet being attached to it. The word 'zero' is chosen here since the stability phenomenon under consideration at the moment is allied to the notion of convergence in the limit as h tends to zero. Zero-stability ensures that those solutions of the difference equation for y_n which arise because the first-order differential equation is being replaced by a higher order difference equation (frequently called *parasitic solutions*) are damped out in the limit as $h \to 0$.

For a one-step method, the polynomial $\rho(\zeta)$ has degree one, and if the method is consistent the only root, ζ_1, is $+1$. Thus a consistent one-step method is necessarily zero-stable. Also, for a consistent zero-stable k-step method, $\sigma(1)$ cannot vanish. For, if it did, then, by (28),

$$\rho(1) = 0 = \rho'(1),$$

which implies that $\rho(\zeta)$ has a double root at $+1$, contrary to the definition of zero-stability.

We now state the fundamental theorem of Dahlquist[35,36] on linear multistep methods.

Theorem 2.1 *The necessary and sufficient conditions for a linear multistep method to be convergent are that it be consistent and zero-stable.*

The proof of this important theorem is beyond the scope of this book; it can be found in Henrici.[67] It is no surprise that the proof is not elementary, since consistency and zero-stability are both simple algebraic properties of the coefficients α_j, β_j, whereas convergence is an analytic property involving a wide class of initial value problems. Qualitatively speaking, consistency controls the magnitude of the local truncation error

† Some authors refer to zero-stability as 'D-stability', short for 'Dahlquist-stability'; the basic theory discussed in this chapter was originally propounded by Dahlquist.[35,36]

committed at each stage of the calculation, while zero-stability controls the manner in which this error is propagated as the calculation proceeds; both are essential if convergence is to be achieved. Convergence is a minimal property which any acceptable linear multistep method must possess. Accordingly we reject out of hand, as having no practical interest, linear multistep methods which are not both consistent and zero-stable.

Example 4 *Illustrate the effect of zero-stability by using the linear multistep method*

$$y_{n+2} - (1 + a)y_{n+1} + ay_n = \tfrac{1}{2}h[(3 - a)f_{n+1} - (1 + a)f_n]$$

with (i) $a = 0$, *(ii)* $a = -5$ *to compute numerical solutions to the initial value problem*

$$y' = 4xy^{\frac{1}{2}}, \qquad y(0) = 1$$

in the interval $0 \leqslant x \leqslant 2$.

The first characteristic polynomial for this method is

$$\rho(\zeta) = \zeta^2 - (1 + a)\zeta + a = (\zeta - 1)(\zeta - a).$$

Thus, when $a = 0$ the method is zero-stable and when $a = -5$ it is zero-unstable. It is easily established that the method has order 2 if $a \neq -5$ and order 3 if $a = -5$.

The theoretical solution of the initial value problem is $y(x) = (1 + x^2)^2$. We shall take $y_0 = 1$ and, for the purposes of this illustration, we shall choose the necessary additional starting value, y_1, to coincide with the theoretical solution; that is, we set $y_1 = (1 + h^2)^2$. The numerical results when $h = 0{\cdot}1$ are given in table 1.

We observe that for the first few steps method (ii), which has order three, gives slightly more accurate results than does method (i), which has order two. However,

Table 1

$h = 0{\cdot}1$		(i)	(ii)
x	Theoretical solution	Numerical solution, $a = 0$	Numerical solution, $a = -5$
0	1·000,000,0	1·000,000,0	1·000,000,0
0·1	1·020,100,0	1·020,100,0	1·020,100,0
0·2	1·081,600,0	1·080,700,0	1·081,200,0
0·3	1·188,100,0	1·185,248,1	1·189,238,5
0·4	1·345,600,0	1·339,629,8	1·338,866,0
0·5	1·562,500,0	1·552,090,0	1·592,993,5
⋮	⋮	⋮	⋮
1·0	4·000,000,0	3·940,690,3	−68·639,804
1·1	4·884,100,0	4·808,219,7	+367·263,92
⋮	⋮	⋮	⋮
2·0	25·000,000	24·632,457	$-6{\cdot}96 \times 10^8$

very quickly the zero-instability of method (ii) manifests itself in the form of a violently growing error, which renders the numerical results completely meaningless. The effect is not lessened by reducing the steplength, as the calculations with $h = 0{\cdot}05$ and $h = 0{\cdot}025$, displayed in table 2, show. As the steplength is reduced, the solution at any fixed station x improves with method (i) and gets rapidly worse with method (ii). This is in agreement with theorem 2.1 since method (i), being consistent and zero-stable, is convergent, while method (ii), being zero-unstable, is divergent.

Table 2

$h = 0{\cdot}05$		(i)	(ii)
x	Theoretical solution	Numerical solution, $a = 0$	Numerical solution, $a = -5$
0	1·000,000,0	1·000,000,0	1·000,000,0
0·1	1·020,100,0	1·020,043,7	1·020,075,0
0·2	1·081,600,0	1·081,239,3	1·081,157,8
0·3	1·188,100,0	1·187,162,0	1·177,715,0
0·4	1·345,600,0	1·343,780,8	1·095,852,1
0·5	1·562,500,0	1·559,454,9	−4·430,547,9
⋮	⋮	⋮	⋮
1·0	4·000,000,0	3·983,891,3	$-5{\cdot}730 \times 10^{7}$
1·1	4·884,100,0	4·863,623,9	$-1{\cdot}432 \times 10^{9}$
⋮	⋮	⋮	⋮
2·0	25·000,000	24·903,668	$-5{\cdot}464 \times 10^{21}$

$h = 0{\cdot}025$		(i)	(ii)
x	Theoretical solution	Numerical solution, $a = 0$	Numerical solution, $a = -5$
0	1·000,000,0	1·000,000,0	1·000,000,0
0·1	1·020,100,0	1·020,077,7	1·020,072,0
0·2	1·081,600,0	1·081,491,5	1·065,009,7
0·3	1·188,100,0	1·187,836,2	−8·915,658,3
0·4	1·345,600,0	1·345,103,6	−6289·8299
0·5	1·562,500,0	1·561,683,2	−3,932,119·5
⋮	⋮	⋮	⋮
1·0	4·000,000,0	3·995,817,9	$-3{\cdot}750 \times 10^{20}$
1·1	4·884,100,0	4·878,799,3	$-2{\cdot}344 \times 10^{23}$
⋮	⋮	⋮	⋮
2·0	25·000,000	24·975,378	$-3{\cdot}411 \times 10^{48}$

Example 5 *Illustrate the effect of inconsistency by using the linear multistep method*

$$y_{n+2} - y_{n+1} = \tfrac{1}{3}h(3f_{n+1} - 2f_n)$$

to compute a numerical solution for the initial value problem of example 4.

For this method,

$$\sum_{j=0}^{k} \alpha_j = 0, \qquad \sum_{j=0}^{k} j\alpha_j = 1, \qquad \sum_{j=0}^{k} \beta_j = \tfrac{1}{3},$$

so that the first of the consistency conditions (26) is satisfied, but the second is not. Since the first characteristic polynomial ρ is given by

$$\rho(\zeta) = \zeta^2 - \zeta = \zeta(\zeta - 1),$$

the method is zero-stable. As in example 4, we take the starting values y_0 and y_1 to coincide with the theoretical solution. Computing with $h = 0{\cdot}1$, $0{\cdot}05$, and $0{\cdot}025$ yields the results given in table 3.

Table 3

x	Theoretical solution	Numerical solution, $h = 0{\cdot}1$	Numerical solution, $h = 0{\cdot}05$	Numerical solution, $h = 0{\cdot}025$
0	1·000,000,0	1·000,000,0	1·000,000,0	1·000,000,0
0·1	1·020,100,0	1·020,100,0	1·015,031,2	1·011,280,6
0·2	1·081,600,0	1·060,500,0	1·045,489,0	1·036,622,5
0·3	1·188,100,0	1·115,951,1	1·090,465,4	1·076,179,4
0·4	1·345,600,0	1·187,794,5	1·150,971,1	1·130,790,3
0·5	1·562,500,0	1·277,661,2	1·228,294,0	1·201,566,2
⋮	⋮	⋮	⋮	⋮
1·0	4·000,000,0	2·075,163,2	1·931,864,0	1·855,935,3
1·1	4·884,100,0	2·324,034,2	2·152,853,2	2·062,348,7
⋮	⋮	⋮	⋮	⋮
2·0	25·000,000	6·803,217,8	6·140,889,8	5·795,792,9

As the steplength decreases, the numerical solution moves further away from the theoretical solution. Since the first of the consistency conditions (26) is satisfied, the calculated solution $\{y_n\}$ will converge to a function $y(x)$ as the steplength tends to zero, but this function $y(x)$ will not be the solution of the initial value problem. We observe that the failure of this inconsistent method is less dramatic than that of the zero-unstable method of example 4. This is because our inconsistent method is zero-stable and thus does not propagate errors in an explosive manner. The inconsistency implies that there is insufficient local accuracy for the numerical solution to remain related to the theoretical solution as the steplength decreases.

Exercises

7. Investigate the zero-stability of Quade's formula, stated in exercise 5 (page 27).

8. Show that the order of the linear multistep method

$$y_{n+2} - y_{n+1} = \frac{h}{12}(4f_{n+2} + 8f_{n+1} - f_n)$$

is zero. By finding the exact solution (satisfying appropriate starting values) of the difference equation which arises when this method is applied to the initial value problem $y' = 1$, $y(0) = 0$, demonstrate that the method is indeed divergent.

9. Show that the order of the linear multistep method

$$y_{n+2} + (b - 1)y_{n+1} - by_n = \tfrac{1}{4}h[(b + 3)f_{n+2} + (3b + 1)f_n]$$

is 2 if $b \neq -1$ and is 3 if $b = -1$. Show that the method is zero-unstable if $b = -1$. Illustrate the resulting divergence of the method with $b = -1$ by applying it to the initial value problem $y' = y$, $y(0) = 1$ and solving exactly the resulting difference equation when the starting values are $y_0 = 1$, $y_1 = 1$. (Note that these values satisfy the restriction on starting values stated in the definition of convergence.)

10. A given linear multistep method is defined by its first and second characteristic polynomials $\rho(\zeta)$ and $\sigma(\zeta)$. Sequences of polynomials $\{\rho_j(\zeta)\}$ and $\{\sigma_j(\zeta)\}, j = 1, 2, 3, \ldots$ are constructed as follows:

$$\rho_1(\zeta) = \rho(\zeta); \qquad \sigma_1(\zeta) = \sigma(\zeta)$$

$$\rho_{j+1}(\zeta) = \zeta\rho_j'(\zeta); \qquad \sigma_{j+1}(\zeta) = \zeta\sigma_j'(\zeta), \qquad j = 1, 2, \ldots.$$

Prove that the given linear multistep method will have order p if and only if

$$\rho_1(1) = 0, \qquad \rho_{j+1}(1) = j\sigma_j(1), \qquad j = 1, 2, \ldots, p \quad \text{and} \quad \rho_{p+2}(1) \neq (p + 1)\sigma_{p+1}(1).$$

Use this technique to verify the answer to exercise 5 (page 27).

2.9 Attainable order of zero-stable methods

For a given stepnumber k, we are usually interested in choosing coefficients α_j, β_j which result in the method (2) having a reasonably high order—if not the highest order possible. In seeking high order, we automatically satisfy the condition of consistency but meet a very real barrier in the condition of zero-stability.

The general k-step method has $2k + 2$ coefficients $\alpha_j, \beta_j, j = 0, 1, \ldots, k$, of which one, α_k is specified to be $+1$. There are thus $2k + 1$ free parameters and this number will be further reduced to $2k$ if we demand that the method be explicit. From (19) it is clear that for the method to have order p, $p + 1$ linear conditions must be satisfied by these parameters. Thus the highest order we can expect from a k-step method is $2k$, if the method is implicit, and $2k - 1$ if it is explicit. However, these maximal orders cannot in general be attained without violating the condition of zero-stability, as the following theorem shows.

Theorem 2.2 *No zero-stable linear multistep method of stepnumber k can have order exceeding $k + 1$ when k is odd, or exceeding $k + 2$ when k is even.*

For a proof of this theorem, the reader is again referred to Henrici.[67]

A zero-stable linear k-step method which has order $k + 2$ is called an *optimal* method. It can be shown that for an optimal method, all the roots of $\rho(\zeta)$ lie on the unit circle. It is of interest to note the unique position that Simpson's rule (12) occupies. It has stepnumber 2 and order 4 and thus is both maximal, since it has order $2k$, and optimal, since it has order $k + 2$. It is the only linear multistep method to possess both these properties. One is tempted to conclude that a zero-stable method with stepnumber 3 is of no interest since its order cannot exceed 4, and this is already achieved by Simpson's rule. However, we shall see in chapter 3 that Simpson's rule suffers certain computational disadvantages which render it unsuitable as a general-purpose method. These disadvantages are indeed shared by all optimal methods and, as a result, methods with odd stepnumber are not as unattractive as theorem 2.2 might suggest.

Example 6 *Find the optimal linear multistep method with stepnumber* 4.

For an optimal method, all the roots of $\rho(\zeta)$ are on the unit circle. Since $\rho(\zeta)$ is a polynomial of degree 4 and has, by consistency, one real root at $+1$, it must have another real root on the unit circle, and this can only be at -1 if the method is to be zero-stable. The two remaining roots must be complex. Hence we have

$$\zeta_1 = +1, \qquad \zeta_2 = -1, \qquad \zeta_3 = \exp(i\phi), \qquad \zeta_4 = \exp(-i\phi), \qquad 0 < \phi < \pi.$$

Hence $\rho(\zeta) = (\zeta - 1)(\zeta + 1)(\zeta - \exp(i\phi))(\zeta - \exp(-i\phi))$ (since the coefficient of ζ^4 is $+1$ by definition). Thus

$$\rho(\zeta) = \zeta^4 - 2\cos\phi\,\zeta^3 + 2\cos\phi\,\zeta - 1.$$

Hence, putting $\cos\phi = \mu$ we obtain

$$\alpha_4 = +1, \qquad \alpha_3 = -2\mu, \qquad \alpha_2 = 0, \qquad \alpha_1 = 2\mu, \qquad \alpha_0 = -1.$$

We now require the method to have order $k + 2 = 6$. In view of the symmetry of the values we have obtained for the α_j, it is convenient to state the order requirement in terms of the coefficients D_q given by (23) rather than in terms of the C_q given by (19).

Putting $t = 2$ in (23) yields

$$D_0 = \alpha_0 + \alpha_1 + \alpha_2 + \alpha_3 + \alpha_4 = 0,$$

$$D_1 = -2\alpha_0 - \alpha_1 + \alpha_3 + 2\alpha_4 - (\beta_0 + \beta_1 + \beta_2 + \beta_3 + \beta_4),$$

$$D_q = \frac{1}{q!}[(-2)^q\alpha_0 + (-1)^q\alpha_1 + \alpha_3 + 2^q\alpha_4]$$

$$-\frac{1}{(q-1)!}[(-2)^{q-1}\beta_0 + (-1)^{q-1}\beta_1 + \beta_3 + 2^{q-1}\beta_4], \qquad q = 2, 3, \ldots.$$

Setting $D_q = 0, q = 2, 3, 4, 5, 6$ appears to give five equations for the four parameters $\beta_0, \beta_1, \beta_3$, and β_4. However, on inserting the values we have obtained for the α_j into these equations we find

$$-2\beta_0 - \beta_1 + \beta_3 + 2\beta_4 = 0,$$

$$2^2\beta_0 + \beta_1 + \beta_3 + 2^2\beta_4 = \tfrac{2}{3}(2^3 - 2\mu),$$

$$-2^3\beta_0 - \beta_1 + \beta_3 + 2^3\beta_4 = 0,$$

$$2^4\beta_0 + \beta_1 + \beta_3 + 2^4\beta_4 = \tfrac{2}{5}(2^5 - 2\mu),$$

$$-2^5\beta_0 - \beta_1 + \beta_3 + 2^5\beta_4 = 0,$$

and it is at once clear that we can satisfy the first, third, and last of these if we choose $\beta_1 = \beta_3, \beta_0 = \beta_4$. The remaining two equations give

$$4\beta_0 + \beta_1 = \tfrac{1}{3}(8 - 2\mu),$$

$$16\beta_0 + \beta_1 = \tfrac{1}{5}(32 - 2\mu),$$

whose solution is

$$\beta_0 = \tfrac{1}{45}(14 + \mu) = \beta_4,$$

$$\beta_1 = \tfrac{1}{45}(64 - 34\mu) = \beta_3.$$

Finally, solving $D_1 = 0$ gives

$$\beta_2 = \tfrac{1}{15}(8 - 38\mu),$$

and we have found the required method. The error constant is

$$D_7 = \frac{1}{7!}(-2^7\alpha_0 - \alpha_1 + \alpha_3 + 2^7\alpha_4) - \frac{1}{6!}(2^6\beta_0 + \beta_1 + \beta_3 + 2^6\beta_4)$$

$$= \frac{2}{7!}(2^7 - 2\mu) - \frac{2}{6!}(2^6\beta_0 + \beta_1) = -\left(\frac{16 + 5\mu}{1890}\right).$$

Since $\mu = \cos\phi, 0 < \phi < \pi$, μ is restricted to the range $-1 < \mu < 1$, so there is no allowable value for μ which will cause D_7 to vanish, that is cause the order of the method to exceed 6; this is in agreement with theorem 2.2. We observe also that there is no allowable value for μ which will cause β_4 to vanish, thus making the method explicit. The error constant is minimized by letting $\mu \to -1$, but when μ is very close to -1 the three roots $\zeta_2, \zeta_3, \zeta_4$ are nearly coincident on the unit circle, that is, nearly in a zero-unstable configuration. Since the value of the error constant fluctuates only by a factor of roughly two as μ ranges over the interval $(-1, +1)$, the choice for μ is not critical, and for simplicity we would probably choose $\mu = 0$ since this causes two coefficients, namely α_1 and α_3, to vanish. Another simplifying choice is $\mu = \frac{4}{19}$, which causes β_2 to vanish; the resulting method turns out to be Quade's method, defined in exercise 5 (page 27).

Exercise

11. Find the range of α for which the linear multistep method

$$y_{n+3} + \alpha(y_{n+2} - y_{n+1}) - y_n = \tfrac{1}{2}(3 + \alpha)h(f_{n+2} + f_{n+1})$$

is zero-stable. Show that there exists a value of α for which the method has order 4 but that if the method is to be zero-stable, its order cannot exceed 2.

2.10 Specification of linear multistep methods

In the days of desk computation, it was common practice to write the right-hand side of a linear multistep method in terms of a power series in a difference operator. A typical example is

$$y_{n+1} - y_n = h(1 - \tfrac{1}{2}\nabla - \tfrac{1}{12}\nabla^2 - \tfrac{1}{24}\nabla^3 - \ldots)f_{n+1}. \tag{29}$$

Truncating the series after two terms gives

$$y_{n+1} - y_n = \tfrac{1}{2}h(f_{n+1} + f_n),$$

which is the Trapezoidal rule (10). Truncating after three terms gives

$$y_{n+1} - y_n = \tfrac{1}{12}h(5f_{n+1} + 8f_n - f_{n-1}),$$

a method which is equivalent to (13).

One reason for expressing a numerical algorithm in a form such as (29) lay in the technique, common in desk computation, of including higher differences of f if they became significantly large at some stage of the calculation. This is equivalent to exchanging the linear multistep method for one with a higher stepnumber. Although this will not affect the zero-stability (which is a function only of the first characteristic polynomial $\rho(\zeta)$), there are, as we shall see in later chapters, other stability phenomena which are functions of the second characteristic polynomial $\sigma(\zeta)$ as well as of the first, and these will be affected if higher differences of f are arbitrarily introduced. Thus the practice can be dangerous unless supported by adequate analysis. In any event, when a digital computer is used, it is much more convenient to compute with a fixed linear multistep method and alter the steplength if a demand for greater accuracy arises at a later stage of the calculation.†

† Recent computer-oriented methods due to Gear and Krogh do, however, make use of families of linear multistep methods with variable stepnumber; these methods will be discussed in section 3.14.

The existence of formulae like (29) has resulted in 'family' names being given to classes of linear multistep methods, of different stepnumber, which share a common form for the first characteristic polynomial $\rho(\zeta)$. Thus methods for which $\rho(\zeta) = \zeta^k - \zeta^{k-1}$ are called *Adams* methods. They have the property that all the spurious roots of ρ are located at the origin; such methods are thus zero-stable. Adams methods which are explicit are called *Adams–Bashforth* methods, while those which are implicit are called *Adams–Moulton* methods. Explicit methods for which $\rho(\zeta) = \zeta^k - \zeta^{k-2}$ are called *Nyström* methods, and implicit methods with the same form for ρ are called *generalized Milne–Simpson* methods; both these families are clearly zero-stable, since they have one spurious root at -1 and the rest at the origin.

Clearly there exist many linear multistep methods which do not belong to any of the families named above. We now specify a selection of linear multistep methods by quoting coefficients α_j, β_j, $j = 0, 1, \ldots, k$ for $k = 1$, 2, 3, 4, giving explicit and implicit methods for each stepnumber. In each case we retain just enough parameters $a, b, c, \ldots$ to enable us to have complete control over the values taken by the spurious roots of $\rho(\zeta)$. These parameters must be chosen so that the spurious roots all lie in a zero-stable configuration. (All of the methods quoted are consistent, so that the principal root of $\rho(\zeta)$ is always $+1$.) The methods have the highest order that can be attained whilst retaining the given number of free parameters. The order, p, and the error constant, C_{p+1}, are quoted in each case.

Explicit methods

$k = 1$:

$$\begin{aligned} \alpha_1 &= 1, & & \\ \alpha_0 &= -1, & \beta_0 &= 1, \\ p &= 1; & C_{p+1} &= \tfrac{1}{2}. \end{aligned}$$

$k = 2$:

$$\begin{aligned} \alpha_2 &= 1, & & \\ \alpha_1 &= -1 - a, & \beta_1 &= \tfrac{1}{2}(3 - a), \\ \alpha_0 &= a, & \beta_0 &= -\tfrac{1}{2}(1 + a), \\ p &= 2; & C_{p+1} &= \tfrac{1}{12}(5 + a). \end{aligned}$$

There exists no value for a which causes the order to exceed 2 and the method to be zero-stable.

$k = 3$:

$$\begin{aligned}
&\alpha_3 = 1, \\
&\alpha_2 = -1 - a, \qquad \beta_2 = \tfrac{1}{12}(23 - 5a - b), \\
&\alpha_1 = a + b, \qquad \beta_1 = \tfrac{1}{3}(-4 - 2a + 2b), \\
&\alpha_0 = -b, \qquad \beta_0 = \tfrac{1}{12}(5 + a + 5b), \\
&p = 3; \qquad C_{p+1} = \tfrac{1}{24}(9 + a + b).
\end{aligned}$$

There exist no values for a and b which cause the order to exceed 3 and the method to be zero-stable.

$k = 4$:

$$\begin{aligned}
&\alpha_4 = 1, \\
&\alpha_3 = -1 - a, \qquad \beta_3 = \tfrac{1}{24}(55 - 9a - b - c), \\
&\alpha_2 = a + b, \qquad \beta_2 = \tfrac{1}{24}(-59 - 19a + 13b + 5c), \\
&\alpha_1 = -b - c, \qquad \beta_1 = \tfrac{1}{24}(37 + 5a + 13b - 19c), \\
&\alpha_0 = c, \qquad \beta_0 = \tfrac{1}{24}(-9 - a - b - 9c), \\
&p = 4; \qquad C_{p+1} = \tfrac{1}{720}(251 + 19a + 11b + 19c).
\end{aligned}$$

There exist no values for a, b, and c which cause the order to exceed 4 and the method to be zero-stable.

Implicit methods

$k = 1$:

$$\begin{aligned}
&\alpha_1 = 1, \qquad \beta_1 = \tfrac{1}{2}, \\
&\alpha_0 = -1, \qquad \beta_0 = \tfrac{1}{2}, \\
&p = 2; \qquad C_{p+1} = -\tfrac{1}{12}.
\end{aligned}$$

$k = 2$:

$$\begin{aligned}
&\alpha_2 = 1, \qquad \beta_2 = \tfrac{1}{12}(5 + a), \\
&\alpha_1 = -1 - a, \qquad \beta_1 = \tfrac{2}{3}(1 - a), \\
&\alpha_0 = a, \qquad \beta_0 = \tfrac{1}{12}(-1 - 5a).
\end{aligned}$$

If $a \neq -1$, then $p = 3$; $C_{p+1} = -\frac{1}{24}(1 + a)$.
If $a = -1$, then $p = 4$; $C_{p+1} = -\frac{1}{90}$.

$k = 3$:

$$\begin{aligned}
\alpha_3 &= 1, & \beta_3 &= \tfrac{1}{24}(9 + a + b), \\
\alpha_2 &= -1 - a, & \beta_2 &= \tfrac{1}{24}(19 - 13a - 5b), \\
\alpha_1 &= a + b, & \beta_1 &= \tfrac{1}{24}(-5 - 13a + 19b), \\
\alpha_0 &= -b, & \beta_0 &= \tfrac{1}{24}(1 + a + 9b), \\
p &= 4; & C_{p+1} &= -\tfrac{1}{720}(19 + 11a + 19b).
\end{aligned}$$

There exist no values for a and b which causes the order to exceed 4 and the method to be zero-stable. (See theorem 2.2.)

$k = 4$:

$$\begin{aligned}
\alpha_4 &= 1, & \beta_4 &= \tfrac{1}{720}(251 + 19a + 11b + 19c), \\
\alpha_3 &= -1 - a, & \beta_3 &= \tfrac{1}{360}(323 - 173a - 37b - 53c), \\
\alpha_2 &= a + b, & \beta_2 &= \tfrac{1}{30}(-11 - 19a + 19b + 11c), \\
\alpha_1 &= -b - c, & \beta_1 &= \tfrac{1}{360}(53 + 37a + 173b - 323c), \\
\alpha_0 &= c, & \beta_0 &= \tfrac{1}{720}(-19 - 11a - 19b - 251c).
\end{aligned}$$

If $27 + 11a + 11b + 27c \neq 0$, then

$$p = 5; C_{p+1} = -\tfrac{1}{1440}(27 + 11a + 11b + 27c).$$

If $27 + 11a + 11b + 27c = 0$, then

$$p = 6; C_{p+1} = -\tfrac{1}{15120}(74 + 10a - 10b - 74c).$$

There exist no values for a, b, and c which cause the order to exceed 6 and the method to be zero-stable. (See theorem 2.2.)

3

Linear multistep methods II: application

3.1 Problems in applying linear multistep methods

Let us suppose that we have selected a particular linear multistep method which is consistent and zero-stable. There are three problems we may have to face before proceeding with the calculation.

(i) In the case when the stepnumber k exceeds 1, how do we find the necessary additional starting values y_μ, $\mu = 1, 2, \ldots, k-1$?
(ii) How do we choose a suitable value for the steplength h?
(iii) In the case when the method is implicit, how do we solve at each step the implicit, and in general non-linear, equation for y_{n+k}?

To these we can add a fourth question which we ought always to ask after the calculation has been completed.

(iv) How accurate is the numerical solution we have obtained?

As we shall see, problem (i) presents little difficulty. Problem (ii), which is closely linked with (iv), constitutes the major problem in the application of linear multistep methods, and one to which a completely satisfactory answer does not exist at the present time. Recall that our chosen method, being consistent and zero-stable, is necessarily convergent; that is, if we are prepared to keep reducing the steplength, then, in the limit as $h \to 0$, the solution generated by the method will converge to the solution of the initial value problem. In practice, of course, we are not prepared to keep on recomputing with smaller mesh lengths and, in general, we would not contemplate performing the calculation for more than one, or perhaps two, fixed values of h. In any event, the property of convergence tells us only what happens *in the limit* as $h \to 0$ and does not give us any assurance that our solution will be acceptable when h is a fixed non-zero number, however small. Discussion of this area will lead us to, among other things, the study of new types of stability phenomena. Problem (iii) will lead us to the study of a new form of linear multistep method, namely the predictor–corrector pair.

3.2 Starting values

Recall the initial value problem

$$y' = f(x, y), \qquad y(a) = \eta.$$

Suppose that we seek a numerical solution of this problem by using, in the main computation, a linear multistep method of order p, whose step-number is $k > 1$. Let us set y_0 equal to η. It is desirable that the additional starting values $y_\mu, \mu = 1, 2, \ldots, k - 1$, should be calculated to an accuracy at least as high as the local accuracy of the main method; that is, we require that

$$y_\mu - y(x_\mu) = O(h^{p+1}), \qquad \mu = 1, 2, \ldots, k - 1. \tag{1}$$

The method used to evaluate the y_μ must itself require no starting values other than y_0; that is, it must be a one-step method. The only one-step methods we have encountered so far, namely Euler's rule and the Trapezoidal rule, which have order one and two respectively, are, in general, inadequate. If enough partial derivatives of $f(x, y)$ with respect to x and y exist, then we may use a truncated Taylor series to estimate y_1 to any required degree of accuracy. Thus, if we set

$$y_1 = y(x_0) + hy^{(1)}(x_0) + \frac{h^2}{2!}y^{(2)}(x_0) + \ldots + \frac{h^p}{p!}y^{(p)}(x_0) \tag{2}$$

then, since $y(x_1) - y_1 = [h^{p+1}/(p + 1)!]y^{(p+1)}(\xi), x_0 < \xi < x_1$, (1) is satisfied. The derivatives in (2) are evaluated by successively differentiating the differential equation. Thus

$$\begin{aligned}
y(x_0) &= y_0, \\
y^{(1)}(x_0) &= [f(x, y)]_{\substack{x=x_0\\y=y_0}}, \\
y^{(2)}(x_0) &= \left[\frac{\partial f}{\partial x} + \frac{\partial f}{\partial y}y'\right]_{\substack{x=x_0\\y=y_0}} = \left[\frac{\partial f}{\partial x} + f\frac{\partial f}{\partial y}\right]_{\substack{x=x_0\\y=y_0}}, \\
y^{(3)}(x_0) &= \left[\frac{\partial^2 f}{\partial x^2} + 2f\frac{\partial^2 f}{\partial x\,\partial y} + f^2\frac{\partial^2 f}{\partial y^2} + \frac{\partial f}{\partial x}\frac{\partial f}{\partial y} + f\left(\frac{\partial f}{\partial y}\right)^2\right]_{\substack{x=x_0\\y=y_0}}. \\
&\;\;\vdots \qquad\qquad\qquad \vdots
\end{aligned} \tag{3}$$

The remaining starting values y_μ, $\mu = 2, 3, \ldots, k - 1$, may be likewise calculated from the truncated Taylor series

$$y_\mu = y(x_{\mu-1}) + hy^{(1)}(x_{\mu-1}) + \frac{h^2}{2!}y^{(2)}(x_{\mu-1}) + \ldots + \frac{h^p}{p!}y^{(p)}(x_{\mu-1}),$$

where $y(x_{\mu-1})$ is taken to be $y_{\mu-1}$, already calculated, and the derivatives are given by (3) where the terms in brackets on the right-hand sides are now evaluated at $x = x_{\mu-1}$, $y = y_{\mu-1}$.

The process described above constitutes the *Taylor algorithm of order p*. It has the obvious weakness that it fails if any one of the necessary partial derivatives fails to exist at the point at which it is to be evaluated. Even when this difficulty does not arise, the evaluation of the total derivatives from (3) can be an excessively tedious business, even for relatively simple functions $f(x, y)$. The reader is invited to convince himself of the truth of this statement by attempting exercise 1.

Exercises

1. Evaluate the total derivatives $y^{(q)}(x_0)$, $q = 1, 2, 3, 4$, for the initial value problem $y' = \sqrt{x^2 + y^2}$, $x_0 = 1$, $y(x_0) = 1$.

2. Express the total derivatives $y^{(q)}(x)$, $q = 1, 2, \ldots$, as functions of x and y only, where $y' = xy^{\frac{1}{2}}$. Show that these derivatives are identically zero for $q \geqslant 5$. What conclusions can be drawn concerning the form of the general solution of the differential equation $y' = xy^{\frac{1}{2}}$?

The use of the Taylor algorithm is not restricted to the finding of additional starting values; it may be regarded as a one-step explicit method which involves higher total derivatives of y. It is natural to ask whether there exist implicit methods of the same type. We thus consider the class of methods

$$\alpha_1 y_{n+1} + \alpha_0 y_n = \sum_{s=1}^{l} h^s(\beta_{s1} y_{n+1}^{(s)} + \beta_{s0} y_n^{(s)}), \qquad l \geqslant 2 \tag{4}$$

where the total derivatives $y_{n+j}^{(s)}$ are computed from (3). For any given l the coefficients α_j, β_{sj}, $j = 0, 1$, $s = 1, 2, \ldots, l$, which give maximum local accuracy may be obtained from an obvious extension of the Taylor expansion technique used in section 2.2 to derive linear multistep methods.

The following methods are obtained, where the quoted local truncation errors are defined in the sense of section 2.7.

$$l = 2: \quad y_{n+1} - y_n = \tfrac{1}{2}h(y^{(1)}_{n+1} + y^{(1)}_n) - \tfrac{1}{12}h^2(y^{(2)}_{n+1} - y^{(2)}_n), \qquad (5i)$$

$$T_{n+1} = h^5 y^{(5)}(x_n)/720 + O(h^6).$$

$$l = 3: \quad y_{n+1} - y_n = \tfrac{1}{2}h(y^{(1)}_{n+1} + y^{(1)}_n) - \tfrac{1}{10}h^2(y^{(2)}_{n+1} - y^{(2)}_n) + \tfrac{1}{120}h^3(y^{(3)}_{n+1} + y^{(3)}_n), \qquad (5ii)$$

$$T_{n+1} = -h^7 y^{(7)}(x_n)/100{,}800 + O(h^8).$$

$$l = 4: \quad y_{n+1} - y_n = \tfrac{1}{2}h(y^{(1)}_{n+1} + y^{(1)}_n) - \tfrac{3}{28}h^2(y^{(2)}_{n+1} - y^{(2)}_n) + \tfrac{1}{84}h^3(y^{(3)}_{n+1} + y^{(3)}_n) - \tfrac{1}{1680}h^4(y^{(4)}_{n+1} - y^{(4)}_n), \qquad (5iii)$$

$$T_{n+1} = h^9 y^{(9)}(x_n)/25{,}401{,}600 + O(h^{10}).$$

These methods, and the Taylor algorithm, are special one-step cases of a more general class of multistep methods containing higher total derivatives of y, the class of *Obrechkoff methods.* We shall discuss this class in section 7.2. Method (5*ii*) is known as *Milne's starting* procedure (Milne[130]). Formulae (5) are capable of producing high accuracy estimates for starting values. For the Taylor algorithm to be of comparable accuracy, higher total derivatives of y are needed. The choice between (5) and the Taylor algorithm depends on the balance, in any particular problem, between the labour in solving the implicit equation in the former and in computing additional higher derivatives in the latter. The following example indicates one circumstance in which (5) is preferable.

Example 1 It is intended to solve the initial value problem

$$y' = x^{\frac{3}{2}} + y, \qquad y(0) = 1,$$

by a linear two-step method, using a steplength of 0·1. *What is the most accurate estimate for the additional starting value* y_1 *that can be obtained from* (*i*) *the Taylor algorithm,* (*ii*) *formulae* (5)?

We start by computing total derivatives of y.

$$y^{(1)} = x^{\frac{3}{2}} + y,$$
$$y^{(2)} = \tfrac{3}{2}x^{\frac{1}{2}} + x^{\frac{3}{2}} + y,$$
$$y^{(3)} = \tfrac{3}{4}x^{-\frac{1}{2}} + \tfrac{3}{2}x^{\frac{1}{2}} + x^{\frac{3}{2}} + y.$$

At the initial point, $x = 0$, $y^{(3)}$ does not exist. Hence we cannot use a Taylor algorithm of order greater than 2. The most accurate estimate we can get for y_1 is then

$$y_1 = y(0) + 0{\cdot}1y^{(1)}(0) + \tfrac{1}{2}(0{\cdot}1)^2 y^{(2)}(0)$$
$$= 1 + 0{\cdot}1 + 0{\cdot}005 = 1{\cdot}105.$$

Since $y^{(3)}$ does not exist at $x = 0$, we cannot use (5*ii*) or (5*iii*). However, from (5*i*)

$$y_1 - y(0) = \tfrac{1}{2}(0{\cdot}1)[y^{(1)}(x_1) + y^{(1)}(x_0)] - \tfrac{1}{12}(0{\cdot}01)[y^{(2)}(x_1) - y^{(2)}(0)]$$

or

$$y_1 - 1 = \tfrac{1}{2}(0{\cdot}1)[(0{\cdot}1)^{\frac{3}{2}} + y_1 + 1] - \tfrac{1}{12}(0{\cdot}01)[\tfrac{3}{2}(0{\cdot}1)^{\frac{1}{2}} + (0{\cdot}1)^{\frac{3}{2}} + y_1 - 1].$$

Solving for y_1, we find $y_1 = 1{\cdot}106{,}39$. The exact value for $y(x_1)$, taken from the theoretical solution to the initial value problem is $y(x_1) = 1{\cdot}106{,}50$.

We point out that the differential equation in this example is linear; if it were non-linear, the labour in solving (5*i*) for y_1 would be much increased.

Finally, additional starting values may be obtained by using one of the *Runge–Kutta methods* which we shall discuss in chapter 4. These methods do not themselves require additional starting values and, for general purposes, probably constitute the most efficient method for generating starting values for linear multistep methods. The block methods, which we shall discuss in section 4.11 are particularly appropriate (Milne,[131] Sarafyan[158]).

3.3† A bound for the local truncation error

If we ignore round-off error, question (iv) of section 3.1 will be answered if we can find a bound for the global truncation error $e_n = y(x_n) - y_n$. (It will, in fact, prove possible to extend the analysis to take some account of round-off error.) The expression for such a bound will be a function of the steplength, and if we stipulate in advance an acceptable level of accuracy by prescribing the magnitude of the error bound, it may be possible to deduce the maximum allowable value for the steplength, thus answering question (ii). We shall see that this is not necessarily the best way to answer (ii). We start by finding a bound for the local truncation error, defined in section 2.7.

Let us assume that the theoretical solution $y(x)$ possesses $p + 1$ continuous derivatives, where p is the order of the linear multistep method. So far, our only measures of local accuracy are the order p and the error constant C_{p+1}. The local truncation error is then

$$\mathscr{L}[y(x_n); h] = C_{p+1}h^{p+1}y^{(p+1)}(x_n) + O(h^{p+2}). \tag{6}$$

This asymptotic formula tells us how the local truncation error behaves in the limit as $h \to 0$ but, because of our ignorance of the magnitude of the $O(h^{p+2})$ term, we cannot deduce a bound for the magnitude of the local truncation error when h is a fixed non-zero constant. A similar problem

† The reader who is willing to accept the broad conclusions of section 3.5 may omit sections 3.3 and 3.4 with no loss of continuity.

exists if we attempt to bound the error committed in approximating the value taken by a sufficiently differentiable function $F(x)$ at $a + h$ by the first $p + 1$ terms in its Taylor expansion about $x = a$. In this case the difficulty is overcome by using one of the several available forms of the remainder for a Taylor series. Thus, using the Lagrange form of the remainder (see, for example, Jones and Jordan,[82] Vol. 1, page 201) we have

$$F(a + h) = F(a) + hF^{(1)}(a) + \dots + \frac{h^p}{p!}F^{(p)}(a) + R_{p+1},$$

where

$$R_{p+1} = \frac{h^{p+1}}{(p + 1)!}F^{(p+1)}(a + \theta h), \qquad 0 < \theta < 1.$$

It is natural to ask whether a similar result holds for the operator $\mathscr{L}$. That is, if $\mathscr{L}$ is of order p, can we write

$$\mathscr{L}[y(x_n); h] = C_{p+1}h^{p+1}y^{(p+1)}(x_n + \theta h), \qquad 0 < \theta < k? \tag{7}$$

We start by making use of a more general form of the remainder R_{p+1} after $p + 1$ terms in a Taylor series (see Jones and Jordan,[82] Vol. 2, page 78); it is

$$R_{p+1} = \frac{1}{p!}\int_0^h (h - t)^p F^{(p+1)}(a + t)\,dt.$$

The Taylor expansion for $y(x_n + jh)$ about x_n may now be written

$$y(x_n + jh) = y(x_n) + jhy^{(1)}(x_n) + \dots + \frac{1}{p!}j^p h^p y^{(p)}(x_n) + \frac{1}{p!}\int_0^{jh} (jh - t)^p y^{(p+1)}(x_n + t)\,dt.$$

Making the substitution $t = hs$, we obtain

$$y(x_n + jh) = y(x_n) + jhy^{(1)}(x_n) + \dots + \frac{1}{p!}j^p h^p y^{(p)}(x_n) + \frac{1}{p!}h^{p+1}\int_0^{j} (j - s)^p y^{(p+1)}(x_n + sh)\,ds.$$

A similar expansion for $y^{(1)}(x_n + jh)$ about x_n, but now truncated after only p terms, yields

$$y^{(1)}(x_n + jh) = y^{(1)}(x_n) + jhy^{(2)}(x_n) + \ldots + \frac{1}{(p-1)!} j^{p-1}h^{p-1}y^{(p)}(x_n)$$
$$+ \frac{1}{(p-1)!} h^p \int_0^j (j-s)^{p-1} y^{(p+1)}(x_n + sh)\, ds.$$

Substituting these expressions into the equation defining the operator $\mathscr{L}$ (equation (17) of chapter 2) and recalling that $\mathscr{L}$ is of order p, we obtain

$$\mathscr{L}[y(x_n); h] = \frac{1}{p!} h^{p+1} \sum_{j=0}^{k} \alpha_j \int_0^j (j-s)^p y^{(p+1)}(x_n + sh)\, ds$$
$$- \frac{1}{(p-1)!} h^{p+1} \sum_{j=0}^{k} \beta_j \int_0^j (j-s)^{p-1} y^{(p+1)}(x_n + sh)\, ds.$$

Employing the notation we introduced in section 2.4, we define

$$z_+ = \begin{cases} z, & \text{if } z \geqslant 0, \\ 0, & \text{if } z < 0. \end{cases}$$

Then

$$\int_0^j (j-s)^p y^{(p+1)}(x_n + sh)\, ds = \int_0^k (j-s)_+^p y^{(p+1)}(x_n + sh)\, ds,$$
$$\int_0^j (j-s)^{p-1} y^{(p+1)}(x_n + sh)\, ds = \int_0^k (j-s)_+^{p-1} y^{(p+1)}(x_n + sh)\, ds,$$

and we obtain

$$\mathscr{L}[y(x_n); h] = \frac{1}{p!} h^{p+1} \int_0^k \sum_{j=0}^{k} [\alpha_j (j-s)_+^p - p\beta_j (j-s)_+^{p-1}] y^{(p+1)}(x_n + sh)\, ds.$$

We now define the *influence function* $G(s)$ of the linear multistep method to be

$$G(s) = \sum_{j=0}^{k} [\alpha_j (j-s)_+^p - p\beta_j (j-s)_+^{p-1}]. \tag{8}$$

Note that $G(s)$ is a function of the coefficients of the method only and does not depend on the function $y(x)$. Since, by consistency, $p \geqslant 1$,

$G(s)$ is a polynomial in each of the subintervals $[0, 1], [1, 2], \ldots, [k-1, k]$. We now have the following expression for the local truncation error:

$$\mathscr{L}[y(x_n); h] = \frac{1}{p!}h^{p+1}\int_0^k G(s)y^{(p+1)}(x_n + sh)\,\mathrm{d}s. \tag{9}$$

This is not yet of the form conjectured in (7). We now consider two separate cases.

Case I $G(s)$ does not change sign in the interval $[0, k]$

In this case we appeal to the generalized mean value theorem for integrals, (see, for example, Jones and Jordan,[82] Vol. 1, page 439) which states that if $\phi(x)$ is continuous and $g(x)$ integrable and of constant sign in the interval $[c, d]$, then there exists $\xi \in (c, d)$ such that

$$\int_c^d \phi(x)g(x)\,\mathrm{d}x = \phi(\xi)\int_c^d g(x)\,\mathrm{d}x.$$

Applying this theorem to the right-hand side of (9) yields the result

$$\mathscr{L}[y(x_n); h] = \frac{1}{p!}h^{p+1}\left(\int_0^k G(s)\,\mathrm{d}s\right)y^{(p+1)}(x_n + \theta h), \qquad 0 < \theta < k. \tag{10}$$

This is now of the form (7). Neither the influence function $G(s)$ nor the error constant C_{p+1} depends on the function $y(x)$. Putting $y(x) = x^{p+1}$ in (10) and in equation 18 of chapter 2 gives, respectively,

$$\mathscr{L}[x_n^{p+1}; h] = \frac{1}{p!}h^{p+1}\left(\int_0^k G(s)\,\mathrm{d}s\right)(p+1)!$$

and

$$\mathscr{L}[x_n^{p+1}; h] = C_{p+1}h^{p+1}(p+1)!,$$

since all derivatives of x^{p+1} of order greater than $p + 1$ vanish identically. Hence

$$\frac{1}{p!}\int_0^k G(s)\,\mathrm{d}s = C_{p+1}, \tag{11}$$

and it follows that the conjecture (7) holds. We can deduce the following bound for the magnitude of the local truncation error at any step within the range of integration $[a, b]$:

$$|\mathscr{L}[y(x_n); h]| \leqslant h^{p+1}GY, \tag{12i}$$

where

$$Y = \max_{x \in [a,b]} |y^{(p+1)}(x)| \tag{12ii}$$

and

$$G = |C_{p+1}| = \frac{1}{p!}\left|\int_0^k G(s)\,ds\right|. \tag{12iii}$$

Case II $G(s)$ changes sign in the interval $[0, k]$

In this case we cannot show that the conjecture (7) holds. However, from (9) we obtain

$$\begin{aligned}|\mathscr{L}[y(x_n); h]| &\leqslant \frac{1}{p!}h^{p+1}\int_0^k |G(s)|\,|y^{(p+1)}(x_n + sh)|\,ds \\ &\leqslant \frac{1}{p!}h^{p+1}Y\int_0^k |G(s)|\,ds.\end{aligned}$$

Thus, once again, we obtain a bound of the form

$$|\mathscr{L}[y(x_n); h]| \leqslant h^{p+1}GY, \tag{12i}$$

where Y is given by (12*ii*) and now

$$G = \frac{1}{p!}\int_0^k |G(s)|\,ds. \tag{12iv}$$

We observe that (12*iv*) holds for all linear multistep methods whether or not the influence function changes sign in $[0, k]$. The expression for G given by (12*iii*) holds only when $G(s)$ has constant sign in $[0, k]$, but in this case

$$\left|\int_0^k G(s)\,ds\right| = \int_0^k |G(s)|\,ds,$$

and (12*iv*) is equivalent to (12*iii*). The bound defined by (12*i*, *ii*, *iv*) is, in all cases, the best that can be obtained from this analysis.

It can be shown that for all Adams–Bashforth and Adams–Moulton methods the influence function does not change sign in $[0, k]$.

Example 2 *Construct the influence function, $G(s)$, for the two-step explicit method*

$$y_{n+2} - (1 + a)y_{n+1} + ay_n = \tfrac{1}{2}h[(3 - a)f_{n+1} - (1 + a)f_n], \qquad a \neq -5.$$

Find the range of a for which $G(s)$ has constant sign, and demonstrate that for this range (11) holds. Find a bound for the local truncation error when this method, with $a = 0$, is applied to the initial value problem of example 4, chapter 2 (page 34).

From section 2.6 we have that if $a \neq -5$, $p = 2$, $C_{p+1} = \frac{1}{12}(5 + a)$. By (8)

$$G(s) = (2 - s)_+^2 - (1 + a)(1 - s)_+^2 + a(0 - s)_+^2$$
$$-2 \cdot \tfrac{1}{2}[(3 - a)(1 - s)_+ - (1 + a)(0 - s)_+].$$

For $s \in [0, 1]$, $(2 - s)_+ = 2 - s$, $(1 - s)_+ = 1 - s$, $(0 - s)_+ = 0$.
For $s \in [1, 2]$, $(2 - s)_+ = 2 - s$, $(1 - s)_+ = 0$, $(0 - s)_+ = 0$.
Hence, for $s \in [0, 1]$,

$$G(s) = (2 - s)^2 - (1 + a)(1 - s)^2 - (3 - a)(1 - s) = -as^2 + (1 + a)s,$$

and for $s \in [1, 2]$, $G(s) = (2 - s)^2$.

Now $-as^2 + (1 + a)s$ vanishes at $s = 0$ and at $s = (1 + a)/a$, and the latter root lies in the interval $(0, 1)$ if and only if $a < -1$. Thus, for $a \geqslant -1$, $G(s)$ is of constant sign in $[0, 1]$ and is, in fact, non-negative there; since it is clearly non-negative in $[1, 2]$, it follows that $G(s)$ does not change sign in $[0, 2]$ provided $a \geqslant -1$. For this range of a

$$\int_0^2 G(s)\,\mathrm{d}s = \int_0^1 [-as^2 + (1 + a)s]\,\mathrm{d}s + \int_1^2 (2 - s)^2\,\mathrm{d}s = \tfrac{1}{6}(5 + a).$$

Hence

$$\frac{1}{p!}\int_0^k G(s)\,\mathrm{d}s = \tfrac{1}{12}(5 + a) = C_{p+1},$$

verifying (11). By (12*iii*) $G = \frac{1}{12}|5 + a| = \frac{1}{12}(5 + a)$ since $a \geqslant -1$. If, instead, we choose to calculate G from (12*iv*), then we obtain

$$2G = \int_0^2 |G(s)|\,\mathrm{d}s = \int_0^{(1+a)/a} |-as^2 + (1 + a)s|\,\mathrm{d}s$$
$$+ \int_{(1+a)/a}^1 |-as^2 + (1 + a)s|\,\mathrm{d}s + \int_1^2 (2 - s)^2\,\mathrm{d}s$$

for *all a*. If $a \geqslant -1$, we find

$$2G = \int_0^{(1+a)/a} [-as^2 + (1 + a)s]\,\mathrm{d}s + \int_{(1+a)/a}^1 [-as^2 + (1 + a)s]\,\mathrm{d}s$$
$$+ \int_1^2 (2 - s)^2\,\mathrm{d}s,$$

and we clearly recover the result $G = \frac{1}{12}(5 + a)$. If, however, $a < -1$, then

$$2G = \int_0^{(1+a)/a} [as^2 - (1 + a)s]\,\mathrm{d}s + \int_{(1+a)/a}^1 [-as^2 + (1 + a)s]\,\mathrm{d}s + \int_1^2 (2 - s)^2\,\mathrm{d}s,$$

which gives

$$G = \tfrac{1}{12}(5 + a) - \tfrac{1}{6}(1 + a)^3/a^2.$$

The differential equation in example 4 of chapter 2 is

$$y' = 4xy^{\frac{1}{2}}.$$

In order to bound $y^{(p+1)}(x) = y^{(3)}(x)$, we differentiate the differential equation totally to obtain

$$y^{(2)} = 4y^{\frac{1}{2}} + 2xy^{-\frac{1}{2}}.4xy^{\frac{1}{2}} = 4y^{\frac{1}{2}} + 8x^2,$$

$$y^{(3)} = 2y^{-\frac{1}{2}}.4xy^{\frac{1}{2}} + 16x = 24x.$$

Hence

$$Y = \max_{x \in [0, b]} |y^{(3)}(x)| = 24b,$$

where b is the right-hand end of the range of integration under consideration. The bound on the local truncation error is thus $2b(5 + a)h^3 = 10bh^3$, when $a = 0$.

If we wish to compare this bound with the actual errors produced in the numerical solutions for the case $a = 0$ given in example 4 of chapter 2, we must recall from section 2.7 that local truncation error coincides with actual truncation error only when no previous truncation errors have been made. Thus we may apply the bound only to the first application of the method, since the starting values y_0 and y_1 were taken to coincide with the theoretical solution. Thus we may take $b = 2h$. Using the numerical results of example 4 of chapter 2, we find the results given in table 4.

Table 4

h	Actual error in y_2	Error bound
0·1	0·0009	0·002
0·05	0·000,056,3	0·000,125

Finally, we observe that in this example it was particularly easy to evaluate Y, partly because repeated total differentiation of the differential equation caused no problem, and partly because the final expression for $y^{(p+1)}(x)$ turned out to be a function of x only. For a general problem we can expect neither of these fortunate contingencies to occur. We shall discuss this point later when we have obtained a bound for the global truncation error.

Exercises

3. If we put $a = -5$ in example 2, the method has order 3. Compute the influence function for this case and show that it does not change sign in $[0, 2]$. Calculate the error constant for this method and verify that (11) holds.

Calculate the bound for the local truncation error when this method is applied to the problem of example 2, and prove that in this particular instance the bound is sharp in the sense that (12*i*) holds with equality. Corroborate this by showing that the bound is equal to the actual local truncation error in the first step when $h = 0{\cdot}1, 0{\cdot}05$. (See example 4 of chapter 2 for the numerical solution in the case when $a = -5$.) Note that the zero-instability of the method does not adversely affect the bound on the *local* truncation error.

4. Construct the influence function for Simpson's rule and show that it does not change sign in $[0, 2]$. Verify that (11) holds.

3.4† A bound for the global truncation error

The derivation of a bound for the global truncation error $e_n = y(x_n) - y_n$ in the case of a general linear multistep method is beyond the scope of this book. It can be found on page 248 of Henrici.[67] However, if we consider only a certain subclass of linear multistep methods, a global error bound can be somewhat more easily derived. Moreover, the bound obtained possesses all the salient features of the bound for the general method. We shall say that the linear multistep method

$$\sum_{j=0}^{k} \alpha_j y_{n+j} = h \sum_{j=0}^{k} \beta_j f_{n+j}$$

is of *class A* if

$$\alpha_k = 1; \qquad \alpha_j \leqslant 0, j = 0, 1, \ldots, k-1; \qquad \sum_{j=0}^{k} \alpha_j = 0.$$

We observe that this class contains the classes of Adams–Bashforth, Adams–Moulton, Nyström and generalized Milne–Simpson methods. It can be shown that methods of class A are necessarily zero-stable (Copson[30]).

We now derive a bound for the global truncation error for the general *explicit* method of class A. The theoretical solution $y(x)$ of the initial value problem satisfies

$$\sum_{j=0}^{k} \alpha_j y(x_{n+j}) - h \sum_{j=0}^{k-1} \beta_j y'(x_{n+j}) = \mathscr{L}[y(x_n); h]$$

or

$$y(x_{n+k}) = \sum_{j=0}^{k-1} [-\alpha_j y(x_{n+j}) + h\beta_j f(x_{n+j}, y(x_{n+j}))] + \mathscr{L}[y(x_n); h].$$

If we assume that the equation defining the numerical solution is solved *exactly* at each stage, then the numerical solution satisfies

$$y_{n+k} = \sum_{j=0}^{k-1} [-\alpha_j y_{n+j} + h\beta_j f(x_{n+j}, y_{n+j})]. \tag{13}$$

Subtracting, we obtain

$$e_{n+k} = \sum_{j=0}^{k-1} \{-\alpha_j e_{n+j} + h\beta_j [f(x_{n+j}, y(x_{n+j})) - f(x_{n+j}, y_{n+j})]\} + \mathscr{L}[y(x_n); h].$$

† The reader who is willing to accept the broad conclusions of section 3.5 may omit this section with no loss of continuity.

Hence

$$|e_{n+k}| \leqslant \sum_{j=0}^{k-1} [|-\alpha_j|\,|e_{n+j}| + h|\beta_j|\,|f(x_{n+j}, y(x_{n+j})) - f(x_{n+j}, y_{n+j})|] + |\mathscr{L}[y(x_n); h]|. \quad (14)$$

It follows from the definition of class A that

$$|-\alpha_j| = -\alpha_j, \qquad j = 0, 1, \ldots, k-1.$$

Also, if L is the Lipschitz constant associated with the initial value problem, then

$$|f(x_{n+j}, y(x_{n+j})) - f(x_{n+j}, y_{n+j})| \leqslant L|y(x_{n+j}) - y_{n+j}| = L|e_{n+j}|.$$

Finally, from (12*i*),

$$|\mathscr{L}[y(x_n); h]| \leqslant h^{p+1}GY$$

where Y and G are given by (12*ii*) and (12*iv*) and p is the order of the method. Substituting these inequalities in (14) gives

$$|e_{n+k}| \leqslant \sum_{j=0}^{k-1} (-\alpha_j + hL|\beta_j|)|e_{n+j}| + h^{p+1}GY, \qquad n = 0, 1, 2, \ldots \quad (15)$$

We now introduce some simplifying notation. Let

$$P_j = -\alpha_j + hL|\beta_j|, \qquad P = \sum_{j=0}^{k-1} P_j, \qquad Q = h^{p+1}GY. \quad (16)$$

Now $\alpha_k = +1$ and $\sum_{j=0}^{k} \alpha_j = 0$. It follows that

$$P = 1 + hLB, \quad (17)$$

where

$$B = \sum_{j=0}^{k-1} |\beta_j|. \quad (18)$$

Finally, let δ be the maximum error in the starting values; that is, let

$$\delta = \max_{\mu = 0, 1, \ldots, k-1} |e_\mu|. \quad (19)$$

In the subsequent analysis we shall make frequent use of the following consequences of (16), (17), (18), and (19):

$$P_j \geqslant 0, \qquad P \geqslant 1, \qquad Q \geqslant 0, \qquad \delta \geqslant 0. \quad (20)$$

Equation (15) may now be written

$$|e_{n+k}| \leqslant \sum_{j=0}^{k-1} P_j|e_{n+j}| + Q, \qquad n = 0, 1, 2, \ldots. \tag{21}$$

Setting $n = 0$ in (21) gives, in view of (19) and (20),

$$|e_k| \leqslant \sum_{j=0}^{k-1} P_j\delta + Q = P\delta + Q. \tag{22}$$

Setting $n = 1$ in (21) gives

$$|e_{k+1}| \leqslant \sum_{j=0}^{k-1} P_j|e_{j+1}| + Q. \tag{23}$$

Now, by (19), $|e_{j+1}| \leqslant \delta$, $j = 0, 1, \ldots, k-2$, and by (22), $|e_k| \leqslant P\delta + Q$. However, by (20), $\delta \leqslant P\delta + Q$, so that we may assert that

$$|e_{j+1}| \leqslant P\delta + Q, \qquad j = 0, 1, \ldots, k-1.$$

Substituting this in (23) yields

$$|e_{k+1}| \leqslant \sum_{j=0}^{k-1} P_j(P\delta + Q) + Q = P^2\delta + (P+1)Q. \tag{24}$$

The pattern is now clear, and we may proceed inductively, making the induction hypothesis

$$|e_{l+k}| \leqslant P^{l+1}\delta + Q\sum_{s=0}^{l} P^s, l = 0, 1, \ldots, m-1. \tag{25}$$

(Equations (22) and (24) verify that this assumption holds for $m = 1, 2$.) By (21),

$$|e_{m+k}| \leqslant \sum_{j=0}^{k-1} P_j|e_{m+j}| + Q. \tag{26}$$

By the induction hypothesis (25), we have

$$\begin{aligned}|e_{m+j}| &\leqslant P^{m-k+j+1}\delta + Q\sum_{s=0}^{m-k+j} P^s, \qquad j = 0, 1, \ldots, k-1,\\ &\leqslant P^m\delta + Q\sum_{s=0}^{m-1} P^s,\end{aligned}$$

in view of (20). Thus, from (26) we obtain

$$|e_{m+k}| \leqslant \sum_{j=0}^{k-1} P_j\left(P^m\delta + Q\sum_{s=0}^{m-1} P^s\right) + Q = P^{m+1}\delta + Q\sum_{s=0}^{m} P^s.$$

Thus (25) holds when $l = m$, and the induction proof is complete. We therefore have

$$|e_{n+k}| \leqslant P^{n+1}\delta + Q \sum_{s=0}^{n} P^s, \qquad n = 0, 1, 2, \dots,$$

$$\leqslant P^{n+k}\delta + Q \sum_{s=0}^{n+k-1} P^s, \qquad n = 0, 1, 2, \dots,$$

since $k \geqslant 1$. Equivalently, we may write

$$|e_n| \leqslant P^n\delta + Q \sum_{s=0}^{n-1} P^s \equiv P^n\delta + Q(P^n - 1)/(P - 1) \qquad n = k, k+1, \dots.$$

The same bound holds when $n = 0, 1, \dots, k-1$, since

$$|e_n| \leqslant \delta \leqslant P^n\delta + Q(P^n - 1)/(P - 1), \qquad n = 0, 1, \dots, k-1,$$

in view of (20). Thus, using (16) and (17), we can assert that the following bound for the global error e_n holds for $n = 0, 1, 2, \dots$:

$$|e_n| \leqslant \delta(1 + hLB)^n + \frac{h^pGY}{LB}[(1 + hLB)^n - 1]. \tag{27}$$

This form of the bound is, however, rather misleading. Consider the case when there are no starting errors—that is, when $\delta = 0$. The right-hand side of (27) is then

$$\frac{h^pGY}{LB}[(1 + hLB)^n - 1] = \frac{h^pGY}{LB}[nhLB + O(h^2)], \qquad h \to 0,$$

which appears to be of the form $Kh^{p+1} + O(h^{p+2})$, K constant. This is not so, for $n \to \infty$ as $h \to 0$, since at any fixed station x_n, $nh = x_n - a$. A less misleading form of the bound can be obtained by observing that $1 + hLB \leqslant \exp(hLB)$, whence $(1 + hLB)^n \leqslant \exp(nhLB) = \exp[LB(x_n - a)]$; substituting this in (27) gives

$$|e_n| \leqslant \delta \exp[LB(x_n - a)] + \frac{h^pGY}{LB}\{\exp[LB(x_n - a)] - 1\}. \tag{28}$$

The second term in the right-hand side of (28) is now seen to be properly of the form $O(h^p)$ as $h \to 0$.

Yet another form can be found by slackening the bound further through using the fact that, for any z, $\exp(z) - 1 < z\exp(z)$. It is

$$|e_n| < [\delta + (x_n - a)h^pGY]\exp[LB(x_n - a)]. \tag{29}$$

In each of the bounds (27), (28), and (29), p is the order of the method, L the Lipschitz constant associated with the initial value problem, and Y, G, B, and δ are given by (12*ii*), (12*iv*) (18), and (19) respectively.

The error bounds we have found can be easily extended to take account of round-off error. Suppose that a round-off error is committed at each application of the method, but that sufficient decimal places are retained to ensure that this error is of order h^{q+1}. Then if we denote by $\tilde{y}_n$ the numerical solution at x_n when round-off error is taken into account, equation (13) is replaced by

$$\tilde{y}_{n+k} = \sum_{j=0}^{k-1} \{-\alpha_j \tilde{y}_{n+j} + h\beta_j f(x_{n+j}, \tilde{y}_{n+j})\} + \theta K_1 h^{q+1},$$

where $|\theta| \leqslant 1$ and K_1 is a positive constant. It is obvious that the above analysis holds if we replace $h^{p+1}GY$ by $h^{p+1}GY + h^{q+1}K_1$, and e_n by $\tilde{e}_n$, where $\tilde{e}_n = y(x_n) - \tilde{y}_n$. We therefore obtain in place of (29) the following bound for the accumulated truncation and round-off error for an explicit linear multistep method of class A:

$$|\tilde{e}_n| < \{\delta + (x_n - a)(h^p GY + h^q K_1)\} \exp\{LB(x_n - a)\}. \tag{30}$$

A bound similar to (30) holds for a *general* (explicit or implicit) zero-stable linear multistep method; its derivation, however, requires somewhat more elaborate mathematical apparatus. The interested reader is referred to page 248 of Henrici.[67] The result, which holds only if $|h\beta_k\alpha_k^{-1}|L < 1$ is

$$|\tilde{e}_n| \leqslant \Gamma^*\{Ak\delta + (x_n - a)(h^p GY + h^q K_1)\} \exp\{\Gamma^* LB(x_n - a)\}, \tag{31}$$

where

$$A = \sum_{j=0}^{k} |\alpha_j|, \qquad B = \sum_{j=0}^{k} |\beta_j|,$$

$$\Gamma^* = \Gamma/(1 - h|\alpha_k^{-1}\beta_k|L),$$

$$\Gamma = \sup_{l=0,1,\ldots} |\gamma_l|,$$

where

$$1/(\alpha_k + \alpha_{k-1}\zeta + \ldots + \alpha_0\zeta^k) = \gamma_0 + \gamma_1\zeta + \gamma_2\zeta^2 + \ldots.$$

3.5 Discussion on error bounds.

There are several general conclusions we can draw from considering the form of the bounds (30) and (31). The influence of the three separate sources of error, namely starting error, local truncation error and local round-off error are represented respectively by the terms δ, h^pGY and h^qK_1.

The term corresponding to starting error is effectively independent of h (completely so in the case of (30)). Note that if we adopt the natural strategy, advocated in section 3.2, of requiring that the accuracy of the starting values be at least as high as the local accuracy of the main method, then we demand that $\delta = O(h^{p+1})$ and the starting errors will certainly not dominate the error bound as $h \to 0$.

The interesting fact about the other two sources of error is that in each case the term in the error bound is a factor h^{-1} greater than the corresponding bound in the local error; thus, the *local* truncation error is bounded by GYh^{p+1} whereas the bound for the *global* truncation error is proportional to GYh^p (if we ignore starting errors). Thus, due to the process of accumulation, *the bound for global error is an order of magnitude greater than that for local error.*

Lastly, we observe that if $q > p$ (a natural choice if it can be achieved), then as $h \to 0$, truncation error, rather than round-off error, will dominate the bound. It is, however, impracticable to demand that the local round-off error be bounded by K_1h^{q+1}, since we normally work with a fixed precision in the computer; it is certainly not possible to increase this precision without limit as $h \to 0$. A more realistic assumption is that the local round-off error is bounded by ε, a small fixed constant. The analysis of the preceding section would then hold with $h^{p+1}GY$ replaced by $h^{p+1}GY + \varepsilon$, rather than by $h^{p+1}GY + h^{q+1}K_1$. In the final bound (31), the term $h^pGY + h^qK_1$ would be replaced by $h^pGY + \varepsilon/h$. Thus, as h is reduced (from some sufficiently large value), the amended bound would initially decrease, but eventually increase when the term ε/h started to dominate the bound. That actual global errors, as opposed to error bounds, behave in just this fashion has been demonstrated experimentally—see, for example, Hull and Creemer.[72]

It also follows from (30) and (31) that if $p \geqslant 1$, $q \geqslant 1$ and $\delta \to 0$ as $h \to 0$, then $\tilde{e}_n \to 0$ as $h \to 0$, $n \to \infty$, $nh = x_n - a$ remaining fixed. If we ignore round-off error (that is, set $K_1 = 0$) then $\tilde{e}_n$ is identical with $e_n = y(x_n) - y_n$, and we have formally proved that consistency ($p \geqslant 1$) implies convergence in the case of explicit methods of class A. (Recall that methods of class A are always zero-stable.) Zero-stability must be assumed in order to establish the more general bound (31) and thus, once again, zero-stability together with consistency implies convergence, in agreement with theorem 2.1.

We now turn to the question of the practical application of these error bounds. The first obstacle consists of finding an estimate for

$Y = \max_{x \in [a,b]} |y^{(p+1)}(x)|$. We have already pointed out that obtaining expressions for the higher total derivatives of y by repeatedly differentiating the differential equation can be tedious, sometimes to the point of impracticability. Even if we succeed, the resulting expression for $y^{(p+1)}(x)$ will, in general, be a function of both x and y, and at the outset we have no knowledge of y except at the initial point. After we have computed a numerical solution, we could, for the purpose of estimating Y, make the approximation $y(x_n) \approx y_n$. (If it proves impossible to express $y^{(p+1)}(x)$ analytically in terms of x and y, another possible approach is to use our knowledge of y_n, $n = 0, 1, 2, \ldots, (b - a)/h$, to estimate $y^{(p+1)}(x)$ by the process of numerical differentiation; such a process is, however, notoriously inaccurate, particularly so when p is large.) Using the error bound in the manner described above thus enables us to answer question (iv) of section 3.1, but not question (ii) (unless, perhaps, by an iterative procedure). However, there may be occasions when we are able to estimate Y before computing a numerical solution. For example, we may be able to prove that $|y^{(p+1)}(x)|$ takes its maximum value at the initial point $x = a$, or perhaps additional external evidence, associated with the physical problem which gives rise to the given initial value problem, allows us to make an *a priori* estimate for Y.

The second practical difficulty—one which always arises with general error bounds in numerical analysis—is that it is possible to find problems for which the error bounds (30) and (31) are excessively conservative. In such a situation, our answer to question (iv) will be misleadingly pessimistic, and, if we attempt to use the bound to answer (ii), the value for h which we shall arrive at will be unnecessarily small. We illustrate the point by the following simple example.

Example 3 Illustrate the pessimistic nature of the error bound (30) in the case when Euler's rule is applied to the initial value problem $y' = \lambda y$, $y(0) = 1$, $\lambda < 0$.

Assume no round-off errors are made; since no additional starting values are required for Euler's rule, we choose $y_0 = 1$ and $\delta = 0$. It is easily established that the following hold for this problem:

$$G = \tfrac{1}{2}, \qquad L = |\lambda|, \qquad B = 1, \qquad p = 1.$$

From (30) we obtain

$$|e_n| < \tfrac{1}{2} x_n h Y \exp(|\lambda| x_n). \tag{32}$$

For this simple problem, a sharper bound can be obtained as follows:

$$y(x_{n+1}) = y(x_n) + h\lambda y(x_n) + T_{n+1},$$

where $T_{n+1} = \frac{1}{2}h^2 y^{(2)}(x_n + \theta h)$, $0 < \theta < 1$.

Since $y^{(2)}(x) = \lambda^2 y(x)$, and y is initially positive and can vanish only where y' vanishes, we can conclude that $y^{(2)}(x) \geqslant 0$ for all $x \geqslant 0$. Hence $0 \leqslant T_{n+1} \leqslant \frac{1}{2}h^2 Y$. Since no round-off errors are made, we have

$$y_{n+1} = y_n + h\lambda y_n$$

and hence

$$e_{n+1} = (1 + h\lambda)e_n + T_{n+1}.$$

Thus

$$\begin{aligned} e_0 &= 0, \\ e_1 &= T_1, \\ e_2 &= (1 + h\lambda)e_1 + T_2 = (1 + h\lambda)T_1 + T_2, \\ &\vdots \qquad\qquad \vdots \\ e_n &= (1 + h\lambda)^{n-1}T_1 + (1 + h\lambda)^{n-2}T_2 + \ldots + T_n. \end{aligned}$$

If $h < -1/\lambda$ then

$$\begin{aligned} e_n &\leqslant \tfrac{1}{2}h^2 Y[(1 + h\lambda)^{n-1} + (1 + h\lambda)^{n-2} + \ldots + 1] \\ &\leqslant \tfrac{1}{2}nh^2 Y = \tfrac{1}{2}x_n h Y. \end{aligned}$$

Thus, for $h < -1/\lambda$, $\lambda < 0$, we have the bound

$$0 \leqslant e_n \leqslant \tfrac{1}{2}x_n h Y. \tag{33}$$

Thus the bound in (32) is greater than that in (33) by a factor $\exp(|\lambda| x_n)$, a factor which may be quite large. Moreover the sharper bound (33) tells us that the accumulated error is always non-negative.

Example 4 Consider the specific case of the problem in example 3, where $\lambda = -5$, $h = 0{\cdot}1$, $n = 0, 1, \ldots, 10$.
(i) Assuming knowledge of the theoretical solution $y(x) = \exp(\lambda x)$, *compare the actual accumulated error at each step with the bounds (32) and (33).*
(ii) Compare the maximum allowable values for h which would be predicted by (32) and (33) if we were to demand that the accumulated error at $x = 1$ *be not greater than* 0·1 *in magnitude.*

(i) When $\lambda = -5$, $-1/\lambda = 0{\cdot}2$, $h = 0{\cdot}1 < -1/\lambda$, and (33) is applicable.
The numerical solution, y_n, satisfies the difference equation

$$y_{n+1} = (1 + h\lambda)y_n, \qquad y_0 = 1,$$

whose solution is $y_n = (1 + h\lambda)^n = (\frac{1}{2})^n$. (Using this solution rather than applying the method step-by-step ensures that no round-off error is propagated.) Thus the actual accumulated error is

$$e_n = \exp(-5x_n) - (\tfrac{1}{2})^n.$$

Also,

$$Y = \max_{x \in [0,\,1]} |25 \exp(-5x)| = 25.$$

Using (33), $|e_n| \leqslant \frac{1}{2}nh^2 25 = \frac{1}{8}n$.
Using (32), $|e_n| < \frac{1}{2}nh^2 25 \exp(5nh) = \frac{1}{8}n \exp(\frac{1}{2}n)$.
These error bounds are compared with actual global truncation errors in table 5.

Table 5

n	Error e_n	Error bound (33)	Error bound (32)
0	0	0	0
1	0·1065	0·125	0·2061
2	0·1179	0·250	0·6796
3	0·0981	0·375	1·6806
4	0·0728	0·500	3·6945
5	0·0508	0·625	7·6141
6	0·0342	0·750	15·0641
7	0·0224	0·875	28·9761
8	0·0144	1·000	54·5982
9	0·0092	1·125	101·2692
10	0·0058	1·250	185·516,25

Although the actual error decreases with n for $n > 2$ both error bounds grow progressively larger, the sharper bound growing linearly and the general bound more than exponentially.
(ii) From (33), $|e_n| \leqslant 0{\cdot}1$ at $x_n = 1$ if $\frac{1}{2}h25 \leqslant 0{\cdot}1$, that is, if $h \leqslant 0{\cdot}008$.
From (32), $|e_n| < 0{\cdot}1$ at $x_n = 1$ if $\frac{1}{2}h25 \exp(5) < 0{\cdot}1$, that is, if $h < 0{\cdot}000{,}054$.
In fact, we have seen that choosing $h = 0{\cdot}1$ gave an error at $x = 1$ of only 0·0058.

Exercise

5. Construct global error bounds for the numerical calculations of example 4 of chapter 2 (page 34) and compare with the actual errors. (Note that when the parameter a has value 0 the method is of class A, but when $a = -5$, it is not.)

Our conclusions from this section are that the importance of the global error bound lies in the theoretical conclusions that can be inferred from its form, rather than in its practical applications. In particular, if the bound for the *local* error is of order h^{p+1} then the bound for the *global* error will be of order h^p. In practical terms, the bound provides an answer, possibly a very pessimistic one, to question (iv); it does not effectively answer question (ii).

We note that in example 4, after an initial growth, the actual global error *decreased* as the computation proceeded. This motivates us to seek some criterion which will ensure that the global error will be damped out as the numerical solution is continued. The situation is similar to one that we discussed in chapter 2, where the rôle of consistency was to limit the size of local inaccuracies, while that of zero-stability was to ensure that these inaccuracies were not propagated in a disastrous manner. In that context, however, we were concerned only with the situation in the limit as $h \to 0$, $n \to \infty$, nh fixed, whereas now we wish to take h as a fixed positive number, and study the propagation of the error as $n \to \infty$. We shall develop a theory, the theory of *weak stability*, which attempts to provide a criterion, involving h, for the global error to be damped out as the computation proceeds. We shall then be able to answer question (ii) by choosing h small enough for this criterion to be satisfied and for the *local* truncation error to be acceptably small.

3.6 Weak stability† theory

Consider the general linear multistep method

$$\sum_{j=0}^{k} \alpha_j y_{n+j} = h \sum_{j=0}^{k} \beta_j f_{n+j}, \tag{34}$$

which, as always in this chapter, we assume to be consistent and zero-stable. The theoretical solution $y(x)$ of the initial value problem satisfies

$$\sum_{j=0}^{k} \alpha_j y(x_{n+j}) = h \sum_{j=0}^{k} \beta_j f(x_{n+j}, y(x_{n+j})) + T_{n+k}, \tag{35}$$

where $T_{n+k} = \mathscr{L}[y(x_n); h]$, the local truncation error. If we denote by $\{\tilde{y}_n\}$ the solution of (34) when a round-off error R_{n+k} is committed at the nth application of the method, then

$$\sum_{j=0}^{k} \alpha_j \tilde{y}_{n+j} = h \sum_{j=0}^{k} \beta_j f(x_{n+j}, \tilde{y}_{n+j}) + R_{n+k}. \tag{36}$$

On subtracting (36) from (35) and defining the global error $\tilde{e}_n$ by $\tilde{e}_n = y(x_n) - \tilde{y}_n$, we find

$$\sum_{j=0}^{k} \alpha_j \tilde{e}_{n+j} = h \sum_{j=0}^{k} \beta_j [f(x_{n+j}, y(x_{n+j})) - f(x_{n+j}, \tilde{y}_{n+j})] + \phi_{n+k},$$

where $\phi_{n+k} = T_{n+k} - R_{n+k}$. If we assume that the partial derivative $\partial f/\partial y$ exists for all $x \in [a, b]$, then, by the mean value theorem, there exists a number ξ_{n+j} lying in the open interval whose end-points are $y(x_{n+j})$

† Sometimes called 'numerical stability' or 'conditional stability'.

and $\tilde{y}_{n+j}$, such that

$$f(x_{n+j}, y(x_{n+j})) - f(x_{n+j}, \tilde{y}_{n+j}) = [y(x_{n+j}) - \tilde{y}_{n+j}]\frac{\partial f(x_{n+j}, \xi_{n+j})}{\partial y}$$

$$= \tilde{e}_{n+j}\frac{\partial f}{\partial y}(x_{n+j}, \xi_{n+j}).$$

Hence

$$\sum_{j=0}^{k} \alpha_j \tilde{e}_{n+j} = h \sum_{j=0}^{k} \beta_j \frac{\partial f(x_{n+j}, \xi_{n+j})}{\partial y}\tilde{e}_{n+j} + \phi_{n+k}.$$

We now make two important simplifying assumptions:

$$\frac{\partial f}{\partial y} = \lambda, \quad \text{constant,} \tag{37i}$$

$$\phi_n = \phi, \quad \text{constant.} \tag{37ii}$$

The equation for $\tilde{e}_n$ now reduces to the *linearized error equation*

$$\sum_{j=0}^{k} (\alpha_j - h\lambda\beta_j)\tilde{e}_{n+j} = \phi. \tag{38}$$

By section 1.7, the general solution of this equation is

$$\tilde{e}_n = \sum_{s=1}^{k} d_s r_s^n - \phi/h\lambda \sum_{j=0}^{k} \beta_j, \tag{39}$$

where the d_s are arbitrary constants and the r_s are the roots, assumed distinct, of the polynomial equation

$$\sum_{j=0}^{k} (\alpha_j - h\lambda\beta_j)r^j = 0. \tag{40}$$

If the first and second characteristic polynomials of the linear multistep method are ρ and σ respectively (see equation (27) of chapter 2), then (40) may be written

$$\pi(r, \bar{h}) = \rho(r) - \bar{h}\sigma(r) = 0, \tag{41}$$

where

$$\bar{h} = h\lambda. \tag{42}$$

The polynomial $\pi(r, \bar{h})$ is frequently referred to as the *characteristic polynomial* of the method. However, since we have already used this phrase to describe $\rho(\zeta)$ and $\sigma(\zeta)$, we shall, instead, call $\pi(r, \bar{h})$ the *stability polynomial* of the method defined by ρ and σ.

We are interested, in this analysis, not in attempting to estimate the magnitude of $\tilde{e}_n$, but in deciding whether or not $\tilde{e}_n$ will grow with n. Thus, from (39) we are motivated to make the following definition:

Definition The linear multistep method (34) is said to be **absolutely stable** *for a given $\bar{h}$ if, for that $\bar{h}$, all the roots r_s of (41) satisfy $|r_s| < 1, s = 1, 2, \dots, k$, and to be* **absolutely unstable** *for that $\bar{h}$ otherwise. An interval (α, β) of the real line is said to be an* **interval of absolute stability** *if the method is absolutely stable for all $\bar{h} \in (\alpha, \beta)$. If the method is absolutely unstable for all $\bar{h}$ it is said to have* **no interval of absolute stability**.

It is easy to show that in the case when the linearized error equation has multiple roots, the condition $|r_s| < 1$ is still sufficient to cause $\tilde{e}_n$ to decay as n increases. This would not have been the case had the definition required that $|r_s| \leqslant 1$.

The interval of absolute stability is determined only by the coefficients of the method. However, the correspondingly largest value of h for which $\tilde{e}_n$ will not increase is, through (42), dependent on λ and hence on the particular differential equation whose solution is sought. For a general non-linear equation, $\partial f/\partial y$ will not be constant, and we choose for λ either a bound or a typical value for $\partial f/\partial y$, possibly holding over a subinterval of the range of integration. If a different value for λ is chosen in a subsequent subinterval, there will be a corresponding change in the maximum allowable value for h.

From (41), we see that when $\bar{h} = 0$ the roots r_s coincide with the zeros ζ_s of the first characteristic polynomial $\rho(\zeta)$, which, by zero-stability, all lie in or on the unit circle. Consistency and zero-stability imply that $\rho(\zeta)$ has a simple zero at $+1$; we have previously labelled this zero ζ_1. Let r_1 be the root of (41) which tends to ζ_1 as $\bar{h} \to 0$. (It can be shown that the roots r_s are continuous functions of $\bar{h}$.) We shall now show that

$$r_1 = \exp(\bar{h}) + O(\bar{h}^{p+1}), \quad \text{as } \bar{h} \to 0, \tag{43}$$

where p is the order of the linear multistep method. By definition of order, $\mathscr{L}[y(x); h] = O(h^{p+1})$, for any sufficiently differentiable function $y(x)$. Choosing $y(x) = \exp(\lambda x)$, we obtain

$$\mathscr{L}[\exp(\lambda x_n); h] = \sum_{j=0}^{k} \{\alpha_j \exp[\lambda(x_n + jh)] - h\beta_j \lambda \exp[\lambda(x_n + jh)]\} = O(\bar{h}^{p+1})$$

or

$$\exp(\lambda x_n) \sum_{j=0}^{k} \{\alpha_j[\exp(\bar{h})]^j - \bar{h}\beta_j[\exp(\bar{h})]^j\} = O(\bar{h}^{p+1}).$$

On dividing by $\exp(\lambda x_n)$, we obtain

$$\pi(\exp(\bar{h}), \bar{h}) \equiv \rho(\exp(\bar{h})) - \bar{h}\sigma(\exp(\bar{h})) = O(\bar{h}^{p+1}).$$

Since the roots of (41) are r_s, $s = 1, 2, \ldots, k$, we may write

$$\pi(r, \bar{h}) \equiv \rho(r) - \bar{h}\sigma(r) \equiv (\alpha_k - \bar{h}\beta_k)(r - r_1)(r - r_2)\ldots(r - r_k),$$

and set $r = \exp(\bar{h})$ to obtain

$$(\exp(\bar{h}) - r_1)(\exp(\bar{h}) - r_2)\ldots(\exp(\bar{h}) - r_k) = O(\bar{h}^{p+1}).$$

As $\bar{h} \to 0$, $\exp(\bar{h}) \to 1$, and $r_s \to \zeta_s$, $s = 1, 2, \ldots, k$. Thus the first factor on the left-hand side tends to zero as $\bar{h} \to 0$; no other factor may do so, since, by zero-stability, ζ_1 $(= +1)$ is a simple zero of $\rho(\zeta)$. Hence we may conclude that $\exp(\bar{h}) - r_1 = O(\bar{h}^{p+1})$, establishing (43).

It follows from (43) that for sufficiently small $\bar{h}$, $r_1 > 1$ whenever $\bar{h} > 0$. *Thus, every consistent zero-stable linear multistep method is absolutely unstable for small positive* $\bar{h}$.†

Having shown that the interval of absolute stability cannot include the positive real line in the neighbourhood of the origin, it is natural to ask whether there always exists some $\bar{h}$ for which the method is absolutely stable. This is not so, and certain linear multistep methods do indeed have no interval of absolute stability; for such methods, growth of error with n is unavoidable.‡ In section 2.9 we defined optimal k-step methods to be zero-stable methods of order $k + 2$. It can be shown that every optimal linear multistep method has no interval of absolute stability. It is easy to see intuitively—and straightforward to prove analytically—that if all the *spurious* zeros of the first characteristic polynomial $\rho(\zeta)$ lie strictly inside the unit circle, then the method has a non-vanishing interval

† Hull and Newbery[77] define a method to be absolutely stable if $|r_s| < 1$, $s = 2, 3, \ldots, k$. In this case the italicized result does not hold, since r_1 does not enter the criterion. The definition we have given has the advantage that it is pertinent to our discussion of stiff systems in chapter 8. For comments on the proliferation of definitions in weak stability theory, see the footnote following the definition of relative stability.

‡ Consistent zero-stable methods which have no interval of absolute stability are sometimes called 'weakly stable', and those with a non-vanishing interval 'strongly stable'. Confusingly, the former class is sometimes said to suffer a 'weak instability'. The situation is further confused by the possibility of defining these terms with respect to *relative stability*, which we shall introduce presently. For these reasons, we prefer to avoid such definitions and reserve the phrase 'weak stability' to act as a general description of the theory we are now developing.

of absolute stability. Again intuitively, we expect that if all the spurious zeros are situated at the origin (that is, if the method is an Adams method) the interval of absolute stability will be substantial.

We now return to the fact, italicized above, that all linear multistep methods are absolutely unstable for small positive $\bar{h}$. This implies that whenever we use the method to integrate numerically a differential equation for which $\partial f/\partial y > 0$, the error will increase as the computation proceeds. This is not as serious a drawback as might be imagined, for recall that the assumption (37*i*) holds strictly only for equations of the type $y' = \lambda y + g(x)$, λ constant. The solution of this equation contains a term $A \exp(\lambda x)$. (We exclude the somewhat pathological case where the initial condition is such that the arbitrary constant A takes the value zero.) The solution of the linearized error equation contains a term r_1^n and, using (43), we may write $r_1^n = (\exp(\bar{h}))^n + O(\bar{h}^p)$. Since $(\exp(\bar{h}))^n = \exp(\lambda(x_n - a))$ it follows that, *if the solution of the error equation is dominated by the term in* r_1^n, then the error grows at a rate similar to that at which the solution grows—a state of affairs which is generally acceptable. The italicized phrase motivates the following alternative definition, which was originally proposed by Hull and Newbery:[77]

Definition† *The linear multistep method (34) is said to be* **relatively stable** *for a given* $\bar{h}$ *if, for that* $\bar{h}$, *the roots* r_s *of (41) satisfy* $|r_s| < |r_1|$, $s = 2, 3, \ldots, k$, *and to be* **relatively unstable** *otherwise. An interval* (α, β) *of the real line is said to be an* **interval of relative stability** *if the method is relatively stable for all* $\bar{h} \in (\alpha, \beta)$.

Linear multistep methods which have no interval of absolute stability may have non-vanishing intervals of relative stability. Consider, for example, Simpson's rule which, being an optimal method, turns out to have no interval of absolute stability. For this method, $\rho(r) = r^2 - 1$, $\sigma(r) = \frac{1}{3}(r^2 + 4r + 1)$ and equation (41) becomes

$$(1 - \tfrac{1}{3}\bar{h})r^2 - \tfrac{4}{3}\bar{h}r - (1 + \tfrac{1}{3}\bar{h}) = 0. \tag{44}$$

It is easily established that the roots r_1 and r_2 of this equation are real for all values of $\bar{h}$. The expressions for these roots as functions of $\bar{h}$ are

† The literature in this area contains many similar but slightly differing definitions. Thus, Ralston[149] requires that $|r_s| \leqslant |r_1|$, $s = 2, 3, \ldots, k$, and that when $|r_s| = |r_1|$, r_s must be a simple root. Klopfenstein and Millman[86] demand that $|r_s| \leqslant \exp(\bar{h})$, $s = 2, 3, \ldots, k$, while Krogh[90] adopts the same definition but requires, in addition, that roots of magnitude $\exp(\bar{h})$ be simple. Crane and Klopfenstein[33] have the same requirement as Krogh, but demand that it hold for $s = 1, 2, \ldots, k$. In view of (43) all these definitions of relative stability are seen to be virtually equivalent for sufficiently small $\bar{h}$, but it can be important, when comparing results from different sources, to be aware of the precise definition being employed. Stetter's[170,171] usage of the terms 'weak' and 'strong' stability differs from ours.

complicated by the appearance of square roots. (In any case, if we were to treat similarly a k-step method with $k \geqslant 3$ it would be impracticable to attempt to find closed analytical expressions for the roots r_s in terms of $\bar{h}$.) However, some insight can be obtained by considering $O(\bar{h})$ approximations to the roots. We already know from (43) that $r_1 = 1 + \bar{h} + O(\bar{h}^2)$. The spurious root of $\rho(\zeta) = 0$ is $\zeta_2 = -1$, so we write $r_2 = -1 + \gamma\bar{h} + O(\bar{h}^2)$. On substituting this value for r in (44), we obtain $\gamma = \frac{1}{3}$. For sufficiently small $\bar{h}$ we can ignore the $O(\bar{h}^2)$ terms and write $r_1 = 1 + \bar{h}$, $r_2 = -1 + \frac{1}{3}\bar{h}$. The positions of these roots relative to the unit circle for $\bar{h} = 0$, $\bar{h} > 0$ and $\bar{h} < 0$ are shown in figure 2.

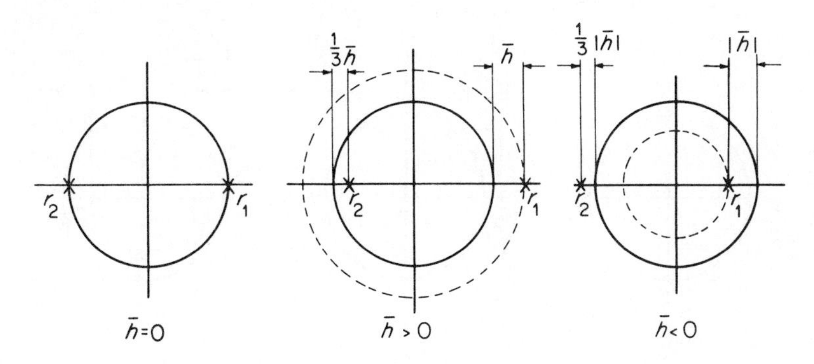

FIGURE 2

It is at once clear that the method has no interval of absolute stability since, if $\bar{h}$ moves away from zero, either positively or negatively, one of the roots will have modulus greater than unity. On the other hand, the interval of relative stability must be of the form $(0, \beta)$, $\beta > 0$, since, for small positive $\bar{h}$, $|r_2| < |r_1|$, while, for small negative $\bar{h}$, $|r_2| > |r_1|$. Using the full expressions for r_1 and r_2 got by solving (44) exactly, it is not hard to show that the interval of relative stability for Simpson's rule is $(0, \infty)$. We cannot, of course, deduce this from our present treatment, which is limited to the case where $\bar{h}$ is small. The implication of this result is that a growing error is unavoidable with Simpson's rule. If $\partial f/\partial y < 0$, it is clearly inadvisable to use this method; if $\partial f/\partial y > 0$, however, the error will not grow relative to the solution. Since for a general problem we do not know in advance the sign of $\partial f/\partial y$ over the range of integration, we conclude that Simpson's rule cannot be recommended as a general procedure. The same comment applies to any optimal method.

It is a mistake to take the view that the sole purpose of introducing the concept of relative stability is to enable us to deal with the case $\bar{h} > 0$.

Consider the following simple example used by Stetter[171] to illustrate the point.

$$y_{n+2} - y_n = \tfrac{1}{2}h(f_{n+1} + 3f_n). \tag{45}$$

This method is consistent and zero-stable. Equation (41) becomes

$$r^2 - \tfrac{1}{2}\bar{h}r - (1 + \tfrac{3}{2}\bar{h}) = 0.$$

Applying the approximate analysis we used for Simpson's rule, we obtain that for sufficiently small $\bar{h}$, $r_1 = 1 + \bar{h}$, $r_2 = -1 - \frac{1}{2}\bar{h}$. The positions of these roots relative to the unit circle are shown in figure 3.

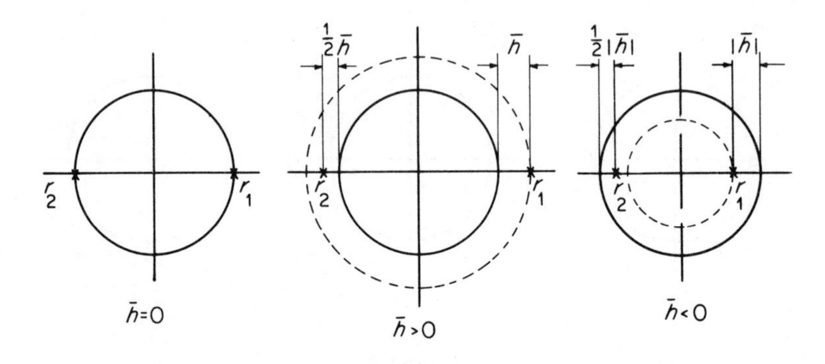

FIGURE 3

It follows that the interval of absolute stability is of the form $(\alpha, 0)$, $\alpha < 0$, while the interval of relative stability takes the form $(0, \beta)$, $\beta > 0$. If we adopt the absolute definition, then we would use the method only for an equation for which $\partial f/\partial y < 0$; in that case the modulus of the error would decay, but less slowly than that of the solution, the former being dominated by a term which behaves like $|r_2|^n$, the latter by one behaving like $|r_1|^n$. If we adopt the relative definition, then we would use the method only when $\partial f/\partial y > 0$ and accept an error which grows in modulus but more slowly than does the solution.

The above example makes us tend to prefer the concept of relative stability to that of absolute stability. However, let us consider a further example. We investigate the relative stability of the consistent zero-stable method

$$y_{n+3} - y_{n+2} + y_{n+1} - y_n = \frac{h}{5}(f_{n+2} - 4f_{n+1} + 13f_n). \tag{46}$$

Equation (41) becomes

$$r^3 - (1 + \tfrac{1}{5}\bar{h})r^2 + (1 + \tfrac{4}{5}\bar{h})r - (1 + \tfrac{13}{5}\bar{h}) = 0. \quad (47)$$

When $\bar{h} = 0$, the roots of this equation are

$$r_1 = 1, \qquad r_2 = i, \qquad r_3 = -i.$$

We know that when $\bar{h} \neq 0$, $r_1 = 1 + \bar{h} + O(\bar{h}^2)$; we can write $O(\bar{h})$ approximations to r_2 and r_3 in the form

$$r_2 = (1 + \gamma\bar{h}) \exp[i(\pi/2 + \theta\bar{h})], \qquad r_3 = (1 + \gamma\bar{h}) \exp[-i(\pi/2 + \theta\bar{h})],$$

since, for sufficiently small $\bar{h}$, r_2 and r_3 will remain complex and, therefore, conjugate. It follows that if we ignore $O(\bar{h}^2)$ terms equation (47) is identical with the equation

$$(r - 1 - \bar{h})\{r - (1 + \gamma\bar{h}) \exp[i(\pi/2 + \theta\bar{h})]\}\{r - (1 + \gamma\bar{h}) \times \exp[-i(\pi/2 + \theta\bar{h})]\} = 0.$$

From consideration of the term independent of r, it follows that

$$-(1 + \bar{h})(1 + \gamma\bar{h})^2 = -(1 + \tfrac{13}{5}\bar{h}).$$

Ignoring the terms in $\bar{h}^2$ and $\bar{h}^3$, we obtain that $\gamma = \frac{4}{5}$. Hence $|r_2| = |r_3| = 1 + \frac{4}{5}\bar{h} + O(\bar{h}^2)$, while $|r_1| = 1 + \bar{h} + O(\bar{h}^2)$. We conclude that the interval of relative stability is of the form $(0, \beta)$, $\beta > 0$. However, if we calculate the *exact* interval of relative stability a somewhat peculiar result emerges.

A tedious calculation involving the discriminant of the cubic (47) (see, for example, Turnbull[177]) shows that for all values of $\bar{h}$ (47) has one real root, r_1, and two complex roots, r_2, r_3. Moreover, except at the point $\bar{h} = -\frac{5}{13}$, where $r_1 = 0$, the function $|r_2|/|r_1|$ ($\equiv |r_3|/|r_1|$ since r_2 and r_3 are complex conjugates) is a continuous function of $\bar{h}$. It follows that the end-points of the interval of relative stability will occur when

$$r_1 = R, \qquad r_2 = R \exp(i\theta), \qquad r_3 = R \exp(-i\theta), \qquad R > 0,$$

or when

$$r_1 = R, \qquad r_2 = -R \exp(i\theta), \qquad r_3 = -R \exp(-i\theta), \qquad R < 0.$$

Thus at values of $\bar{h}$ which correspond to end-points of the interval of relative stability, (47) must be identical with the equation

$$(r - R)[r \mp R \exp(i\theta)][r \mp R \exp(-i\theta)] = 0,$$

or

$$r^3 - (1 \pm 2\cos\theta)Rr^2 + (1 \pm 2\cos\theta)R^2 r - R^3 = 0,$$

where we take the + signs when $R > 0$ and the − signs when $R < 0$. Comparing with (47), we have

$$1 + \bar{h}/5 = (1 \pm 2\cos\theta)R; \qquad 1 + 4\bar{h}/5 = (1 \pm 2\cos\theta)R^2;$$
$$1 + 13\bar{h}/5 = R^3.$$

Thus a necessary condition for $\bar{h}^*$ to be an end-point of the interval of relative stability is that $\bar{h}^*$ be a root of the equation

$$(1 + 4\bar{h}/5)^3 = (1 + 13\bar{h}/5)(1 + \bar{h}/5)^3.$$

This quartic equation in $\bar{h}$ turns out to have four real roots, namely 0, $(-5 \pm 15\sqrt{3})/13$, 10; moreover, we find that all four of them do indeed represent end-points of intervals of relative stability! Specifically, we find the following

$$\bar{h}^* = 0; \qquad R = 1, \qquad \theta = \pm\pi/2,$$

$$\bar{h}^* = (-5 + 15\sqrt{3})/13; \qquad R = \sqrt{3}, \qquad \theta = \pi \pm \cos^{-1}[(2\sqrt{3} - 5)/13],$$

$$\bar{h}^* = (-5 - 15\sqrt{3})/13; \qquad R = -\sqrt{3}, \qquad \theta = \pi \pm \cos^{-1}[(2\sqrt{3} + 5)/13],$$

$$\bar{h}^* = 10; \qquad R = 3, \qquad \theta = \pm\pi/2.$$

As $\bar{h} \to \pm\infty$ it follows from (47) that $|r_1| \to \infty$ while r_2 and r_3 tend to the roots of $\sigma(r) = 0$, that is, to finite complex numbers. We now have enough information to sketch the form of the graph of $|r_2|/|r_1|$ against $\bar{h}$; the result is shown in figure 4. It follows that the method is relatively stable if

$$\bar{h} \in (-\infty, (-5 - 15\sqrt{3})/13),$$
$$\bar{h} \in (0, (-5 + 15\sqrt{3})/13),$$

or

$$\bar{h} \in (10, \infty),$$

and is relatively unstable otherwise.

The first conclusion we can draw from this example is that the interval of relative stability, hitherto tacitly assumed to be a single interval, can, in fact, be the union of a number of disjoint intervals. This is not a point of

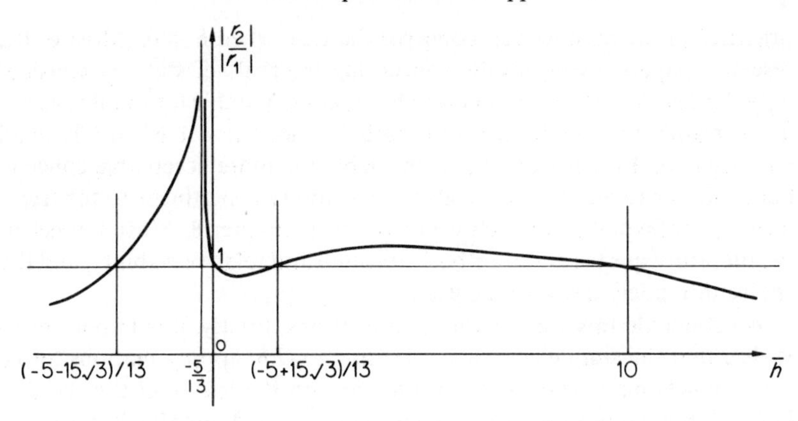

FIGURE 4

great practical significance, however, since we normally operate with a small value for h (in order to keep the local truncation error acceptably small) and are frequently interested in recomputing with a smaller value for h. Our primary interest, therefore, concerns intervals of stability in the immediate neighbourhood of the origin. One practical point does emerge, however. Suppose we have no knowledge whatsoever of the interval of stability (absolute or relative). Knowing λ, we could select a value for h, compute the roots of (41), and ascertain from the appropriate definition that we are operating inside an interval of stability. If we subsequently reduce h, we cannot conclude, without further investigation, that we continue to operate in an interval of stability.

We might conclude from figure 4 that it would be disastrous to compute with $\bar{h}$ in the neighbourhood of $-\frac{5}{13}$. This is not the case, since with $\bar{h} = -\frac{5}{13}$, $r_1 = 0$, $|r_2| = |r_3| = \frac{9}{13}$, and the error does not grow with n. Relative stability (as we have defined it) makes us unduly pessimistic in this example, just as absolute stability made us unduly optimistic for the method (45). Note that the statement that $r_1 = 0$ at $\bar{h} = -\frac{5}{13}$ does not contradict the statement (43) that $r_1 = \exp(\bar{h}) + O(\bar{h}^{p+1})$; the latter is an asymptotic statement which holds only as $\bar{h} \to 0$. Had we adopted one of the definitions of relative stability which replace $|r_1|$ by $\exp \bar{h}$ then, for this somewhat bizarre example, figure 4 would look very different; in particular, we would not be led to expect a calamity at $\bar{h} = -\frac{5}{13}$. The analysis, on the other hand, would be very much more complicated.

In summarizing the results of the above discussion, we can say that for small positive $\bar{h}$, all consistent zero-stable methods are unstable in the absolute sense; relative stability may be appropriate here. The main practical problem, however, concerns the case when $\bar{h} < 0$. (Most of the research papers giving results concerning intervals of stability consider only the case $\bar{h} < 0$.) A method which is relatively stable for small negative $\bar{h}$ is certainly absolutely stable for such $\bar{h}$, since $|r_1| < 1$ when $\bar{h}$ is small and negative. Relative stability is probably the more acceptable concept, but is harder to analyse, while absolute stability is pertinent to the treatment of stiff equations which we consider in chapter 8. Most sources in the literature quote results for both absolute and relative stability, and the choice ultimately lies with the user.

We conclude this section with two cautions. Firstly, it is important to realize that the simplified analysis leading to $O(\bar{h})$ approximations to the roots of (41) may yield useful information on the *nature* of the stability interval, but cannot be used to determine its *size*. A simple illustration is afforded by considering intervals of absolute stability for Euler's rule and the Trapezoidal rule. Both are one-step methods and thus (41) has only one root r_1. It is easily established that for Euler's rule $r_1 = 1 + \bar{h}$ and for the Trapezoidal rule $r_1 = (1 + \frac{1}{2}\bar{h})/(1 - \frac{1}{2}\bar{h})$; both expressions are exact. It follows that the interval of absolute stability for Euler's rule is $(-2, 0)$, while that for the Trapezoidal rule is $(-\infty, 0)$. On the other hand, if we accept $O(\bar{h})$ approximations, then $r_1 = 1 + \bar{h}$ for both methods. (Note that our definition of relative stability is not applicable to one-step methods.)

Secondly, it is unwise to attempt to draw too sharp conclusions from weak stability theory, since it is very much subject to the substantial hypothesis (37*i*) that $\partial f/\partial y$ is constant. For a highly non-linear problem the theory gives only a rough guide to the maximum allowable value for h, and it can be dangerous to choose h close to this maximum.

Example 5 Consider the method

$$y_{n+2} - (1 + a)y_{n+1} + ay_n = \frac{h}{12}[(5 + a)f_{n+2} + 8(1 - a)f_{n+1} - (1 + 5a)f_n],$$

where $-1 \leqslant a < 1$.

(*i*) *Prove that the interval of absolute stability is* $(6(a + 1)/(a - 1), 0)$ *and that the interval of relative stability is* $(3(a + 1)/2(a - 1), \infty)$.

(*ii*) *Illustrate in the case* $a = -0{\cdot}9$, *by computing solutions to the initial value problem* $y' = -20y$, $y(0) = 1$.

(i) The condition $-1 \leqslant a < 1$ guarantees zero-stability. The method has order 3 if $a \neq -1$, and order 4 if $a = -1$ (Simpson's rule). The stability polynomial is

$$\pi(r, \bar{h}) = \left[1 - \frac{\bar{h}}{12}(5 + a)\right]r^2 - \left[(1 + a) + \frac{2\bar{h}}{3}(1 - a)\right]r + \left[a + \frac{\bar{h}}{12}(1 + 5a)\right] = 0. \quad (48)$$

The discriminant, Δ, of this quadratic may be written

$$\Delta = (1 - a)^2 + \bar{h}(1 - a^2) + \tfrac{1}{12}\bar{h}^2(7 - 2a + 7a^2).$$

Regarding Δ as a quadratic in $\bar{h}$ we find that the discriminant of *this* quadratic is $-\frac{4}{3}(1 - a)^4 < 0$ for all a.

Hence Δ is strictly positive for all values of $\bar{h}$ and the equation (48) has real distinct roots r_1, r_2 for all $\bar{h}$, for all a.

Absolute stability. We already know that one end of the interval of absolute stability corresponds to (48) having a root $+1$. Since the roots are always real, the other end of the interval must correspond to (48) having a root -1. Putting $r = -1$ in (48) gives $\bar{h} = 6(a + 1)/(a - 1) < 0$. Hence the interval of absolute stability is $(6(a + 1)/(a - 1), 0)$. Note that it vanishes when $a = -1$ (Simpson's rule) and grows infinitely large as a approaches $+1$; but note that as a gets arbitrarily close to $+1$, zero-instability threatens.

Relative Stability. Since r_1 and r_2 are always real, the end-points of the interval of relative stability are given by $r_1 = r_2$ and $r_1 = -r_2$. We have shown that the former never occurs, that is, the interval extends to $+\infty$; the latter occurs when the coefficient of r in (48) vanishes, that is when $\bar{h} = 3(a + 1)/2(a - 1)$. The range of relative stability is thus $(3(a + 1)/2(a - 1), \infty)$.

(ii) For $y' = -20y$, $\partial f/\partial y = \lambda = -20$.

Putting $a = -0{\cdot}9$ in the results of (i), we see that the method will be absolutely stable if $-0{\cdot}316 < -20h < 0$; that is, if $h < 0{\cdot}016$; it will be relatively stable if $-0{\cdot}079 < -20h < \infty$, that is, if $h < 0{\cdot}003{,}95$. The initial value problem is solved numerically for $h = 0{\cdot}01, 0{\cdot}02, 0{\cdot}04$; the absolute values of the errors are shown in table 6 and compared with the theoretical solution. Starting values are taken from the theoretical solution.

Table 6

x	Theoretical solution	Absolute value of error in numerical solution		
		$h = 0{\cdot}01$	$h = 0{\cdot}02$	$h = 0{\cdot}04$
0·0	1·000,000,00	0	0	0
0·2	0·018,315,64	27×10^{-8}	12×10^{-8}	0·000,259,74
0·4	0·000,335,46	11×10^{-8}	27×10^{-8}	0·000,430,32
0·6	0·000,006,14	7×10^{-8}	35×10^{-8}	0·000,933,81
0·8	0·000,000,11	3×10^{-8}	46×10^{-8}	0·002,014,26
1·0	0·000,000,00	6×10^{-8}	63×10^{-8}	0·004,345,24

For $h = 0{\cdot}01$, the errors are seen to decay (making allowance for round-off error), while for $h = 0{\cdot}02$ and $h = 0{\cdot}04$, they increase. Only the first of these values for h is within the interval permitted by absolute stability. All three values are greater than the maximum value for h permitted by relative stability; observe that even in the case $h = 0{\cdot}01$ where the errors decay they do so less rapidly than the solution. We see from this example what a severe requirement relative stability can be.

Example 6 *Demonstrate that Simpson's rule is unstable in both absolute and relative senses when $\partial f/\partial y < 0$, by using it to compute a numerical solution to the initial value problem $y' = -10(y-1)^2$, $y(0) = 2$.*

Since $\partial f/\partial y$ is $-20(y-1)$, it is clearly negative in the neighbourhood of the origin. (The theoretical solution is $y = 1 + 1/(1+10x)$ and thus $\partial f/\partial y$ is in fact negative for all $x \geqslant 0$.) The numerical solution displayed in table 7 is computed with $h = 0{\cdot}1$, the starting values being taken from the theoretical solution.

Table 7

x	Theoretical solution	Solution by Simpson's rule
0	2·000,000	2·000,000
0·1	1·500,000	1·500,000
0·2	1·333,333	1·302,776
0·3	1·250,000	1·270,115
0·4	1·200,000	1·165,775
⋮	⋮	⋮
3·8	1·025,641	0·867,153
3·9	1·025,000	0·953,325
4·0	1·024,390	0·850,962
⋮	⋮	⋮
4·8	1·020,408	0·040,686
4·9	1·020,000	−5·990,968
5·0	1·019,608	−394·086

In example 5, we were operating with a linear equation close to the ends of the interval of absolute stability; the instabilities invoked were therefore numerically mild. The present example demonstrates that the growth of error due to weak instability can be sufficiently violent to swamp the genuine solution, and corroborates our finding that optimal methods cannot be recommended as general purpose methods.

Note that the method is still convergent. If we were to continue to reduce the steplength, then the point x_n at which the numerical solution differs from the theoretical by a given amount (or fraction) would keep moving to the right. In the limit as $h \to 0$, $n \to \infty$, the numerical solution would converge to the theoretical solution but for any fixed positive h, no matter how small, the absolute value of the numerical solution tends to infinity as n tends to infinity.

Exercises

6. Show that the method $y_{n+1} - y_n = hf_{n+1}$ is absolutely stable for all $\bar{h} \notin [0, 2]$.

7. Repeat the analysis of example 5 for the two-step explicit method,

$$y_{n+2} - (1+a)y_{n+1} + ay_n = \tfrac{1}{2}h[(3-a)f_{n+1} - (1+a)f_n], \qquad -1 \leqslant a < 1.$$

8. The method

$$y_{n+3} + \alpha(y_{n+2} - y_{n+1}) - y_n = \tfrac{1}{2}(3 + \alpha)h(f_{n+2} + f_{n+1})$$

is zero-stable if $-3 < \alpha < 1$. (See exercise 11 of chapter 2, page 40.) By finding $O(\bar{h})$ approximations to the roots of the stability polynomial, $\pi(r, \bar{h})$, show that, for all α in the above range, the method has no interval of absolute stability, and an interval of relative stability of the form $(0, \beta)$, $\beta > 0$.

3.7 General methods for finding intervals of absolute and relative stability

In the previous section we succeeded in finding *precise* intervals of absolute or relative stability for specific methods only by using special algebraic properties of polynomials of low degree; the devices we employed do not extend to deal with the case of a general k-step method. In this section we consider some general methods for establishing intervals of stability.

The first and most direct method is the *root locus method.* This consists of repeatedly solving the polynomial equation (41) for a range of values of $\bar{h}$ in the neighbourhood of the origin. Any standard numerical method, such as Newton–Raphson iteration, may be employed for the approximate solution of (41). A plot of $|r_s|$, $s = 1, 2, \ldots, k$ against $\bar{h}$ then allows us to deduce intervals of stability in the neighbourhood of the origin. The literature contains root locus plots for a number of methods, but these are mainly predictor–corrector pairs—the most practical form of linear multistep method. After we have discussed such methods, some results from the literature concerning intervals of stability will be quoted.

Example 7 Use the root locus method to find the intervals of absolute and relative stability for method (45) (page 70).

For this method, $\pi(r, \bar{h}) = r^2 - \tfrac{1}{2}\bar{h}r - (1 + \tfrac{3}{2}\bar{h}) = 0$. The roots r_1 and r_2 of this equation are found for $\bar{h} = -2{\cdot}0(0{\cdot}2)(1{\cdot}0)$, and figure 5 shows a plot of $|r_1|$ and $|r_2|$ against $\bar{h}$. The interval of absolute stability is seen to be $(-1{\cdot}33, 0)$, and the interval of relative stability to be of the form $(0, \beta)$, where $\beta > 1$; in fact, we can easily show that $\beta = \infty$.

This is in agreement with the qualitative results which we obtained earlier for this example. The graph of $\exp(\bar{h})$ is also shown in figure 5, and it is clear that if we adopt a definition of relative stability which demands that $|r_2| < \exp(\bar{h})$, then we again find that $(0, \beta)$ is an interval of relative stability. Note that there also exists an interval $(-0{\cdot}8, -0{\cdot}55)$ for which $|r_2| < \exp(\bar{h})$. We again emphasize that the various alternative definitions of relative stability are virtually equivalent only in some neighbourhood of the origin.

We shall call the second method we consider the *Schur criterion.* In fact, several criteria based on theorems of Schur[159] have been proposed—

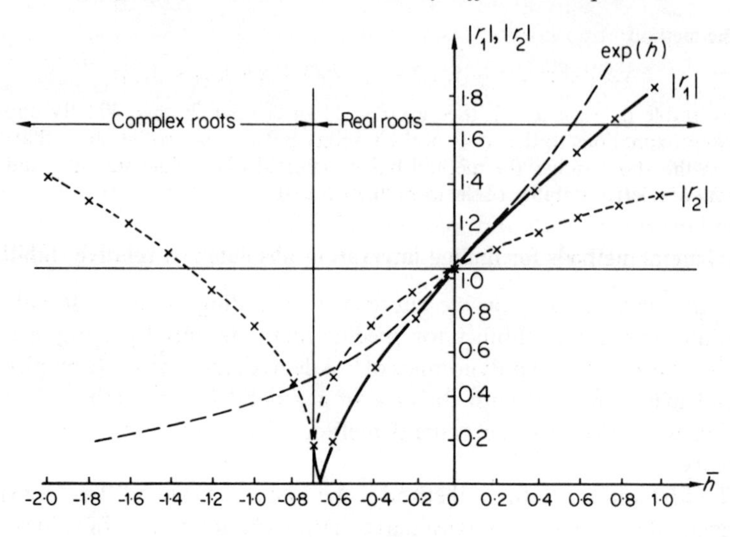

FIGURE 5

notably the *Wilf stability criterion* (Wilf,[182] Emanuel[40]). A useful source of criteria for the location of roots of polynomials arising in various branches of numerical analysis is a recent paper by Miller,[128] whose approach we follow here. Such an approach has the advantage of extending readily to the case of a system of differential equations, which, as we shall see in chapter 8, corresponds to the coefficients of the stability polynomial taking complex values. Accordingly, we shall state the criterion for a general kth degree polynomial, with complex coefficients,

$$\phi(r) = c_k r^k + c_{k-1} r^{k-1} + \ldots + c_1 r + c_0, \tag{49}$$

where $c_k \neq 0$, $c_0 \neq 0$. The polynomial $\phi(r)$ is said to be a *Schur polynomial* if its roots r_s satisfy $|r_s| < 1$, $s = 1, 2, \ldots, k$. Define the polynomials

$$\hat{\phi}(r) = c_0^* r^k + c_1^* r^{k-1} + \ldots + c_{k-1}^* r + c_k^*, \tag{50}$$

where c_j^* is the complex conjugate of c_j, and

$$\phi_1(r) = \frac{1}{r}[\hat{\phi}(0)\phi(r) - \phi(0)\hat{\phi}(r)]. \tag{51}$$

Clearly $\phi_1(r)$ has degree at most $k - 1$. Then, by a theorem of Schur,[159] $\phi(r)$ is a Schur polynomial if and only if $|\hat{\phi}(0)| > |\phi(0)|$ and $\phi_1(r)$ is a Schur polynomial.

Clearly, the interval (α, β) is an interval of absolute stability if, for all $\bar{h} \in (\alpha, \beta)$, the (real) stability polynomial $\pi(r, \bar{h})$, defined by (41), is a Schur polynomial. Writing $\pi(r, \bar{h})$ for $\phi(r)$ in (49), we can construct $\hat{\pi}(r, \bar{h})$ and $\pi_1(r, \bar{h})$ from (50) and (51). The first condition $|\hat{\pi}(0, \bar{h})| > |\pi(0, \bar{h})|$ yields our first inequality in $\bar{h}$, while the second condition may be tested by writing $\pi_1(r, \bar{h})$ for $\phi(r)$ in (49) and repeating the process, thereby obtaining a second inequality for $\bar{h}$, and so on. At each stage, the degree of the polynomial under test is reduced by one, so that eventually we merely have to state a criterion for a polynomial of degree one to be a Schur polynomial, and, obviously, this can easily be done.

We cannot use the Schur criterion directly to determine intervals of relative stability in the sense previously defined. However, if we adopt a definition of relative stability which requires that $|r_s| < \exp(\bar{h})$, $s = 1, 2, \ldots, k$, then it is technically possible to use the Schur criterion. Substituting $r = R\exp(\bar{h})$ into (41) gives a polynomial equation in R,

$$\rho(R\exp(\bar{h})) - \bar{h}\sigma(R\exp(\bar{h})) = 0.$$

If the roots of this equation are R_s, $s = 1, 2, \ldots, k$, then the Schur criterion gives necessary and sufficient conditions for $|R_s| < 1$, $s = 1, 2, \ldots, k$, that is, for $|r_s| < \exp(\bar{h})$, $s = 1, 2, \ldots, k$. The resulting inequalities, polynomial in $\bar{h}$ in the case of absolute stability, are now transcendental in $\bar{h}$, and can be awkward to handle.

Example 8 Use the Schur criterion to investigate the weak stability of method (45) (page 70).

The stability polynomial is $\pi(r, \bar{h}) = r^2 - \frac{1}{2}\bar{h}r - (1 + \frac{3}{2}\bar{h})$. Thus,

$$\hat{\pi}(r, \bar{h}) = -(1 + \tfrac{3}{2}\bar{h})r^2 - \tfrac{1}{2}\bar{h}r + 1$$

and the condition $|\hat{\pi}(0, \bar{h})| > |\pi(0, \bar{h})|$ is satisfied if $|1 + \frac{3}{2}\bar{h}| < 1$, that is, if $\bar{h} \in (-\frac{4}{3}, 0)$. By (51),

$$\begin{aligned}\pi_1(r, \bar{h}) = 1/r[r^2 - \tfrac{1}{2}\bar{h}r - (1 + \tfrac{3}{2}\bar{h}) + (1 + \tfrac{3}{2}\bar{h})\{-(1 + \tfrac{3}{2}\bar{h})r^2 - \tfrac{1}{2}\bar{h}r + 1\}]\\ = -\tfrac{1}{2}\bar{h}(2 + \tfrac{3}{2}\bar{h})(3r + 1),\end{aligned}$$

which has its only root at $-\frac{1}{3}$, and is therefore a Schur polynomial. It follows that the interval of absolute stability is $(-\frac{4}{3}, 0)$, in agreement with the root locus plot of figure 5.

Using the device described above to find, for this example, the interval of relative stability, given by the requirement that $|r_s| < \exp(\bar{h})$, $s = 1, 2, \ldots, k$, we consider the polynomial $\phi(R, \bar{h}) = \exp(2\bar{h})R^2 - \frac{1}{2}\bar{h}\exp(\bar{h})R - (1 + \frac{3}{2}\bar{h})$. Then,

$$\hat{\phi}(R, \bar{h}) = -(1 + \tfrac{3}{2}\bar{h})R^2 - \tfrac{1}{2}\bar{h}\exp(\bar{h})R + \exp(2\bar{h}),$$

and we obtain the first condition $|1 + \frac{3}{2}\bar{h}| < |\exp(2\bar{h})|$. After a little manipulation, we find that the first-degree polynomial $\phi_1(R, \bar{h})$ is Schur if and only if

$$|\{\tfrac{1}{2}\bar{h}\exp(\bar{h})\}/\{\exp(2\bar{h}) - 1 - \tfrac{3}{2}\bar{h}\}| < 1.$$

A full solution of this pair of simultaneous inequalities for $\bar{h}$ involves considerable computation, but, on expanding the exponentials in powers of $\bar{h}$, it becomes clear that both inequalities are satisfied for all positive $\bar{h}$ and that the second is not satisfied for small negative $\bar{h}$. We conclude that $(0, \infty)$ is an interval of relative stability.

An alternative to the Schur criterion, advocated by Fraboul,[46] consists of applying a transformation which maps the interior of the unit circle into the left-hand half-plane, and then appealing to the well-known *Routh–Hurwitz criterion*, which gives necessary and sufficient conditions for the roots of a polynomial to have negative real parts. The appropriate transformation is $r = (1 + z)/(1 - z)$; this maps the circle $|r| = 1$ into the imaginary axis $\operatorname{Re} z = 0$, the interior of the circle into the half-plane

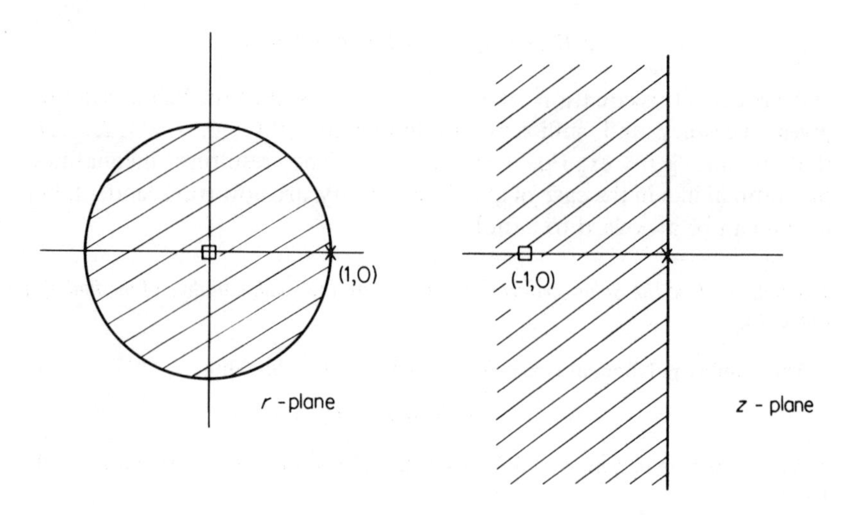

FIGURE 6

$\operatorname{Re} z < 0$, and the point $r = 1$ into $z = 0$. (See figure 6.) Under this transformation, equation (41) becomes

$$\rho((1 + z)/(1 - z)) - \bar{h}\sigma((1 + z)/(1 - z)) = 0.$$

On multiplying through by $(1 - z)^k$, this becomes a polynomial equation of degree k, which we write

$$a_0 z^k + a_1 z^{k-1} + \ldots + a_k = 0, \tag{52}$$

where we may assume, without loss of generality, that $a_0 > 0$. (Note that the convention we adopt for labelling the coefficients of (52) differs from that used in (49); this is done in order that we may apply the Routh–Hurwitz criterion in its conventional form.) The necessary and sufficient condition for the roots of (52) to lie in the half-plane Re $z < 0$, that is, for the roots of $\pi(r, \bar{h}) = 0$ to lie within the circle $|r| < 1$, is that all the leading principal minors of Q be positive, where Q is the $k \times k$ matrix defined by

$$Q = \begin{bmatrix} a_1 & a_3 & a_5 & \dots & a_{2k-1} \\ a_0 & a_2 & a_4 & \dots & a_{2k-2} \\ 0 & a_1 & a_3 & \dots & a_{2k-3} \\ 0 & a_0 & a_2 & \dots & a_{2k-4} \\ \vdots & \vdots & \vdots & & \vdots \\ 0 & 0 & 0 & & a_k \end{bmatrix},$$

and where $a_j = 0$ if $j > k$. (See, for example, Sanchez.[157]) It can be shown that this condition implies $a_j > 0$, $j = 0, 1, \dots, k$. Thus the positivity of the coefficients in (52) is a necessary but not sufficient condition for absolute stability. For $k = 2, 3, 4$ the necessary and sufficient conditions for absolute stability given by this criterion are as follows:

$k = 2$: $a_0 > 0, a_1 > 0, a_2 > 0$.

$k = 3$: $a_0 > 0, a_1 > 0, a_2 > 0, a_3 > 0, a_1 a_2 - a_3 a_0 > 0$.

$k = 4$: $a_0 > 0, a_1 > 0, a_2 > 0, a_3 > 0, a_4 > 0, a_1 a_2 a_3 - a_0 a_3^2 - a_4 a_1^2 > 0$.

Compared with the Schur criterion, this approach has the disadvantage that preliminary manipulation is necessary to obtain equation (52). On the other hand, for small values of k, it is easier to write down inequalities for $\bar{h}$ from the Routh–Hurwitz criterion than it is from the Schur criterion. Hall[63] uses a modified form of this approach to determine intervals of absolute stability for certain predictor–corrector pairs.

Example 9 Use the Routh–Hurwitz criterion to investigate the weak stability of method (45) (page 70).

The stability polynomial is $\pi(r, \bar{h}) = r^2 - \frac{1}{2}\bar{h}r - (1 + \frac{3}{2}\bar{h})$, giving, on transformation,

$$(1 - z)^2\left[\left(\frac{1 + z}{1 - z}\right)^2 - \tfrac{1}{2}\bar{h}\left(\frac{1 + z}{1 - z}\right) - (1 + \tfrac{3}{2}\bar{h})\right] = a_0 z^2 + a_1 z + a_2,$$

where $a_0 = -\bar{h}, a_1 = 4 + 3\bar{h}, a_2 = -2\bar{h}$.

The Routh–Hurwitz criterion is clearly satisfied if and only if $\bar{h} \in (-\frac{4}{3}, 0)$, which is the required interval of absolute stability.

If we investigate relative stability as given by the requirement $|r_s| < \exp(\bar{h})$, $s = 1, 2, \ldots, k$, we obtain

$$(1 - z)^2\left[\left(\frac{1+z}{1-z}\right)^2 \exp(2\bar{h}) - \tfrac{1}{2}\bar{h}\left(\frac{1+z}{1-z}\right)\exp(\bar{h}) - (1 + \tfrac{3}{2}\bar{h})\right] = a_0 z^2 + a_1 z + a_2,$$

where, for relative stability,

$$a_0 = \exp(2\bar{h}) + \tfrac{1}{2}\bar{h}\exp(\bar{h}) - 1 - \tfrac{3}{2}\bar{h} > 0,$$

$$a_1 = 2\exp(2\bar{h}) + 2 + 3\bar{h} > 0,$$

$$a_2 = \exp(2\bar{h}) - \tfrac{1}{2}\bar{h}\exp(\bar{h}) - 1 - \tfrac{3}{2}\bar{h} > 0.$$

Proceeding as in example 7, we find once again that $(0, \infty)$ is an interval of relative stability.

Finally, we consider a method for finding intervals of absolute stability which requires neither the computation of the roots of a polynomial nor the solving of simultaneous inequalities; we shall call it the *boundary locus method*. The roots r_s of the stability polynomial are, in general, complex numbers; for the moment let us regard $\bar{h}$ also as a complex number. Then, instead of defining an *interval* of absolute stability to be an interval of the real $\bar{h}$-line such that the roots of $\pi(r, \bar{h}) = 0$ lie within the unit circle whenever $\bar{h}$ lies in the interior of the interval, we define a *region* of absolute stability to be a region of the complex $\bar{h}$-plane such that the roots of $\pi(r, \bar{h}) = 0$ lie within the unit circle whenever $\bar{h}$ lies in the interior of the region. Let us call the region $\mathscr{R}$ and its boundary $\partial\mathscr{R}$. Since the roots of $\pi(r, \bar{h}) = 0$ are continuous functions of $\bar{h}$, $\bar{h}$ will lie on $\partial\mathscr{R}$ when one of the roots of $\pi(r, \bar{h}) = 0$ lies on the boundary of the unit circle, that is, when $\pi(\exp(i\theta), \bar{h}) \equiv \rho(\exp(i\theta)) - \bar{h}\sigma(\exp(i\theta)) = 0$. It follows that the locus of $\partial\mathscr{R}$ is given by $\bar{h}(\theta) = \rho(\exp(i\theta))/\sigma(\exp(i\theta))$, which we can plot in the complex $\bar{h}$-plane. For real $\bar{h}$, the end-points of the interval of absolute stability will be given by the points at which $\partial\mathscr{R}$ cuts the real axis. Note that this procedure tells us only the end-points of the interval or intervals of absolute stability. In order to determine which of the various possible intervals defined by these end-points are intervals of absolute stability, it is necessary to compute the roots at a number of appropriate spot values of $\bar{h}$.† We shall refer to the boundary locus method again in chapter 8, when we discuss systems of ordinary differential equations. We shall see that for systems of equations, the parameter $\bar{h}$ does in fact assume complex

† Frequently it will be possible to deduce which intervals are intervals of absolute stability by using the fact, proved in section 3.6, that every consistent zero-stable linear multistep method is absolutely unstable for small positive $\bar{h}$.

values, and that the whole of the region $\mathscr{R}$ of absolute stability, not only its intercept on the real axis, is then meaningful. The device we used to extend the Schur criterion to deal with relative stability, as defined by $|r_s| < \exp(\bar{h})$, $s = 1, 2, \ldots, k$, fails similarly to extend the boundary locus method. This is because the method depends on the fact that the equation $\pi(r, \bar{h}) \equiv \rho(r) - \bar{h}\sigma(r) = 0$ can be readily solved for $\bar{h}$; the corresponding equation for relative stability, $\rho(R|\exp(\bar{h})|) - \bar{h}\sigma(R|\exp(\bar{h})|) = 0$, is non-linear in $\bar{h}$ and it is not, in general, possible to obtain a solution for $\bar{h}$ in closed form. (Numerical solutions are, of course, feasible; this possibility is investigated in section 3.11.)

Example 10 Use the boundary locus method to establish the interval of absolute stability of method (45) (page 70).

For this method, $\rho(r) = r^2 - 1$, $\sigma(r) = \frac{1}{2}(r + 3)$;

$$\begin{aligned}\bar{h}(\theta) = \rho(\exp(i\theta))/\sigma(\exp(i\theta)) &= 2[\exp(2i\theta) - 1]/[\exp(i\theta) + 3]\\ &= [3(\cos 2\theta - 1) + i(3\sin 2\theta + 2\sin\theta)]/(5 + 3\cos\theta).\end{aligned}$$

This is the locus of $\partial\mathscr{R}$, and it crosses the real axis where $\sin\theta = 0$ or $3\cos\theta = -1$, that is, at $\theta = 0, \pi, \pi \pm \cos^{-1}(\frac{1}{3})$. At $\theta = 0, \pi$, $\bar{h}(\theta) = 0$, while at $\theta = \pi \pm \cos^{-1}(\frac{1}{3})$, $\bar{h}(\theta) = -\frac{4}{3}$. The end-points of the interval of absolute stability are thus $-\frac{4}{3}$ and 0. At the spot value $\bar{h} = -\frac{2}{3}$, the equation $\rho(r) - \bar{h}\sigma(r) = 0$ reduces to $r^2 + \frac{1}{3}r = 0$, whose roots are clearly within the unit circle. Alternatively, we can observe that the method is necessarily absolutely unstable for small positive $\bar{h}$. We conclude that the interval of absolute stability is $(-\frac{4}{3}, 0)$.

Exercises

9. Find the interval of absolute stability of (i) the two-step Adams–Bashforth method $y_{n+2} - y_{n+1} = \frac{1}{2}h(3f_{n+1} - f_n)$, and (ii) the three-step Adams–Moulton method $y_{n+3} - y_{n+2} = \frac{1}{24}h(9f_{n+3} + 19f_{n+2} - 5f_{n+1} + f_n)$, using each of the four methods described above.

Use the Schur or the Routh–Hurwitz criterion to establish intervals of relative stability for (i) and (ii), and compare the results with the intervals of relative stability given by the root locus method.

10. Use the boundary locus method to find the interval of absolute stability of the method of example 5 (page 74).

11. (i) Deduce from the boundary locus method that if the product

$$\rho(\exp(i\theta))\sigma(\exp(-i\theta))$$

is purely imaginary, then the end-points of the interval of absolute stability are all located at the origin, and that the method defined by ρ and σ is therefore absolutely stable either for no $\bar{h}$ or for all $\bar{h} < 0$.

(ii) Call the linear multistep method

$$\sum_{j=0}^{k} \alpha_j y_{n+j} = h \sum_{j=0}^{k} \beta_j f_{n+j}$$

symmetric if $\alpha_j = -\alpha_{k-j}, \beta_j = \beta_{k-j}, j = 0, 1, \ldots, k$. By evaluating the coefficients of $\cos l\theta, l = 0, 1, \ldots, k$ in the product $\rho(\exp(i\theta))\sigma(\exp(-i\theta))$ show that all symmetric methods are absolutely stable either for no $\bar{h}$ or for all $\bar{h} < 0$.

(iii) Observe that the Trapezoidal rule, Simpson's rule, and the method of exercise 8 (page 77) are all symmetric methods for which we have already found intervals of absolute stability. Verify that these previous results do not contradict (ii).

(iv) Quade's method (exercise 5 of chapter 2, page 27) is symmetric. By finding the zeros, at a convenient spot value of $\bar{h}$, of the stability polynomial for this method, deduce that the method has no interval of absolute stability.

12. Using (i) the Schur criterion and (ii) the Routh–Hurwitz criterion, sketch the region D of the a–b plane such that for all points (a, b) inside D the roots of the quadratic $r^2 - ar + b = 0$ lie within the unit circle. Use the result to verify that there are no values of a and b which cause the explicit and implicit three-step methods quoted in section 2.10 to have orders greater than three and four respectively, and to remain zero-stable.

13. Using the boundary locus method, compare the intervals of absolute stability of the most accurate k-step (explicit) Adams–Bashforth methods with those of the most accurate k-step (implicit) Adams–Moulton methods, in the cases $k = 1, 2, 3$. (See section 2.10 for lists of coefficients.) Observe that for each k, the implicit method possesses the larger interval of absolute stability, and that, for both explicit and implicit methods, the interval of absolute stability diminishes as k (and therefore the order) increases.

3.8 Comparison of explicit and implicit linear multistep methods

One of the four questions we posed at the beginning of this chapter has not yet been tackled, namely that of how best to solve, at each step of the computation, the implicit equation defining y_{n+k} in the case when the linear multistep method is implicit. Before answering this question, it is natural to ask why we wish to use implicit methods at all, since no such problem arises with explicit methods. It is clear from the methods quoted in section 2.10 that, for a given stepnumber k, the highest attainable order for a zero-stable method is less in the case of an explicit method than in the case of an implicit one; but we have seen that there is no serious difficulty in obtaining additional starting values and thus there is no apparent reason why we should not prefer an explicit method of higher stepnumber to an implicit method of the same order, but with lower stepnumber. Implicit methods, however, possess advantages over explicit methods other than that of higher order for given stepnumber. As an illustration, we compare explicit Adams–Bashforth methods with implicit Adams–Moulton methods for stepnumber $k = 1, 2, 3, 4$; in every case, the method considered has the highest attainable order. The coefficients can be obtained from the general methods quoted in section 2.10 by setting each of the parameters a, b, c to zero. The results of this comparison are shown in table 8, where p is the

Table 8

Adams–Bashforth (explicit)				
k	1	2	3	4
p	1	2	3	4
C_{p+1}	$\frac{1}{2}$	$\frac{5}{12}$	$\frac{3}{8}$	$\frac{251}{720}$
α	-2	-1	$-\frac{6}{11}$	$-\frac{3}{10}$
Adams–Moulton (implicit)				
k	1	2	3	4
p	2	3	4	5
C_{p+1}	$-\frac{1}{12}$	$-\frac{1}{24}$	$-\frac{19}{720}$	$-\frac{3}{160}$
α	$-\infty$	-6	-3	$-\frac{90}{49}$

order, C_{p+1} the error constant, and $(\alpha, 0)$ the interval of absolute stability. If we compare explicit and implicit methods of the same order, it is clear that the latter, besides having error constants of smaller absolute value, possess a considerable advantage in the size of the interval of absolute stability. Thus, for example, if we wish to use a fourth-order Adams method, we have the choice between the explicit four-step method and the implicit three-step; the latter has an interval of absolute stability which is ten times greater than that of the former and, moreover, its error constant is smaller by a factor of approximately $\frac{1}{13}$. These considerations, which are typical of more general comparisons between explicit and implicit methods, so favour implicit methods that explicit linear multistep methods are seldom used on their own; they do, however, play an important ancillary rôle in predictor–corrector pairs.

3.9 Predictor–corrector methods

Let us suppose that we intend using an implicit linear k-step method to solve our initial value problem. At each step we must solve for y_{n+k} the equation

$$y_{n+k} + \sum_{j=0}^{k-1} \alpha_j y_{n+j} = h\beta_k f(x_{n+k}, y_{n+k}) + h \sum_{j=0}^{k-1} \beta_j f_{n+j}, \tag{53}$$

where $y_{n+j}, f_{n+j}, j = 0, 1, \ldots, k-1$, are known; in general, this equation is non-linear. In section 2.1 we showed that a unique solution for y_{n+k} exists and can be approached arbitrarily closely by the iteration†

$$y_{n+k}^{[s+1]} + \sum_{j=0}^{k-1} \alpha_j y_{n+j} = h\beta_k f(x_{n+k}, y_{n+k}^{[s]}) + h \sum_{j=0}^{k-1} \beta_j f_{n+j}, \qquad s = 0, 1, \ldots, \tag{54}$$

where $y_{n+k}^{[0]}$ is arbitrary, provided that the condition

$$h < 1/L|\beta_k| \tag{55}$$

is satisfied. (Normally, the limit on the acceptable size of steplength h is determined by considerations other than (55). The exception arises in the case of a differential equation with a very large Lipschitz constant; we shall consider such cases in the sections in chapter 8 on stiff equations.)

Each step of the iteration (54) clearly involves an evaluation of $f(x_{n+k}, y_{n+k}^{[s]})$. We are thus concerned to keep to a minimum the number of times the iteration (54) is applied—particularly so when the evaluation of f at given values of its arguments is time-consuming. (Such an evaluation may call several subroutines.) We would therefore like to make the initial guess $y_{n+k}^{[0]}$ as accurate as we can. This we do by using a separate *explicit* method to estimate y_{n+k}, and taking this predicted value to be the initial guess $y_{n+k}^{[0]}$. The explicit method is called the *predictor*, and the implicit method (53) the *corrector*.

We can now proceed in one of two different ways. The first consists of continuing the iteration (54) until the iterates have converged (in practice, until some criterion such as $|y_{n+k}^{[s+1]} - y_{n+k}^{[s]}| < \varepsilon$, where ε is a pre-assigned tolerance, of the order of the local round-off error, say, is satisfied). We then regard the value $y_{n+k}^{[s+1]}$ so obtained as an acceptable approximation to the exact solution y_{n+k} of (53). Since each iteration corresponds to one application of the corrector, we shall call this mode of operation of the predictor–corrector method *correcting to convergence*. In this mode, we cannot tell in advance how many iterations will be necessary, that is, how many function evaluations will be required at each step. On the other hand, the accepted value $y_{n+k}^{[s+1]}$ being independent of the initial guess $y_{n+k}^{[0]}$, the local truncation error and the weak stability characteristics of the overall method are precisely those of the corrector alone; the properties of the predictor are of no importance. In particular, h must be chosen so that $\bar{h}$ lies within an interval of absolute or relative stability of the corrector;

† Iteration superscripts are in square brackets; superscripts denoting total derivatives are in round brackets.

no harm will be done if this value of $\bar{h}$ does not lie within a stability interval of the predictor (or even if the predictor is not zero-stable).

In the alternative approach, which is once again motivated by the desire to restrict the number of function evaluations per step, we stipulate in advance the number of times, m, we will apply the corrector at each step. It is no longer true that the local truncation error and weak stability characteristics of the overall method are those of the corrector alone, and the situation must be carefully analysed; this will be done in later sections. In the present section, we shall define these new modes of operation precisely, introducing a standard notation, initiated by Hull and Creemer,[72] which not only neatly describes the particular mode in which a predictor–corrector pair is applied, but also tells us immediately how many function evaluations per step are required.

Let P indicate an application of the predictor, C a *single* application of the corrector, and E an evaluation of f in terms of known values of its arguments. Suppose that we compute $y_{n+k}^{[0]}$ from the predictor, evaluate $f_{n+k}^{[0]} \equiv f(x_{n+k}, y_{n+k}^{[0]})$, and apply the corrector once to get $y_{n+k}^{[1]}$; the calculation so far is denoted by PEC. A further evaluation of $f_{n+k}^{[1]} \equiv f(x_{n+k}, y_{n+k}^{[1]})$ followed by a second application of the corrector yields $y_{n+k}^{[2]}$, and the calculation is now denoted by $PECEC$, or $P(EC)^2$. Applying the corrector m times is similarly denoted by $P(EC)^m$. Since m is fixed, we accept $y_{n+k}^{[m]}$ as the numerical solution at x_{n+k}. At this stage, the last computed value we have for f_{n+k} is $f_{n+k}^{[m-1]} \equiv f(x_{n+k}, y_{n+k}^{[m-1]})$, and we have a further decision to make, namely, whether or not to evaluate $f_{n+k}^{[m]} \equiv f(x_{n+k}, y_{n+k}^{[m]})$. If this final evaluation is made, we denote the mode by $P(EC)^mE$, and if not, by $P(EC)^m$. This choice clearly affects the next step of the calculation, since both predicted and corrected values for y_{n+k+1} will depend on whether f_{n+k} is taken to be $f_{n+k}^{[m]}$ or $f_{n+k}^{[m-1]}$. Note that, for a given m, both $P(EC)^mE$ and $P(EC)^m$ modes apply the corrector the same number of times; but the former calls for one more function evaluation per step than the latter. We shall see in section 3.11 that the weak stability characteristics of $P(EC)^mE$ modes are quite different from those of $P(EC)^m$ modes.

We now define the above modes precisely. It will turn out to be advantageous if the predictor and the corrector are separately of the same order, and this requirement may well make it necessary for the stepnumber of the predictor to be greater than that of the corrector. (See, for example, table 8.) The notationally simplest way to deal with this contingency is to let both predictor and corrector have the same stepnumber k, but in the case of the corrector, to relax the condition, stated in section 2.1, that not both α_0 and β_0 shall vanish. Let the linear multistep method used as predictor be defined by the characteristic polynomials

$$\rho^*(\zeta) = \sum_{j=0}^{k} \alpha_j^* \zeta^j, \qquad \alpha_k^* = 1, \qquad \sigma^*(\zeta) = \sum_{j=0}^{k-1} \beta_j^* \zeta^j \tag{56}$$

and that used as corrector by

$$\rho(\zeta) = \sum_{j=0}^{k} \alpha_j \zeta^j, \qquad \alpha_k = 1, \qquad \sigma(\zeta) = \sum_{j=0}^{k} \beta_j \zeta^j. \tag{57}$$

Then the modes $P(EC)^m E$ and $P(EC)^m$ described above are formally defined as follows for $m = 1, 2, \ldots$:

$P(EC)^m E$:

$$\left.\begin{aligned} y_{n+k}^{[0]} + \sum_{j=0}^{k-1} \alpha_j^* y_{n+j}^{[m]} &= h \sum_{j=0}^{k-1} \beta_j^* f_{n+j}^{[m]}, \\ f_{n+k}^{[s]} &= f(x_{n+k}, y_{n+k}^{[s]}), \\ y_{n+k}^{[s+1]} + \sum_{j=0}^{k-1} \alpha_j y_{n+j}^{[m]} &= h\beta_k f_{n+k}^{[s]} + h \sum_{j=0}^{k-1} \beta_j f_{n+j}^{[m]}, \\ f_{n+k}^{[m]} &= f(x_{n+k}, y_{n+k}^{[m]}). \end{aligned}\right\} s = 0, 1, \ldots, m-1, \tag{58}$$

$P(EC)^m$:

$$\left.\begin{aligned} y_{n+k}^{[0]} + \sum_{j=0}^{k-1} \alpha_j^* y_{n+j}^{[m]} &= h \sum_{j=0}^{k-1} \beta_j^* f_{n+j}^{[m-1]}, \\ f_{n+k}^{[s]} &= f(x_{n+k}, y_{n+k}^{[s]}), \\ y_{n+k}^{[s+1]} + \sum_{j=0}^{k-1} \alpha_j y_{n+j}^{[m]} &= h\beta_k f_{n+k}^{[s]} + h \sum_{j=0}^{k-1} \beta_j f_{n+j}^{[m-1]}, \end{aligned}\right\} s = 0, 1, \ldots, m-1. \tag{59}$$

Note that as $m \to \infty$, the results of computing with either of the above modes will tend to those given by the mode of correcting to convergence. In practice, it is unusual to use a mode for which m is greater than 2.

3.10 The local truncation error of predictor–corrector methods: Milne's device

Let the predictor P and the corrector C defined by (56) and (57) have associated linear difference operators $\mathscr{L}^*$ and $\mathscr{L}$, orders p^* and p, and error constants $C_{p^*+1}^*$ and C_{p+1} respectively. We shall derive expressions for the principal local truncation error at x_{n+k} of the predictor–corrector method defined by P and C in either $P(EC)^mE$ or $P(EC)^m$ mode, under the usual localizing assumption that all numerical solutions at previous points $x_{n+j}, j = 0, 1, \ldots, k-1$, are exact. We shall also assume that the theoretical solution $y(x)$ of the initial value problem is sufficiently differentiable for the following to hold:

$$\begin{aligned} \mathscr{L}^*[y(x); h] &= C^*_{p^*+1}h^{p^*+1}y^{(p^*+1)}(x) + O(h^{p^*+2}), \\ \mathscr{L}[y(x); h] &= C_{p+1}h^{p+1}y^{(p+1)}(x) + O(h^{p+2}). \end{aligned} \tag{60}$$

Applying equation (25) of chapter 2 (page 28) to the predictor, we obtain

$$y(x_{n+k}) - y^{[0]}_{n+k} = C^*_{p^*+1}h^{p^*+1}y^{(p^*+1)}(x_n) + O(h^{p^*+2}). \tag{61}$$

For the corrector, we repeat the analysis leading to equation (25) of chapter 2 to obtain

$$\sum_{j=0}^{k} \alpha_j y(x_{n+j}) = h \sum_{j=0}^{k} \beta_j f(x_{n+j}, y(x_{n+j})) + \mathscr{L}[y(x_n); h],$$

and

$$y^{[s+1]}_{n+k} + \sum_{j=0}^{k-1} \alpha_j y^{[m]}_{n+j} = h\beta_k f(x_{n+k}, y^{[s]}_{n+k}) + h \sum_{j=0}^{k-1} \beta_j f(x_{n+j}, y^{[m-t]}_{n+j}),$$
$$s = 0, 1, \ldots, m-1,$$

where, by (58) and (59), $t = 0$ if the predictor–corrector method is used in $P(EC)^mE$ mode, and $t = 1$ if it is used in $P(EC)^m$ mode. On subtracting, we have, by virtue of the localizing assumption,

$$\begin{aligned} y(x_{n+k}) - y^{[s+1]}_{n+k} &= h\beta_k[f(x_{n+k}, y(x_{n+k})) - f(x_{n+k}, y^{[s]}_{n+k})] + \mathscr{L}[y(x_n); h] \\ &= h\beta_k \frac{\partial f(x_{n+k}, \eta_{n+k,s})}{\partial y}[y(x_{n+k}) - y^{[s]}_{n+k}] + \mathscr{L}[y(x_n); h], \end{aligned}$$
$$s = 0, 1, \ldots, m-1, \tag{62}$$

for some $\eta_{n+k,s}$ within the interval whose end-points are $y^{[s]}_{n+k}$ and $y(x_{n+k})$.

Consider the case when $p^* \geqslant p$. We may substitute from (61) in the right-hand side of (62) with $s = 0$ to obtain, in view of (60),

$$y(x_{n+k}) - y^{[1]}_{n+k} = C_{p+1}h^{p+1}y^{(p+1)}(x_n) + O(h^{p+2}).$$

This expression for $y(x_{n+k}) - y^{[1]}_{n+k}$ can now be substituted into the right-hand side of (62) with $s = 1$. On successive substitution in (62) with $s = 2, 3, \ldots, m-1$, we find that

$$y(x_{n+k}) - y^{[m]}_{n+k} = C_{p+1}h^{p+1}y^{(p+1)}(x_n) + O(h^{p+2}), \qquad m = 1, 2, \ldots.$$

Thus, if $p^* \geqslant p$, then for $m \geqslant 1$ the principal local truncation error of the predictor–corrector method is that of the corrector alone; note that this holds for both $P(EC)^mE$ and $P(EC)^m$ modes.

Consider now the case $p^* = p - 1$. Substitution of (61) into (62) with $s = 0$ yields

$$y(x_{n+k}) - y_{n+k}^{[1]} = \left[\beta_k \frac{\partial f}{\partial y} C_p^* y^{(p)}(x_n) + C_{p+1} y^{(p+1)}(x_n)\right] h^{p+1} + O(h^{p+2}).$$

Thus, when $m = 1$, the principal local truncation error of the predictor–corrector algorithm is of the same order as that of the corrector, but is not identical with it. However, on successive substitution in (62), we find that for $m \geqslant 2$,

$$y(x_{n+k}) - y_{n+k}^{[m]} = C_{p+1} h^{p+1} y^{(p+1)}(x_n) + O(h^{p+2}),$$

and the principal local truncation error again becomes that of the corrector.

Consider now the case $p^* = p - 2$. Then substitution of (61) into (62) with $s = 0$ yields

$$y(x_{n+k}) - y_{n+k}^{[1]} = \beta_k \frac{\partial f}{\partial y} C_{p-1}^* h^p y^{(p-1)}(x_n) + O(h^{p+1}).$$

Hence, for $m = 1$, the principal local truncation error is one order less than that of the corrector. Substituting again in (62) gives

$$y(x_{n+k}) - y_{n+k}^{[2]} = \left[\left(\beta_k \frac{\partial f}{\partial y}\right)^2 C_{p-1}^* y^{(p-1)}(x_n) + C_{p+1} y^{(p+1)}(x_n)\right] h^{p+1} + O(h^{p+2}),$$

and it follows that for $m = 2$ the principal local truncation error is of the same order as the corrector, but is not identical with it. On further successive substitution in (62) we find that for $m \geqslant 3$,

$$y(x_{n+k}) - y_{n+k}^{[m]} = C_{p+1} h^{p+1} y^{(p+1)}(x_n) + O(h^{p+2}),$$

and once again the principal local truncation error is that of the corrector.

The argument clearly extends to general p^* and we can summarize the result as follows. Let the predictor–corrector method for which the predictor has order p^* and the corrector has order p be applied in $P(EC)^mE$ or $P(EC)^m$ mode, where p^*, p, m are integers and $p^* \geqslant 0$, $p \geqslant 1$, $m \geqslant 1$. Then, if $p^* \geqslant p$, the principal local truncation error of the algorithm is that of the corrector alone. If $p^* = p - q, 0 < q \leqslant p$, then the principal local truncation error of the algorithm is

(i) that of the corrector alone, when $m \geqslant q + 1$,
(ii) of the same order as that of the corrector, but not identical with it, when $m = q$,
(iii) of the form $Kh^{(p-q+m+1)} + O(h^{p-q+m+2})$, when $m \leqslant q - 1$.

Note that in the mode of correcting to convergence the principal local truncation error will be that of the corrector alone, no matter what the order of the predictor.

There is clearly little to be gained by choosing a predictor–corrector method for which $p^* > p$; it would normally have an unnecessarily large stepnumber, and the higher accuracy of the predictor would not be reflected in the local truncation error of the overall method. The discussion above might suggest that it would be advantageous to choose a method for which $p^* = p - m \geqslant 0$. However it turns out that when $p^* = p$ it is possible to estimate the principal local truncation error of the predictor–corrector method (which, as we have seen, then coincides with that of the corrector) *without estimating higher derivatives of* $y(x)$. This technique was originated by W. E. Milne, and we shall refer to it as *Milne's device.*

Assume that $p^* = p$; then, as we have seen above,

$$C_{p+1}h^{p+1}y^{(p+1)}(x_n) = y(x_{n+k}) - y_{n+k}^{[m]} + O(h^{p+2}).$$

Also,

$$C_{p+1}^*h^{p+1}y^{(p+1)}(x_n) = y(x_{n+k}) - y_{n+k}^{[0]} + O(h^{p+2}),$$

from (61). On subtracting, we obtain

$$(C_{p+1}^* - C_{p+1})h^{p+1}y^{(p+1)}(x_n) = y_{n+k}^{[m]} - y_{n+k}^{[0]} + O(h^{p+2}).$$

It follows that the principal local truncation error of the method is given by

$$C_{p+1}h^{p+1}y^{(p+1)}(x_n) = \frac{C_{p+1}}{C_{p+1}^* - C_{p+1}}(y_{n+k}^{[m]} - y_{n+k}^{[0]}), \tag{63}$$

since, in the definition of principal local truncation error, $O(h^{p+2})$ terms are ignored. Note that the estimate for the principal local truncation error given by (63) is dependent on the localizing assumption. In practice, the values of $y_{n+j}^{[m]}, j = 0, 1, \ldots, k - 1$, will not be exact, and we can regard (63) as giving only an approximation to the principal local truncation error. However, Henrici[67] (page 257) shows that, in the mode of correcting to convergence, (63) holds without the localizing assumption, provided that the maximum starting error δ, defined by (19), satisfies $\delta = O(h^q)$ with $q > p$, and that $\alpha_j^* = \alpha_j, j = 0, 1, \ldots, k$. In particular, this last condition is satisfied when the predictor is an Adams–Bashforth method, and the corrector an Adams–Moulton corrector.

The main use to which the error estimate (63) is put is as an aid in determining an appropriate value for the steplength h; this aspect will be discussed in section 3.12. Alternatively, as proposed originally by Hamming,[65] (63) can be used to improve the local accuracy of both predicted and corrected values. From (63) we can easily deduce an estimate for the principal local truncation error of the predictor; it is

$$C_{p+1}^* h^{p+1} y^{(p+1)}(x_n) = \frac{C_{p+1}^*}{C_{p+1}^* - C_{p+1}}(y_{n+k}^{[m]} - y_{n+k}^{[0]}).$$

We cannot use this estimate to improve the predicted value since, at the prediction stage, $y_{n+k}^{[m]}$ is not yet known. However,

$$\begin{aligned} C_{p+1}^* h^{p+1} y^{(p+1)}(x_n) &= C_{p+1}^* h^{p+1} y^{(p+1)}(x_{n-1}) + O(h^{p+2}) \\ &= \frac{C_{p+1}^*}{C_{p+1}^* - C_{p+1}}(y_{n+k-1}^{[m]} - y_{n+k-1}^{[0]}) + O(h^{p+2}). \end{aligned}$$

Recalling the sense in which local truncation error is defined (see equation (25) of chapter 2), we replace $y_{n+k}^{[0]}$ by the *modified* value $\hat{y}_{n+k}^{[0]}$, where

$$\hat{y}_{n+k}^{[0]} = y_{n+k}^{[0]} + \frac{C_{p+1}^*}{C_{p+1}^* - C_{p+1}}(y_{n+k-1}^{[m]} - y_{n+k-1}^{[0]}). \tag{64}$$

We describe this step as a *modifier* and denote its application by M. We may also sacrifice the error estimate afforded by (63), using it instead to improve the corrected value $y_{n+k}^{[m]}$, which we replace by the modified value $\hat{y}_{n+k}^{[m]}$, where

$$\hat{y}_{n+k}^{[m]} = y_{n+k}^{[m]} + \frac{C_{p+1}}{C_{p+1}^* - C_{p+1}}(y_{n+k}^{[m]} - y_{n+k}^{[0]}). \tag{65}$$

One or both of the modifiers (64), (65) can be incorporated into either of the modes $P(EC)^m E$, $P(EC)^m$, with obvious notational changes. Note that the first modifier is applied immediately after the prediction stage and the second after the final correction, so that the resulting modes may be denoted by $PM(EC)^m ME$, $PM(EC)^m M$, respectively. Modifiers may also be incorporated into the mode of correcting to convergence. It is important, however, to realize that for all modes the use of a modifier after the final correction robs us of the possibility of using (63) for step-control purposes. If we wish to have more accurate corrected values, it is probably preferable to increase the order of the method, even at the cost of an increased number of additional starting values. (See Ralston.[147] An extensive study of the effect of modifiers has also been made by Crane.[32]) In recent years, predictor–corrector algorithms discussed in the literature have not, in general, made use of modifiers; nevertheless, Hamming's method (see example 11) remains popular.

Example 11 Consider the predictor, P, and the two correctors, $C^{(1)}$, $C^{(2)}$, defined as follows, by their characteristic polynomials:

$$P: \quad \rho^*(\zeta) = \zeta^4 - 1; \qquad \sigma^*(\zeta) = \tfrac{4}{3}(2\zeta^3 - \zeta^2 + 2\zeta), \tag{66i}$$

$$C^{(1)}: \quad \rho_1(\zeta) = \zeta^2 - 1; \qquad \sigma_1(\zeta) = \tfrac{1}{3}(\zeta^2 + 4\zeta + 1), \tag{66ii}$$

$$C^{(2)}: \quad \rho_2(\zeta) = \zeta^3 - \tfrac{9}{8}\zeta^2 + \tfrac{1}{8}; \qquad \sigma_2(\zeta) = \tfrac{3}{8}(\zeta^3 + 2\zeta^2 - \zeta). \tag{66iii}$$

Use Milne's device to construct an error estimate for a predictor–corrector method which uses (a) P and $C^{(1)}$, and (b) P and $C^{(2)}$. State formally the algorithm which uses (c) P and $C^{(1)}$ in PECE mode, and (d) P and $C^{(2)}$ in PMECME mode. (The predictor–corrector pair in (c) are due to Milne,[129] and methods employing them in various modes are each referred to by different authors as **Milne's method***; (d) is* **Hamming's method**.[65])

Using the coefficients listed in section 2.10, we find that P, $C^{(1)}$, and $C^{(2)}$ all have order 4 (thus making Milne's device applicable) and error constants $C_5^* = \frac{14}{45}$, $C_5^{(1)} = -\frac{1}{90}$, and $C_5^{(2)} = -\frac{1}{40}$, respectively. Substituting in (63), we find the following error estimates, applicable when the particular predictor–corrector pair is employed in any mode:

(a) $$C_5^{(1)}h^5 y^{(5)}(x_{n+k}) \approx -\tfrac{1}{29}(y_{n+k}^{[m]} - y_{n+k}^{[0]}),$$

(b) $$C_5^{(2)}h^5 y^{(5)}(x_{n+k}) \approx -\tfrac{9}{121}(y_{n+k}^{[m]} - y_{n+k}^{[0]}).$$

In order to state the algorithm (c) precisely, we first observe that the stepnumber of the predictor is 4, while that of the corrector is 2. According to our convention, we make both stepnumbers 4 by rewriting $C^{(1)}$ in the equivalent form

$$y_{n+4} - y_{n+2} = \frac{h}{3}(f_{n+4} + 4f_{n+3} + f_{n+2}).$$

The algorithm (c) is then

$$P: \quad y_{n+4}^{[0]} - y_n^{[1]} = \frac{4h}{3}(2f_{n+3}^{[1]} - f_{n+2}^{[1]} + 2f_{n+1}^{[1]}),$$

$$E: \quad f_{n+4}^{[0]} = f(x_{n+4}, y_{n+4}^{[0]}),$$

$$C: \quad y_{n+4}^{[1]} - y_{n+2}^{[1]} = \frac{h}{3}(f_{n+4}^{[0]} + 4f_{n+3}^{[1]} + f_{n+2}^{[1]}),$$

$$E: \quad f_{n+4}^{[1]} = f(x_{n+4}, y_{n+4}^{[1]}).$$

The modifiers necessary for (d) are readily computed from (64) and (65), giving the following algorithm:

$$
\begin{aligned}
&P: && y^{[0]}_{n+4} - \hat{y}^{[1]}_{n} = \frac{4h}{3}(2\hat{f}^{[1]}_{n+3} - \hat{f}^{[1]}_{n+2} + 2\hat{f}^{[1]}_{n+1}),\\
&M: && \hat{y}^{[0]}_{n+4} = y^{[0]}_{n+4} + \tfrac{112}{121}(y^{[1]}_{n+3} - y^{[0]}_{n+3}),\\
&E: && \hat{f}^{[0]}_{n+4} = f(x_{n+4}, \hat{y}^{[0]}_{n+4}),\\
&C: && y^{[1]}_{n+4} - \tfrac{9}{8}\hat{y}^{[1]}_{n+3} + \tfrac{1}{8}\hat{y}^{[1]}_{n+1} = \frac{3h}{8}(\hat{f}^{[0]}_{n+4} + 2\hat{f}^{[1]}_{n+3} - \hat{f}^{[1]}_{n+2}),\\
&M: && \hat{y}^{[1]}_{n+4} = y^{[1]}_{n+4} - \tfrac{9}{121}(y^{[1]}_{n+4} - y^{[0]}_{n+4}),\\
&E: && \hat{f}^{[1]}_{n+4} = f(x_{n+4}, \hat{y}^{[1]}_{n+4}).
\end{aligned}
$$

Note that the principal local truncation error of algorithm (c) is estimated by (a), with $k = 4$; we have no readily computable error estimate for algorithm (d).

Example 12 Compare the accuracies of the numerical solutions obtained when the predictor P and the corrector $C^{(2)}$ of example 11, combined in (i) the mode of correcting to convergence, (ii) PECE mode, and (iii) PMECE mode are used to solve the initial value problem $y' = -10(y-1)^2$, $y(0) = 2$ in the range $0 \leqslant x \leqslant 0{\cdot}2$ using a steplength of 0·01. Using Milne's device, calculate estimates, at every second step, for the principal local truncation error for each of the above modes, and compare with the actual errors.

The necessary starting values are taken to coincide with the theoretical solution $y(x) = 1 + 1/(1 + 10x)$. In the mode of correcting to convergence, the iteration is continued until $|y^{[s+1]}_{n+k} - y^{[s]}_{n+k}| < 10^{-9}$. For all three modes, the formula for the error estimate is that obtained in example 11(b). (See exercise 15 for justification in the case of the *PMECE* mode.) Actual errors and error estimates at every second step are displayed in table 9.

Table 9
Errors $\times 10^5$

	Correcting to convergence		*PECE*		*PMECE*	
x	Actual error	Estimated error	Actual error	Estimated error	Actual error	Estimated error
0·04	0·68	1·02	1·41	1.07	1·41	1·07
0·06	1·38	0·50	3·01	0·65	1·88	0·54
0·08	1·58	0·28	3·66	0·44	1·85	0·27
0·10	1·54	0·15	3·66	0·25	1·68	0·13
0·12	1·41	0·08	3·39	0·13	1·49	0·07
0·14	1·26	0·04	3·04	0·07	1·31	0·04
0·16	1·12	0·02	2·69	0·04	1·15	0·02
0·18	0·99	0·01	2·38	0·02	1·02	0·01
0·20	0·88	0·01	2·11	0·01	0·90	0·01

Comparing first actual errors of the three modes, it is clear that the loss of accuracy through correcting only once is small in this example, (very small indeed, in the case of the *PMECE* mode). On comparing actual and estimated errors, one might conclude from a casual glance that Milne's device has resulted in gross underestimates of error. This is not the case, for the actual and estimated errors are different in kind, the former being *global* or accumulated errors, while the latter are *local* errors. Only at the first computed step, that is, at $x = 0{\cdot}04$, are both errors local, and here the agreement is satisfactory, bearing in mind that Milne's device estimates only the *principal* local truncation error. The subsequent decrease in the estimated local error as the computation proceeds suggests that the actual local error decreases. The eventual decrease in the actual global error indicates that this is the case. If the only criterion governing the choice of steplength were that the local truncation error be everywhere less than the initial local error, then the estimated local errors would indicate that an increase in steplength would have been permissible. That we have not made such an increase has resulted in the actual global error being of the same order as the initial actual local error, whereas, from section 3.5, we can expect global errors to be up to an order of magnitude greater than local errors. (In general, weak stability characteristics of the method will also affect the choice of steplength; clearly in this example the results show no sign of numerical instability.)

Exercises

14. Let the predictor have order p^* and the corrector have order p. Show that an estimate of the principal local truncation error similar to, but not identical with (63) can be formulated when $p^* > p$, but not when $p^* < p$.

15. Let the predictor and corrector have orders p^* and p respectively. Show that if $p^* = p$ the principal local truncation error of the predictor–corrector pair in either $PM(EC)^mE$ or $PM(EC)^m$ mode, where $m \geqslant 1$, is identical with that of the corrector alone. Deduce that when $p^* = p$ Milne's device is applicable to these modes.

16. Construct formulae which estimate the principal local truncation errors of the predictor–corrector methods which use
(i) Adams–Bashforth predictor with Adams–Moulton corrector, of orders 2, 3, and 4.
(ii) Nyström predictor with generalized Milne–Simpson corrector both of order 4. Investigate similar pairs of orders 2 and 3.
(Make use of the coefficients and error constants quoted in section 2.10.)

17. Formally define the 4-step algorithms which use the fourth-order Adams–Bashforth–Moulton predictor–corrector pair in (i) *PEC*, (ii) *PECE*, (iii) *PMEC*, and (iv) *PMECE* mode.

18. Find a fourth-order predictor which, when used with the corrector $C^{(2)}$ defined in example 11, will satisfy the requirement $\alpha_j^* = \alpha_j, j = 0, 1, \ldots, 4$. (See the remarks following equation (63).) Use this predictor along with $C^{(2)}$ to repeat the calculations of example 12.

3.11 Weak stability of predictor–corrector methods

In early papers on predictor–corrector methods, it was tacitly assumed that, even when the corrector was applied only for a fixed number of times at each step, the weak stability characteristics of the overall method were those of the corrector alone; the error introduced by not correcting to convergence was regarded as round-off error. It was first pointed out by Chase[26] that if a mode other than that of correcting to convergence is employed, the predictor, as well as the corrector, influences the weak stability of the method.

Recall from section 3.6 that the weak stability characteristics of the corrector alone are determined by the roots of the stability polynomial

$$\pi(r, \bar{h}) = \rho(r) - \bar{h}\sigma(r),$$

where $\rho(r)$ and $\sigma(r)$ are the first and second characteristic polynomials of the corrector and $\bar{h} = h\partial f/\partial y$, assumed constant. Let the first and second characteristic polynomials of the predictor be $\rho^*(r)$ and $\sigma^*(r)$. We now derive an expression for the stability polynomial of the general *PECE* mode in terms of $\rho(r)$, $\sigma(r)$, $\rho^*(r)$, $\sigma^*(r)$ and $\bar{h}$. Let $\tilde{y}_{n+k}^{[0]}$ and $\tilde{y}_{n+k}^{[1]}$ be the approximations to $y(x_{n+k})$ given by the predictor and corrector, respectively, when round-off error is allowed for. Then, from (58) with $m = 1$, we obtain

$$\tilde{y}_{n+k}^{[0]} + \sum_{j=0}^{k-1} \alpha_j^* \tilde{y}_{n+j}^{[1]} = h \sum_{j=0}^{k-1} \beta_j^* f(x_{n+j}, \tilde{y}_{n+j}^{[1]}) + R_n^*,$$

$$\sum_{j=0}^{k} \alpha_j \tilde{y}_{n+j}^{[1]} = h\beta_k f(x_{n+k}, \tilde{y}_{n+k}^{[0]}) + h \sum_{j=0}^{k-1} \beta_j f(x_{n+j}, \tilde{y}_{n+j}^{[1]}) + R_n,$$

where R_n^* and R_n are the local round-off errors committed at the predictor and corrector stages respectively. The theoretical solution $y(x)$ of the initial value problem satisfies

$$\sum_{j=0}^{k} \alpha_j^* y(x_{n+j}) = h \sum_{j=0}^{k-1} \beta_j^* f(x_{n+j}, y(x_{n+j})) + T_n^*$$

and

$$\sum_{j=0}^{k} \alpha_j y(x_{n+j}) = h \sum_{j=0}^{k} \beta_j f(x_{n+j}, y(x_{n+j})) + T_n,$$

where T_n^* and T_n are the local truncation errors of the predictor and corrector respectively. Defining the global errors $\tilde{e}_n^{[0]}$ and $\tilde{e}_n^{[1]}$ by

$$\tilde{e}_n^{[0]} = y(x_n) - \tilde{y}_n^{[0]}, \qquad \tilde{e}_n^{[1]} = y(x_n) - \tilde{y}_n^{[1]},$$

we may proceed exactly as in section 3.6, (making the same assumptions that $\partial f/\partial y = \lambda$, $T^* - R^*$ and $T_n - R_n$ are all constants) to obtain the following pair of linearized error equations:

$$\tilde{e}^{[0]}_{n+k} + \sum_{j=0}^{k-1} \alpha^*_j \tilde{e}^{[1]}_{n+j} = \bar{h} \sum_{j=0}^{k-1} \beta^*_j \tilde{e}^{[1]}_{n+j} + \text{constant},$$

$$\sum_{j=0}^{k} \alpha_j \tilde{e}^{[1]}_{n+j} = \bar{h}\beta_k \tilde{e}^{[0]}_{n+k} + \bar{h} \sum_{j=0}^{k-1} \beta_j \tilde{e}^{[1]}_{n+j} + \text{constant},$$

where $\bar{h} = h\lambda$. On eliminating $\tilde{e}^{[0]}_{n+k}$ we obtain

$$\sum_{j=0}^{k} \alpha_j \tilde{e}^{[1]}_{n+j} - \bar{h} \sum_{j=0}^{k-1} \beta_j \tilde{e}^{[1]}_{n+j} = -\bar{h}\beta_k \left[\sum_{j=0}^{k-1} \alpha^*_j \tilde{e}^{[1]}_{n+j} - \bar{h} \sum_{j=0}^{k-1} \beta^*_j \tilde{e}^{[1]}_{n+j} \right] + \text{constant}.$$

Adding $-\bar{h}\beta_k \tilde{e}^{[1]}_{n+k}$ to both sides (and recalling that $\alpha^*_k = 1, \beta^*_k = 0$), yields

$$\sum_{j=0}^{k} (\alpha_j - \bar{h}\beta_j) \tilde{e}^{[1]}_{n+j} = -\bar{h}\beta_k \sum_{j=0}^{k} (\alpha^*_j - \bar{h}\beta^*_j) \tilde{e}^{[1]}_{n+j} + \text{constant}.$$

The solution for $\tilde{e}^{[1]}_n$, the global error in the *corrected* value, thus has the form

$$\tilde{e}^{[1]}_n = \sum_{s=1}^{k} d_s r_s^n + \text{constant},$$

where the d_s are arbitrary constants and the r_s are the roots (assumed distinct) of the polynomial equation

$$\rho(r) - \bar{h}\sigma(r) + \bar{h}\beta_k[\rho^*(r) - \bar{h}\sigma^*(r)] = 0.$$

Thus, intervals of absolute and relative stability may be defined precisely as before, except that the roots r_s are no longer the zeros of the stability polynomial (41) but are the zeros of the new stability polynomial

$$\pi_{PECE}(r, \bar{h}) = \rho(r) - \bar{h}\sigma(r) + \bar{h}\beta_k[\rho^*(r) - \bar{h}\sigma^*(r)].$$

The analysis can be extended to give the following stability polynomial for the general $P(EC)^m E$ mode:

$$\pi_{P(EC)^m E}(r, \bar{h}) = \rho(r) - \bar{h}\sigma(r) + M_m(\bar{h})[\rho^*(r) - \bar{h}\sigma^*(r)], \tag{67}$$

where

$$M_m(\bar{h}) = (\bar{h}\beta_k)^m(1 - \bar{h}\beta_k)/[1 - (\bar{h}\beta_k)^m], \qquad m = 1, 2, \ldots. \tag{68}$$

A similar, but somewhat more complicated analysis, leads to the following expression for the stability polynomial of the general $P(EC)^m$ mode:

$$\pi_{P(EC)^m}(r, \bar{h}) = \beta_k r^k[\rho(r) - \bar{h}\sigma(r)] + M_m(\bar{h})[\rho^*(r)\sigma(r) - \rho(r)\sigma^*(r)]. \quad (69)$$

Note that, if we take $L = |\partial f/\partial y|$, the condition (55) (page 86) becomes

$$|\bar{h}\beta_k| < 1. \quad (70)$$

If (70) is satisfied, then $M_m(\bar{h}) \to 0$ as $m \to \infty$. It follows that $\pi_{P(EC)^m E} \to \pi$, $\pi_{P(EC)^m} \to \beta_k r^k \pi$ as $m \to \infty$, where π is the stability polynomial of the corrector alone. This corroborates our earlier statement that, in the mode of correcting to convergence, the weak stability characteristics are those of the corrector alone, provided (55) is satisfied.

In section 3.6 we showed that every consistent zero-stable linear multistep method is absolutely unstable for small positive $\bar{h}$. This was deduced from the fact that the stability polynomial $\pi(r, \bar{h})$ had a zero r_1 which satisfied

$$r_1 = \exp(\bar{h}) + O(\bar{h}^{p+1}),$$

where p was the order of the method. The proof of this last result hinged on the fact that

$$\pi(\exp(\bar{h}), \bar{h}) = O(\bar{h}^{p+1}).$$

Now, using the results of section 3.6, we can state that if the predictor and corrector both have order p, then

$$\rho^*(\exp(\bar{h})) - \bar{h}\sigma^*(\exp(\bar{h})) = O(\bar{h}^{p+1}) \quad (71)$$

and

$$\rho(\exp(\bar{h})) - \bar{h}\sigma(\exp(\bar{h})) = O(\bar{h}^{p+1}). \quad (72)$$

Multiplying (71) by $\sigma(\exp(\bar{h}))$ and (72) by $\sigma^*(\exp(\bar{h}))$ and subtracting we obtain

$$\rho^*(\exp(\bar{h}))\sigma(\exp(\bar{h})) - \rho(\exp(\bar{h}))\sigma^*(\exp(\bar{h})) = O(\bar{h}^{p+1}).$$

It follows that

$$\pi_{P(EC)^m E}(\exp(\bar{h}), \bar{h}) = O(\bar{h}^{p+1}), \pi_{P(EC)^m}(\exp(\bar{h}), \bar{h}) = O(\bar{h}^{p+1}),$$

since $M_m(\bar{h}) = O(\bar{h}^m)$, $m = 1, 2, \ldots$. We conclude, as in section 3.6, that the stability polynomial for the general $P(EC)^m E$ or $P(EC)^m$ mode possesses a root $r_1 = \exp(\bar{h}) + O(\bar{h}^{p+1})$, and that the corresponding methods are absolutely unstable for small positive $\bar{h}$.

Of the four methods we described in section 3.7 for finding intervals of absolute and relative stability, the root-locus method, the Schur criterion and the Routh–Hurwitz criterion, may be applied without modification to the new stability polynomials. Since the coefficients of the new stability polynomials are now rational functions of $\bar{h}$, the inequalities which result from the Schur criterion or the Routh–Hurwitz criterion will in general be much more complicated than in the case of linear multistep methods. Moreover, in the case of $P(EC)^m$ modes, the stability polynomial now has degree $2k$ (although it may have a factor r^q, since we do not demand that both α_0 and β_0 do not vanish). In certain cases, particularly when intervals of relative stability are sought, the root-locus method may be the only practicable possibility. The fourth method, the boundary-locus method which, in the case of a linear multistep method made use of the fact that the stability polynomial was linear in $\bar{h}$, now needs modification.

Let $\hat{\pi}(r, \bar{h})$ represent the stability polynomial of a general $P(EC)^mE$ or $P(EC)^m$ mode. The boundary $\partial\mathscr{R}$ of the region of absolute stability in the complex $\bar{h}$-plane, is the locus of the points $\bar{h}$ for which $r = \exp(i\theta)$. It is thus obtained by solving for $\bar{h}$ the equation

$$\hat{\pi}(\exp(i\theta), \bar{h}) = 0.$$

Since $\hat{\pi}$ is no longer linear in $\bar{h}$ it is not, in general, possible to express the solution in closed analytic form, as was the case for a linear multistep method. However, we may solve it numerically for $\bar{h}$, for each of a number of discrete values of θ (typically $\theta = 0°, 30°, 60°, \ldots$) and fit a curve through the resulting points in the $\bar{h}$-plane. This will approximate to $\partial\mathscr{R}$ and the end-points of the interval of absolute stability are given approximately by the points where the curve cuts the real axis. A similar technique applied to the equation

$$\hat{\pi}(|\exp(\bar{h})|\exp(i\theta), \bar{h}) = 0$$

will yield the approximate boundary of the region of relative stability defined by the requirement $|r_s| < |\exp(\bar{h})|$, $s = 1, 2, \ldots, k$.

Example 13 Using the root locus method, Chase[26] *has shown that the interval of absolute stability of Milne's predictor–corrector pair in PECE mode (see example 11, page 93) is* $(-0{\cdot}8, -0{\cdot}3)$. *Verify this result by using this algorithm to solve the initial value problem* $y' = -y$, $y(0) = 1$.

For this problem, $\partial f/\partial y = -1$ and thus $\bar{h} = -h$. Numerical solutions are computed in the interval $0 \leqslant x \leqslant 100$, using steplengths $h = 0{\cdot}1, 0{\cdot}3, 0{\cdot}5, 0{\cdot}8$, and $1{\cdot}0$. The magnitudes of the errors are displayed in table 10, and clearly corroborate Chase's result that the method is absolutely stable if $-0{\cdot}8 < \bar{h} < -0{\cdot}3$. Note that if $\bar{h} = -0{\cdot}8, -0{\cdot}3$, there is a persistent error which is approximately constant.

Table 10

x	$h = 0{\cdot}1$	$h = 0{\cdot}3$	$h = 0{\cdot}5$	$h = 0{\cdot}8$	$h = 1{\cdot}0$
0	0	0	0	0	0
20	5×10^{-6}	2×10^{-5}	1×10^{-8}	2×10^{-3}	4×10^{-2}
40	7×10^{-4}	2×10^{-5}	$<10^{-10}$	8×10^{-4}	2×10^{-1}
60	8×10^{-2}	2×10^{-5}	$<10^{-10}$	4×10^{-4}	2
80	1×10	2×10^{-5}	$<10^{-10}$	2×10^{-4}	9
100	1×10^{3}	2×10^{-5}	$<10^{-10}$	7×10^{-5}	6×10

Example 14 Repeat example 13, replacing the linear initial value problem by the non-linear problem

$$y' = -5xy^2 + 5/x - 1/x^2, \qquad y(1) = 1.$$

The theoretical solution of this problem is $y(x) = 1/x$, whence

$$\partial f/\partial y = -10xy = -10,$$

on substituting the theoretical solution for y. Thus $\bar{h} = -10h$, and in order to test Chase's result we compute with $h = 0{\cdot}01, 0{\cdot}03, 0{\cdot}05, 0{\cdot}08$, and $0{\cdot}10$. The magnitudes of the errors are displayed in table 11. Once again, these results corroborate that the algorithm is absolutely stable for $-0{\cdot}8 < \bar{h} < -0{\cdot}3$. Once again there is a persistent error when $\bar{h} = -0{\cdot}8, -0{\cdot}3$. For $\bar{h}$ outside the interval of absolute stability, the

Table 11

x	$h = 0{\cdot}01$	$h = 0{\cdot}03$	$h = 0{\cdot}05$	$h = 0{\cdot}08$	$h = 0{\cdot}10$
1	0	0	0	0	0
21	1×10^{-1}	3×10^{-8}	$<10^{-10}$	5×10^{-10}	4×10^{-2}
41	5×10^{-2}	3×10^{-8}	$<10^{-10}$	$<10^{-10}$	3×10^{-2}
61	3×10^{-2}	3×10^{-8}	$<10^{-10}$	$<10^{-10}$	1×10^{-2}
81	2×10^{-2}	3×10^{-8}	$<10^{-10}$	$<10^{-10}$	2×10^{-2}

errors do not show the growth we might have expected; this can be ascribed to the non-linearity of the differential equation. Nevertheless, the errors for $h = 0{\cdot}01$ and $h = 0{\cdot}10$ are large for a fourth-order method. A curious feature of this example is that the solution given by $h = 0{\cdot}10$ is as accurate as that given by $h = 0{\cdot}01$. It should be noted that the differential equation, though non-linear, has the special property that $\partial f/\partial y$ is constant on the theoretical solution, but not on neighbouring integral curves.

Exercises

19. Show that the algorithm which uses the predictor P and the corrector $C^{(2)}$ of example 11 (page 93) in *PECE* mode has an interval of absolute stability of

$(-0{\cdot}5, 0)$ (Chase[26]). Verify this result by calculating solutions to the initial value problems of examples 12 and 13.

20. Consider the following predictor due to Stetter:[170]

$$y_{n+2} + 4y_{n+1} - 5y_n = h(4f_{n+1} + 2f_n).$$

Form the stability polynomial for the *PECE* algorithm which uses this predictor with Simpson's rule as corrector. Hence show that for small negative $\bar{h}$ the algorithm is relatively stable according to the definition $|r_s| < |r_1|$, $s = 2, 3, \ldots, k$. (Note that this is an interesting result in view of the fact that Simpson's rule is itself relatively unstable for all $\bar{h} < 0$.)

3.12 Step-control policy for predictor-corrector methods

At the beginning of this chapter, we drew attention to the most difficult problem in the application of linear multistep methods, namely that of choosing an appropriate value for the steplength. Let us recapitulate the progress that has been made towards solving this problem. In section 3.5 we concluded that a bound for the global truncation error does not, in general, provide an adequate basis for choosing h. In section 3.6 we turned instead to the idea of finding an interval for h which ensures that the global truncation error does not grow in a certain sense. It is still essential to choose h such that the local truncation error is acceptably small; but, from section 3.3, we see that the application of a bound for the local truncation error is bedevilled by the practical difficulty of finding a bound for $|y^{(p+1)}(x)|$. However, if we use a predictor–corrector method in an appropriate mode, then we can avoid estimating higher derivatives by using Milne's device to estimate the principal local truncation error (which we then accept as an approximation to the true local truncation error). We are then in a position to apply the following *step-control policy*: the steplength h will be chosen such that (i) the principal local truncation error as estimated by Milne's device remains at each step less than a pre-assigned tolerance, (ii) $\bar{h}$ lies inside an interval of absolute or relative stability, and (iii) the condition (55) is satisfied. It will be convenient to refer to (i) as *error-control*, (ii) as *stability-control*, and (iii) as *convergence-control*. ('Convergence' in this context refers to the convergence of the iteration (54), not to the convergence of the method in the sense of chapter 2.) The step-control policy may call for a decrease, or permit an increase, in h as the computation proceeds.

In order to implement stability-control it is necessary to estimate $\partial f/\partial y$ at each step, so that $\bar{h} = h(\partial f/\partial y)$ may be calculated. A device, proposed by Nordsieck,[137] allows us easily to compute an estimate for $\bar{h}$ from two successive applications of the corrector. Let us assume that we are using either a $P(EC)^mE$ or a $P(EC)^m$ mode, where $m \geqslant 2$. Consider the last two applications of the corrector in either of the above modes, which by (58) and (59) may be written

$$y_{n+k}^{[m-1]} + \sum_{j=0}^{k-1} \alpha_j y_{n+j}^{[m]} = h\beta_k f_{n+k}^{[m-2]} + h \sum_{j=0}^{k-1} \beta_j f_{n+j}^{[m-t]},$$

$$y_{n+k}^{[m]} + \sum_{j=0}^{k-1} \alpha_j y_{n+j}^{[m]} = h\beta_k f_{n+k}^{[m-1]} + h \sum_{j=0}^{k-1} \beta_j f_{n+j}^{[m-t]},$$

where $t = 0$ for the $P(EC)^mE$ mode, and $t = 1$ for the $P(EC)^m$ mode. On subtracting, and applying the mean value theorem, we obtain

$$y_{n+k}^{[m]} - y_{n+k}^{[m-1]} = h\beta_k \frac{\partial f(x_{n+k}, \eta_{n+k}^{[m-1]})}{\partial y}(y_{n+k}^{[m-1]} - y_{n+k}^{[m-2]}),$$

where $\eta_{n+k}^{[m-1]}$ is an interior point of the open interval whose end-points are $y_{n+k}^{[m-1]}$ and $y_{n+k}^{[m-2]}$. Hence, at each step we obtain, at low computational cost, the following local estimate for $\bar{h}$:

$$\bar{h} = h\frac{\partial f}{\partial y} \approx (y_{n+k}^{[m]} - y_{n+k}^{[m-1]})/\beta_k(y_{n+k}^{[m-1]} - y_{n+k}^{[m-2]}). \tag{73i}$$

Note that this estimate is applicable only if $m \geqslant 2$. If $m = 1$, the following alternative may be used to estimate $\bar{h}$:

$$\bar{h} = h\frac{\partial f}{\partial y} \approx h[f(x_{n+k}, y_{n+k}^{[1]}) - f(x_{n+k}, y_{n+k}^{[0]})]/(y_{n+k}^{[1]} - y_{n+k}^{[0]}). \tag{73ii}$$

(It is emphasized that both the estimates (73) are applicable only when a single differential equation is to be solved.)

In this section we have discussed the principles on which a policy for choosing and changing the steplength may be founded. How such a change can be effected in practice will be discussed in section 3.14, after we have reached some conclusions on the choice of appropriate predictor–corrector methods.

3.13 Choice of predictor–corrector methods

The literature in this subject presents the user with such a variety of algorithms that it is difficult to decide which to adopt for a specific purpose. We shall tackle this problem by first discussing the relative merits of different modes, and then briefly reviewing some of the specific methods which have been proposed.

In many cases the major computational cost of a predictor–corrector algorithm is the evaluation of the function $f(x, y)$. (A recent report by

Hull, Enright, Fellen, and Sedgwick,[73] however, shows that overhead costs frequently dominate function evaluation costs.) Nevertheless, most existing algorithms call for at most two function evaluations per step, and accordingly we restrict this discussion to the modes *PEC*, *PECE*, $P(EC)^2$, possibly with the addition of modifiers. Since the use of Milne's device is an essential part of the step-control policy outlined in section 3.12, we shall assume that predictor and corrector have the same order, and, in general, preclude the use of a modifier after the final correction.

From the point of view of truncation error, it is hard to make any general comparison between these modes, since, as we have seen, all have the same principal local truncation error (though they will not, of course, have the same actual truncation error).

When comparing weak stability characteristics of the above modes, it is necessary first to point out that only the $P(EC)^2$ mode permits us to use (73*i*) to estimate $\bar{h}$. In the case of the *PECE* mode, (73*ii*) may be used instead; no additional computation is required, since $f^{[0]}_{n+k}$ and $f^{[1]}_{n+k}$ have to be evaluated in any case. In the *PEC* mode, $f^{[1]}_{n+k}$ is not evaluated, and additional computation is necessary if (73*ii*) is to be used; however, it is not usually essential to recompute an estimate for $\bar{h}$ at every step, and occasional evaluation of $f^{[1]}_{n+k}$ for the purposes of applying (73*ii*) is normally adequate.

The literature contains much quantitative information on stability intervals for specific predictor–corrector pairs in specific modes. An interesting comparison is made by Brown, Riley, and Bennett[9] who compute intervals of absolute stability for the fourth-order Adams–Bashforth–Moulton pair (see (77)) in various modes; their results are summarized below.

$$
\begin{array}{ll}
\text{Correcting to convergence:} & (-3{\cdot}00, 0), \\
PEC: & (-0{\cdot}16, 0), \\
PECE: & (-1{\cdot}25, 0), \\
P(EC)^2: & (-0{\cdot}90, 0), \\
PM(ECM)^2: & (-0{\cdot}66, 0), \\
P(ECM)^2: & (-0{\cdot}95, 0).
\end{array}
\tag{74}
$$

We observe from these results that the modes which correct a fixed number of times possess substantially poorer weak stability characteristics than does the mode of correcting to convergence. However, this is not always the case. (See example 13, page 99, and exercise 20, page 101). The last two modes in (74) employ modifiers after the final correction, and are effectively of fifth order. We have seen that, in general, stability intervals diminish as order increases, so that the modified modes in (74) are seen to possess quite substantial intervals of absolute stability. Again, this trend does not hold in general (see exercise 21, page 105). There is insufficient evidence in the literature to enable us to draw any firm conclusions on the general effect on weak stability of incorporating modifiers. Finally, we observe from (74) that the stability interval of the *PEC* mode is considerably smaller than those of the *PECE* or $P(EC)^2$ modes. This observation does appear to extend to more general cases. Thus, when the corrector is specified to be the fourth-order Adams–Moulton method, Klopfenstein and Millman[86] have synthesized a predictor (see (79)) which results in the *PEC* mode having an interval of absolute stability of optimal length, while Crane and Klopfenstein[33] have done likewise (see (78)) for the *PECE* mode. The resulting interval for the latter is approximately three times that for the former.†

That PEC modes in general may be expected to have poor weak stability characteristics can be shown as follows. From (69), the stability polynomial π_{PEC} is given by

$$\frac{1}{\beta_k}\pi_{PEC}(r, \bar{h}) = r^k\rho(r) - \bar{h}[r^k\sigma(r) + \rho(r)\sigma^*(r) - \rho^*(r)\sigma(r)]. \tag{75}$$

Denote by $\tilde{P}$ the linear multistep method whose first and second characteristic polynomials are $\tilde{\rho}(r)$ and $\tilde{\sigma}(r)$ respectively, where

$$\tilde{\rho}(r) = r^k\rho(r); \;\; \tilde{\sigma}(r) = [r^k - \rho^*(r)]\sigma(r) + \rho(r)\sigma^*(r). \tag{76}$$

Note that ρ and σ are polynomials of degree k in r, while $r^k - \rho^*$ and σ^* have degree $k - 1$, since the predictor P is explicit and $\alpha_k^* = 1$. It follows that $\tilde{\rho}$ has degree $2k$, while $\tilde{\sigma}$ has degree $2k - 1$; in other words, the method $\tilde{P}$ is *explicit*. Note also that since α_0 and β_0 may both vanish, it is possible that $\tilde{\rho}$ and $\tilde{\sigma}$ have a common factor r^q. Thus the stepnumber of $\tilde{P}$ is less than or equal to $2k$. It is possible to show (Lambert[101]) that if the predictor P and the corrector C both have order p, then $\tilde{P}$ also has order p; moreover, $\tilde{P}$ and C then have the same error constant. Now the stability polynomial of the method $\tilde{P}$ is

$$\pi_{\tilde{P}}(r, \bar{h}) = \tilde{\rho}(r) - \bar{h}\tilde{\sigma}(r) = \frac{1}{\beta_k}\pi_{PEC}(r, \bar{h}),$$

from (75). It follows that the interval of stability (absolute or relative) of the general pth order *PEC* method is identical with that of the pth order

† Hull and Creemer[72] test a family of $P(EC)^m$ algorithms on a wide class of problems. They conclude that the $P(EC)^2$ mode is best, and is on average superior to *PEC*.

explicit linear multistep method $\tilde{P}$; moreover $\tilde{P}$ and C have the same principal local truncation error. Since we have already decided not to advocate the use of explicit linear multistep methods largely on account of their inadequate weak stability properties, there would appear to be little argument in favour of the *PEC* mode. This analysis can be extended to other modes. Thus, to any given $P(EC)^2$ algorithm, there corresponds a $\tilde{P}ECE$ algorithm with identical principal local truncation error and identical weak stability properties. The converse is not true, and we conclude that in seeking a predictor–corrector algorithm which costs two function evaluations per step and which is in some sense 'best' with respect to both principal local truncation error and weak stability properties, it would appear that we ought to search among *PECE* algorithms only. Such a 'best' algorithm may, or may not, be replaceable by a $P(EC)^2$ algorithm having the same principal local truncation error and weak stability properties; where such a replacement is possible, the step-number will be reduced.

Thus, for general purposes, we come down in favour of the *PECE* mode. The exception arises when the initial value problem is such that $\partial f/\partial y$ is everywhere sufficiently small for error-control rather than stability-control to dictate the choice of steplength; in such a case, the *PEC* mode is appropriate. (Note, for example, that if stability-control dictates the choice of steplength, the optimal Klopfenstein—Millman *PEC* algorithm will require roughly 50% more function evaluations over a given range than will the optimal Crane–Klopfenstein *PECE* algorithm.) The only virtue of the $P(EC)^2$ mode would appear to be that it permits the use of (73*i*) to estimate $\bar{h}$.

Exercises

21. Derive the stability polynomial for the mode *PMECME*. Use the root locus method to find the interval of absolute stability of (a) Milne's, (b) Hamming's predictor–corrector pair (example 11, page 93) in *PMECME* mode. Compare with the results for the same pairs in *PECE* mode (example 13, page 99, exercise 19, page 100) to see that the addition of modifiers almost halves the stability interval for Milne's pair, and almost doubles it for Hamming's.

22. We have demonstrated above the equivalence of the stability interval of a given *PEC* method to that of a related explicit method $\tilde{P}$. It does not follow that the methods are computationally equivalent. Consider the following five equations selected from the application of a two-step *PEC* method at a number of consecutive steps:

$$y^{[0]}_{n+q+2} + \alpha^*_1 y^{[1]}_{n+q+1} + \alpha^*_0 y^{[1]}_{n+q} = h(\beta^*_1 f^{[0]}_{n+q+1} + \beta^*_0 f^{[0]}_{n+q}), \qquad q = 0, 1, 2,$$

$$y^{[1]}_{n+q+2} + \alpha_1 y^{[1]}_{n+q+1} + \alpha_0 y^{[1]}_{n+q} = h(\beta_2 f^{[0]}_{n+q+2} + \beta_1 f^{[0]}_{n+q+1} + \beta_0 f^{[0]}_{n+q}), \qquad q = 0, 1.$$

Deduce that the *predicted* values $y_{n+j}^{[0]}$ generated by the *PEC* method satisfy the method $\tilde{P}$. In what way does this fact corroborate our result on the equivalence of the stability intervals?

We now consider the choice of linear multistep methods to act as predictor and corrector. The classic example of a predictor–corrector is Milne's method[129] (example 11, page 93). However, its inadequate weak stability properties (example 13, page 99) make it unsuitable as a general purpose algorithm. Although a smoothing process developed by Milne and Reynolds[132,133] can be applied to control the instability, it is natural to seek a replacement for Simpson's rule as corrector. Hamming's[65] choice (example 11, page 93) was motivated by a variety of factors, some of which are strictly relevant only to desk computation; moreover, the interval of absolute stability in *PECE* mode is only $(-0{\cdot}5, 0)$. In a series of papers, Hull and Newbery[75,76,77] make a systematic study of the dependence of both truncation error and weak stability characteristics of the corrector on the position of the spurious zeros of $\rho(\zeta)$. The broad conclusion to be drawn from this study is that Adams–Moulton correctors are as good as any, for general purposes. Although Hull and Newbery's investigation assumed that the corrector would be iterated to convergence, *PECE* algorithms which employ Adams–Moulton correctors are frequently advocated. In particular, the following fourth-order Adams–Bashforth–Moulton pair is popular:

$$\text{Predictor:}\quad y_{n+4} - y_{n+3} = \frac{h}{24}(55f_{n+3} - 59f_{n+2} + 37f_{n+1} - 9f_n), \quad (77i)$$

$$\text{Corrector:}\quad y_{n+4} - y_{n+3} = \frac{h}{24}(9f_{n+4} + 19f_{n+3} - 5f_{n+2} + f_{n+1}), \quad (77ii)$$

$$\text{Error estimate:}\quad C_5 h^5 y^{(5)}(x_n) \approx -\tfrac{19}{270}(y_{n+4}^{[1]} - y_{n+4}^{[0]}),$$

Interval of absolute stability in *PECE* mode: $(-1{\cdot}25, 0)$.

Crane and Klopfenstein[33] have derived a fourth-order predictor which when used with (77*ii*) as corrector in *PECE* mode, gives an extended interval of absolute stability. Their algorithm is defined as follows:

Predictor:

$$y_{n+4} - 1{\cdot}547652y_{n+3} + 1{\cdot}867503y_{n+2} - 2{\cdot}017204y_{n+1} + 0{\cdot}697353y_n$$
$$= h(2{\cdot}002247f_{n+3} - 2{\cdot}031690f_{n+2} + 1{\cdot}818609f_{n+1} - 0{\cdot}714320f_n), \quad (78)$$

Corrector: (77*ii*),

Error estimate: $C_5h^5y^{(5)}(x_n) = -(y_{n+4}^{[1]} - y_{n+4}^{[0]})/16{\cdot}219{,}66$,

Interval of absolute stability in *PECE* mode: $(-2{\cdot}48, 0)$.

The Crane–Klopfenstein algorithm clearly requires more storage locations than does the fourth-order Adams–Bashforth–Moulton (the additional numbers to be stored at each step being $y_{n+3}^{[1]}, y_{n+2}^{[1]}$, and $y_{n+1}^{[1]}$). Such additional storage will increase as the order (and therefore the stepnumber) of algorithms with extended intervals of absolute stability increases, and can become significant when a large system of simultaneous differential equations is to be solved. In order to reduce storage requirements, Krogh[90] has developed predictors with $\alpha_j^* = 0$, $j = 0, 1, \ldots, k-3$, which, when used with Adams–Moulton correctors in *PECE* mode, have large intervals of absolute stability. Such predictors, together with Adams–Bashforth predictors and Adams–Moulton correctors of orders 4, 5, 6, 7, and 8 can be found in Krogh's paper, which also contains much useful information on stability regions (absolute and relative) of algorithms using these formulae.† Using the results of Krogh's paper, we list in table 12 the approximate intervals of absolute stability

Table 12

Order	AM	ABM	KAM
4	$(-3, 0)$	$(-1{\cdot}3, 0)$	$(-1.8, 0)$
5	$(-1{\cdot}8, 0)$	$(-1{\cdot}0, 0)$	$(-1{\cdot}4, 0)$
6	$(-1{\cdot}2, 0)$	$(-0{\cdot}7, 0)$	$(-1.0, 0)$
7	$(-0{\cdot}8, 0)$	$(-0{\cdot}5, 0)$	$(-0{\cdot}8, 0)$
8	$(-0{\cdot}5, 0)$	$(-0{\cdot}4, 0)$	$(-0{\cdot}6, 0)$

for the Adams–Moulton corrector iterated to convergence (AM), the Adams–Bashforth–Moulton pair in *PECE* mode (ABM), and Krogh's predictor with Adams–Moulton corrector in *PECE* mode (KAM), for orders 4 to 8. The interesting point about this table is that as order increases the 'stability gap' between the mode of correcting to convergence and the *PECE* modes closes; indeed, for order 8, the KAM algorithm has better stability than the mode of correcting to convergence. This suggests that for greater efficiency we should contemplate the use of algorithms of fairly high order—a point of view supported by earlier numerical experiments made by Hull and Creemer.[72]

† Useful information on intervals of relative stability for a large selection of methods is also given by Rodabaugh.[153]

The specific methods we have advocated above all use the *PECE* mode. When the problem is such that a *PEC* mode would be economical, we advocate the following algorithm of Klopfenstein and Millman,[86] which possesses an interval of absolute stability which is large compared with those of other *PEC* algorithms:

$$\text{Predictor:}\quad y_{n+4} + 0{\cdot}29y_{n+3} + 15{\cdot}39y_{n+2} - 12{\cdot}13y_{n+1} - 4{\cdot}55y_n = h(2{\cdot}27f_{n+3} + 6{\cdot}65f_{n+2} + 13{\cdot}91f_{n+1} + 0{\cdot}69f_n), \tag{79}$$

Corrector: (77*ii*),

Error estimate: $C_5h^5y^{(5)}(x_n) \approx -(y^{[1]}_{n+4} - y^{[0]}_{n+4})/18{\cdot}0274$,

Interval of absolute stability in *PEC* mode: $(-0{\cdot}78, 0)$.

Finally, we briefly mention a recent development due to Stetter[172] who considers a new mode denoted by $P(EC)^mLE$, where the L stage consists of forming a linear combination of all previously obtained estimates for y_{n+k}. Thus, the $P(EC)^mE$ mode defined by (58) is modified by inserting after the last application of the corrector the step $\tilde{y}^{[m]}_{n+k} = \sum_{s=0}^{m} \gamma_s y^{[s]}_{n+k}$, where $\sum_{s=0}^{m} \gamma_s = 1$. The final evaluation step is $\tilde{f}^{[m]}_{n+k} = f(x_{n+k}, \tilde{y}^{[m]}_{n+k})$, and in the remainder of (58) $y^{[m]}_{n+j}$ and $f^{[m]}_{n+j}$ are replaced by $\tilde{y}^{[m]}_{n+j}$ and $\tilde{f}^{[m]}_{n+j}$ respectively, $j = 0, 1, 2, \dots, (k-1)$. Stetter shows how the constants γ_s may be chosen to produce algorithms with increased intervals of absolute stability. For example, the second-order pair consisting of $y_{n+2} - y_{n+1} = \frac{1}{2}h(3f_{n+1} - f_n)$ as predictor and $y_{n+2} - y_{n+1} = \frac{1}{2}h(f_{n+2} + f_{n+1})$ as corrector, used in *PECLE* mode with $\gamma_0 = 0{\cdot}65$ and $\gamma_1 = 0{\cdot}35$, has an interval of absolute stability of almost $(-6, 0)$.

3.14 Implementation of predictor–corrector methods: Gear's method

The step-control policy described in section 3.12 may, at a certain stage of the computation, call for a reduction or permit an increase in the steplength. We now consider how such changes can be implemented computationally. In order to fix ideas, let us assume that the algorithm we are using has stepnumber four. When the steplength is held constant the new value y_{n+4} is evaluated in terms of previously calculated numerical values for y and f at x_{n+j}, $j = 0, 1, 2, 3$; let us call these values "back values". If however the steplength is changed before we calculate the new value y_{n+4}, the necessary back values may or may not have been already computed. Thus, for example, if the change consists of doubling the steplength (a common practice when the step-control permits an increase) no additional computation is necessary, since the appropriate back values are now the computed values for y and f at $x_{n+3}, x_{n+1}, x_{n-1}$, and x_{n-3}, all of which values have previously been calculated (but not necessarily stored; thus a programming complication may arise). If on the other hand the steplength is halved, the appropriate back values consist of the values of y and f at $x_{n+3}, x_{n+\frac{5}{2}}, x_{n+2}$, and $x_{n+\frac{3}{2}}$, and we have no computed values for y and f at $x_{n+\frac{5}{2}}, x_{n+\frac{3}{2}}$. We now describe briefly some alternative methods for finding such back values.

Firstly, we may regard the problem as a special case of the problem of finding starting values, which we described in section 3.2. Any of the methods described there, namely the Taylor algorithm, one-step Obrechkoff methods or Runge–Kutta methods, is applicable. The last are probably the easiest to implement from the programming point of view.

The second approach is to find back values for y by any of the standard interpolation formulae, which can be found in elementary texts on numerical analysis. The error in the interpolation formula should, of course, be of the same order as the local truncation error of the method used to solve the initial value problem. The back values for f are then found by function evaluation. It is of interest to note that when the method consists of an Adams–Bashforth–Moulton predictor–corrector pair, the back values consist of y_{n+3}—which is always available—and values of f only. Thus the interpolation may be performed directly in terms of previously computed f-values, and function evaluations thereby saved.

Thirdly, Ceschino[25] has derived formulae of Adams type for use when the steplength is changed from h to ωh. In particular, he has produced a set of three explicit and two implicit fourth-order formulae all of which revert to Adams–Bashforth and Adams–Moulton methods respectively, when $\omega = 1$. These formulae are such that the logical structure of the predictor–corrector mode is preserved through the change of steplength.

Finally, an approach originally due to Nordsieck[137] may be employed. The essence of the device is that, instead of storing a number of back values of y and f, we store derivatives *at a single point* of the local polynomial interpolant which represents the solution. The original Nordsieck methods are equivalent to Adams–Moulton methods corrected to convergence. (This equivalence is further investigated by Osborne[141] who shows that any linear multistep method of appropriate order can be written as a Nordsieck method.) We shall describe here a variation (due to Gear,[51,54]) of the Nordsieck device which produces a one-step predictor–corrector algorithm equivalent to an Adams–Bashforth–Moulton pair in $P(EC)^m$ mode. If such a pair has stepnumber k then it is easily shown that the order p is k, and that the stepnumber of the corrector is one less than that of the predictor. Thus, from (59) we have

$$P:\quad y_{n+k}^{[0]} = y_{n+k-1}^{[m]} + h\sum_{j=0}^{k-1}\beta_j^* f_{n+j}^{[m-1]}, \tag{80i}$$

$$C:\quad y_{n+k}^{[s+1]} = y_{n+k-1}^{[m]} + h\beta_k f_{n+k}^{[s]} + h\sum_{j=0}^{k-1}\beta_j f_{n+j}^{[m-1]}, \qquad s = 0, 1, \ldots, m-1, \tag{80ii}$$

with $\beta_0 = 0$. It follows that

$$y_{n+k}^{[s+1]} - y_{n+k}^{[s]} = h\beta_k(f_{n+k}^{[s]} - f_{n+k}^{[s-1]}), \qquad s = 1, 2, \ldots, m-1 \tag{81}$$

and

$$y_{n+k}^{[1]} - y_{n+k}^{[0]} = h\beta_k\left(f_{n+k}^{[0]} - \sum_{j=0}^{k-1}\frac{\beta_j^* - \beta_j}{\beta_k}f_{n+j}^{[m-1]}\right). \tag{82}$$

Let us define $\delta_j^* = (\beta_j^* - \beta_j)/\beta_k$, $j = 0, 1, \ldots, k-1$ and $d_{n+k} = \sum_{j=0}^{k-1}\delta_j^* f_{n+j}^{[m-1]}$. Then (82) may be written

$$y_{n+k}^{[1]} - y_{n+k}^{[0]} = h\beta_k(f_{n+k}^{[0]} - d_{n+k}). \tag{83}$$

Comparing (83) with (81), we see that d_{n+k} is playing the rôle of $f_{n+k}^{[-1]}$, were such a quantity defined: thus we may regard d_{n+k} as a 'predicted' value for f_{n+k}. Define the $(k+1)$-vector $\mathbf{y}_{n+k}^{[s]}$ by

$$\mathbf{y}_{n+k}^{[s]} = \begin{cases} [y_{n+k}^{[0]}, hd_{n+k}, hf_{n+k-1}^{[m-1]}, hf_{n+k-2}^{[m-1]}, \ldots, hf_{n+1}^{[m-1]}]^T, & \text{if } s = 0,\\ [y_{n+k}^{[s]}, hf_{n+k}^{[s-1]}, hf_{n+k-1}^{[m-1]}, hf_{n+k-2}^{[m-1]}, \ldots, hf_{n+1}^{[m-1]}]^T, & \text{if } s = 1, 2, \ldots, m. \end{cases} \tag{84}$$

It follows that (80) can be written in the form

$$\begin{aligned} P:&\quad \mathbf{y}_{n+k}^{[0]} = B\mathbf{y}_{n+k-1}^{[m]},\\ C:&\quad \mathbf{y}_{n+k}^{[s+1]} = \mathbf{y}_{n+k}^{[s]} + F(\mathbf{y}_{n+k}^{[s]})\mathbf{c}, \qquad s = 0, 1, \ldots, m-1, \end{aligned} \tag{85}$$

where B is a $(k+1)\times(k+1)$ matrix, $\mathbf{c}$ a $(k+1)$-vector, and F a scalar, given by

$$B = \begin{bmatrix} 1 & \beta_{k-1}^* & \beta_{k-2}^* & \ldots & \beta_1^* & \beta_0^*\\ 0 & \delta_{k-1}^* & \delta_{k-2}^* & \ldots & \delta_1^* & \delta_0^*\\ 0 & 1 & 0 & \ldots & 0 & 0\\ 0 & 0 & 1 & \ldots & 0 & 0\\ \vdots & \vdots & \vdots & & \vdots & \vdots\\ 0 & 0 & 0 & \ldots & 1 & 0 \end{bmatrix}, \qquad \mathbf{c} = \begin{bmatrix} \beta_k\\ 1\\ 0\\ 0\\ \vdots\\ 0 \end{bmatrix},$$

$$F(\mathbf{y}_{n+k}^{[s]}) = \begin{cases} h(f_{n+k}^{[0]} - d_{n+k}) \equiv h[f(x_{n+k}, y_{n+k}^{[0]}) - d_{n+k}], & \text{if } s = 0, \\ h(f_{n+k}^{[s]} - f_{n+k}^{[s-1]}) \equiv h[f(x_{n+k}, y_{n+k}^{[s]}) - f_{n+k}^{[s-1]}], & \text{if } s = 1, 2, \ldots, m-1. \end{cases} \tag{86}$$

Note that B and $\mathbf{c}$ depend only on the coefficients in (80) and are independent of h. We now have, in (85), a one-step form of the predictor–corrector method (80). However, if we attempt to change the steplength, the earlier difficulties are still present, since the vector of back values $\mathbf{y}_{n+k-1}^{[m]}$ contains information computed at a number of different points. Recall that in section 2.4 we saw that a linear multistep method could be derived by eliminating the coefficients of a polynomial $I(x)$ which locally represents the solution. Let us illustrate by considering the case $k = 3$. Then the predictor (80*i*) can be so derived from the polynomial $I(x) = ax^3 + bx^2 + cx + d$ by letting

$$\begin{aligned} &I(x_{n+3}) = y_{n+3}^{[0]}, \qquad I(x_{n+2}) = y_{n+2}^{[m]}, \\ &I'(x_{n+2}) = f_{n+2}^{[m-1]}, \qquad I'(x_{n+1}) = f_{n+1}^{[m-1]}, \qquad I'(x_n) = f_n^{[m-1]}. \end{aligned} \tag{87}$$

(Elimination of the four coefficients a, b, c, and d between these five equations will yield the 3-step Adams–Bashforth method.) Clearly, specification of the back vector

$$\mathbf{y}_{n+2}^{[m]} = [y_{n+2}^{[m]}, hf_{n+2}^{[m-1]}, hf_{n+1}^{[m-1]}, hf_n^{[m-1]}]^T$$

determines $I(x)$ uniquely; alternatively, we could determine $I(x)$ by specifying its value and those of its first three derivatives *at a single point*, x_{n+2}. Thus essentially the same information as is carried in the vector $\mathbf{y}_{n+2}^{[m]}$ is contained in the vector $\mathbf{z}_{n+2}^{[m]}$, where

$$\mathbf{z}_{n+2}^{[m]} = \left[I(x_{n+2}), hI'(x_{n+2}), \frac{h^2}{2!}I^{(2)}(x_{n+2}), \frac{h^3}{3!}I^{(3)}(x_{n+2}) \right]^T.$$

Indeed, a straightforward calculation involving the equations (87) shows that

$$\mathbf{z}_{n+2}^{[m]} = Q\mathbf{y}_{n+2}^{[m]}, \qquad Q = \begin{bmatrix} 1 & 0 & 0 & 0 \\ 0 & 1 & 0 & 0 \\ 0 & \frac{3}{4} & -1 & \frac{1}{4} \\ 0 & \frac{1}{6} & -\frac{1}{3} & \frac{1}{6} \end{bmatrix}. \tag{88}$$

Note that the scaling of the components of $\mathbf{z}_{n+2}^{[m]}$ by powers of h has resulted in Q being independent of h.

For general k, all relevant back information can be carried in the vector

$$\mathbf{z}_{n+k-1}^{[m]} = \left[I(x_{n+k-1}), hI'(x_{n+k-1}), \ldots, \frac{h^k}{k!} I^{(k)}(x_{n+k-1}) \right]^T, \tag{89}$$

where $\mathbf{z}_{n+k-1}^{[m]} = Q\mathbf{y}_{n+k-1}^{[m]}$, Q being a $(k+1) \times (k+1)$ matrix depending only on the coefficients of the predictor (80i). Note that the first and second components of $\mathbf{z}_{n+k-1}^{[m]}$ are always identical with the first and second components, respectively, of $\mathbf{y}_{n+k-1}^{[m]}$: hence the first two rows of the matrix Q always have the form $[1\,0\,0\,0\ldots0]$ and $[0\,1\,0\,0\ldots0]$ respectively. Finally, with this Q, we apply the transformations $\mathbf{z}_{n+j}^{[s]} = Q\mathbf{y}_{n+j}^{[s]}$, $s = 0, 1, \ldots, m, j = 0, 1, \ldots, k$, to (85). Note that for any $\mathbf{v}$, $F(\mathbf{v})$, as defined by (86) depends only on the first two components of $\mathbf{v}$; by the form of Q, these components remain unaltered by the transformation $\mathbf{v} \to Q\mathbf{v}$, and hence $F(\mathbf{v}) = F(Q\mathbf{v})$. We thus obtain the one-step method

$$\begin{aligned} P:&\quad \mathbf{z}_{n+k}^{[0]} = QBQ^{-1}\mathbf{z}_{n+k-1}^{[m]} \\ C:&\quad \mathbf{z}_{n+k}^{[s+1]} = \mathbf{z}_{n+k}^{[s]} + F(\mathbf{z}_{n+k}^{[s]})\mathbf{l}, \qquad s = 0, 1, \ldots, m-1, \end{aligned} \tag{90}$$

where $\mathbf{l} = Q\mathbf{c}$. This form of (80) has the advantage that the vector of back values $\mathbf{z}_{n+k-1}^{[m]}$ defined by (89) now contains only information calculated at the single point x_{n+k-1}. If we wish to change steplength from h to αh, all we have to do is to multiply the ith component of $\mathbf{z}_{n+k-1}^{[m]}$ by α^i, $i = 0, 1, \ldots, k$.

Gear's method[54,55] incorporates into a sophisticated programme Adams–Bashforth–Moulton methods in the form (90). Methods of order one to seven are available, and the programme can automatically change order as well as steplength, the strategy being such as to minimize the computational effort required for the local error to be less than a stipulated bound. The programme also offers the user the option of replacing the Adams–Bashforth–Moulton pairs by special predictor–corrector pairs appropriate for the solution of a stiff system of differential equations; this aspect will be discussed in section 8.9. Variable order is also incorporated in *Krogh's method*,[93,94,95] in which orders from one to thirteen are available. This method, like Gear's, is self-starting and automatically controls both order and steplength. Both contain several control processes, some of them heuristic, and it is impracticable to reproduce them here. In a recent sophisticated test of numerical methods for initial value problems, Hull, Enright, Fellen, and Sedgwick[73] compare variable-order predictor–corrector methods, Runge–Kutta methods (see chapter 4), and extrapolation methods (see chapter 6) on a wide selection of test problems, and conclude that when function evaluations are relatively expensive (quantitatively, when each function evaluation costs more than roughly twenty-five arithmetic operations per component), variable-order methods are best, that of Krogh being slightly preferable to that of Gear (presumably as a result of the wider range of order available in the former).

Exercises

23. Verify that the 3-step Adams–Bashforth–Moulton method can be derived from (87). Verify that (88) holds.

24. In the case $k = 3$, use the coefficients given in section 2.10 to find the matrix B. Hence show that the matrix QBQ^{-1} in (90) becomes the 4×4 Pascal matrix

$$\begin{bmatrix} 1 & 1 & 1 & 1 \\ 0 & 1 & 2 & 3 \\ 0 & 0 & 1 & 3 \\ 0 & 0 & 0 & 1 \end{bmatrix}.$$

(This result extends to a general k. It can be utilized in a programme which allows the prediction stage of (90) to be performed without multiplication; see Gear.[51])

25. Show that the predictor–corrector pair (80), used in $P(EC)^mE$ mode, can also be reduced to a form similar to (90), by suitably modifying the notation and adding to (85) one further iteration with $\mathbf{c}$ replaced by $[0, 1, 0, \ldots, 0]^T$.

4

Runge–Kutta methods

4.1 Introduction

Recall the initial value problem

$$y' = f(x, y), \qquad y(a) = \eta.$$

Of all computational methods for the numerical solution of this problem, the easiest to implement is Euler's rule,

$$y_{n+1} - y_n = hf(x_n, y_n) \equiv hf_n.$$

It is explicit and, being a one-step method, it requires no additional starting values and readily permits a change of steplength during the computation. Its low order, of course, makes it of limited practical value. Linear multistep methods achieve higher order by sacrificing the one-step nature of the algorithm, whilst retaining linearity with respect to y_{n+j}, f_{n+j}, $j = 0, 1, \ldots, k$. Higher order can also be achieved by sacrificing linearity, but preserving the one-step nature of the algorithm. This is the philosophy behind the methods first proposed by Runge[156] and subsequently developed by Kutta[99] and Heun.[70] Runge–Kutta methods thus retain the advantages of one-step methods but, due to the loss of linearity, error analysis is considerably more difficult than in the case of linear multistep methods. Traditionally, Runge–Kutta methods are all explicit, although, recently, implicit Runge–Kutta methods, which have improved weak stability characteristics, have been considered. We shall use the phrase 'Runge–Kutta method' to mean 'explicit Runge–Kutta method'. Thus a Runge–Kutta method may be regarded as a particular case of the general explicit one-step method

$$y_{n+1} - y_n = h\phi(x_n, y_n, h). \tag{1}$$

4.2 Order and convergence of the general explicit one-step method

The fact that the general method (1) makes no mention of the function $f(x, y)$, which defines the differential equation, makes it impossible to define the order of the method independently of the differential equation, as was the case with linear multistep methods.

Definition *The method (1) is said to have* **order** *p if p is the largest integer for which*

$$y(x + h) - y(x) - h\phi(x, y(x), h) = O(h^{p+1}) \tag{2}$$

holds, where $y(x)$ is the theoretical solution of the initial value problem.

Definition *The method (1) is said to be* **consistent** *with the initial value problem if*

$$\phi(x, y, 0) \equiv f(x, y). \tag{3}$$

If the method (1) is consistent with the initial value problem (in future we shall simply say 'consistent'), then

$$\begin{aligned} y(x + h) - y(x) - h\phi(x, y(x), h) &= hy'(x) - h\phi(x, y(x), 0) + O(h^2) \\ &= O(h^2), \end{aligned}$$

since $y'(x) = f(x, y(x)) = \phi(x, y(x), 0)$, by (3). Thus a consistent method has order at least one.

The only linear multistep method which falls within the class (1) is Euler's rule which we obtain by setting

$$\phi(x, y, h) = \phi_E(x, y, h) \equiv f(x, y).$$

(The subscript E denotes 'Euler'.) The consistency condition (3) is then obviously satisfied and a simple calculation shows that the order, according to (2), is one. Thus the definitions of order and consistency given above do not contradict those of chapter 2.

The Taylor algorithm of order p (see section 3.2) also falls within the class (1) and is obtained by setting

$$\phi(x, y, h) = \phi_T(x, y, h) \equiv f(x, y) + \frac{h}{2!} f^{(1)}(x, y) + \dots + \frac{h^{p-1}}{p!} f^{(p-1)}(x, y), \tag{4}$$

where

$$f^{(q)}(x, y) \equiv \frac{d^q}{dx^q} f(x, y), \qquad q = 1, 2, \dots, (p - 1).$$

(The subscript T denotes 'Taylor'.) Once again, the new definition of order (2) does not contradict our previous usage.

The following theorem, whose proof may be found in Henrici[67] (page 71), states necessary and sufficient conditions for the method (1) to be convergent in the sense of the definition of section 2.5.

Theorem 4.1 (i) *Let the function $\phi(x, y, h)$ be continuous jointly as a function of its three arguments, in the region $\mathscr{D}$ defined by $x \in [a, b]$, $y \in (-\infty, \infty)$, $h \in [0, h_0]$, $h_0 > 0$.*

(ii) *Let $\phi(x, y, h)$ satisfy a Lipschitz condition of the form*

$$|\phi(x, y^*, h) - \phi(x, y, h)| \leqslant M|y^* - y|$$

for all points (x, y^*, h), (x, y, h) in $\mathscr{D}$.

Then the method (1) is convergent if and only if it is consistent.

For all the methods we shall consider, conditions (i) and (ii) are satisfied if $f(x, y)$ satisfies the conditions stated in theorem 1.1. For such methods consistency is necessary and sufficient for convergence. Note that there is no requirement corresponding to zero-stability, since no parasitic solutions can arise with a one-step method.

4.3 Derivation of classical Runge–Kutta methods

By 'classical' Runge–Kutta methods, we mean methods which were first derived in the pre-computer era. The choice of coefficients in these methods was largely motivated by the need to produce methods convenient for desk computation. Such methods are not necessarily the most suitable for use on an automatic computer. Only methods of order less than or equal to four are considered in this section.

The general *R-stage Runge–Kutta method* is defined by

$$y_{n+1} - y_n = h\phi(x_n, y_n, h), \tag{5i}$$

$$\left.\begin{aligned} \phi(x, y, h) &= \sum_{r=1}^{R} c_r k_r, \\ k_1 &= f(x, y), \\ k_r &= f\left(x + ha_r, y + h\sum_{s=1}^{r-1} b_{rs}k_s\right), \qquad r = 2, 3, \ldots, R, \end{aligned}\right\} \tag{5(ii)}$$

$$a_r = \sum_{s=1}^{r-1} b_{rs}, \qquad r = 2, 3, \ldots, R. \tag{5iii}$$

Note that an R-stage Runge–Kutta method involves R function evaluations per step. Each of the functions $k_r(x, y, h)$, $r = 1, 2, \dots, R$, may be interpreted as an approximation to the derivative $y'(x)$, and the function $\phi(x, y, h)$ as a weighted mean of these approximations. Note also that consistency demands that $\sum_{r=1}^{R} c_r = 1$. If we can choose values for the constants c_r, a_r, b_{rs} such that the expansion of the function $\phi(x, y, h)$ defined by (5*ii*) in powers of h differs from the expansion for $\phi_T(x, y, h)$ given by (4) only in the pth and higher powers of h, then the method clearly has order p. (Note that in (4) we are assuming that $y(x) \in C^p[a, b]$.)

There is a great deal of tedious manipulation involved in deriving Runge–Kutta methods of higher order; accordingly, we shall derive only methods of order up to three and quote some well known methods of order four.

Introducing the shortened notation

$$f = f(x, y), \qquad f_x = \frac{\partial f(x, y)}{\partial x}, \qquad f_{xx} = \frac{\partial^2 f(x, y)}{\partial x^2}, \qquad f_{xy} = \frac{\partial^2 f}{\partial x \partial y}, \text{ etc.}$$

and using the expressions for the total derivatives of f which we found in section 3.2, we may write the expansion (4) in the form

$$\phi_T(x, y, h) = f + \tfrac{1}{2}hF + \tfrac{1}{6}h^2(Ff_y + G) + O(h^3), \tag{6}$$

where

$$F = f_x + ff_y, \qquad G = f_{xx} + 2ff_{xy} + f^2 f_{yy}. \tag{7}$$

It turns out that it is possible to achieve third order with $R = 3$, and thus we need derive expansions only for the functions k_1, k_2, and k_3, where, using (5*ii*) and (5*iii*)

$$\begin{aligned} k_1 &= f(x, y) = f, \\ k_2 &= f(x + ha_2, y + ha_2 k_1), \\ k_3 &= f(x + ha_3, y + h(a_3 - b_{32})k_1 + hb_{32}k_2). \end{aligned} \tag{8}$$

Expanding k_2 as a Taylor series about the point (x, y) we obtain

$$k_2 = f + ha_2(f_x + k_1 f_y) + \tfrac{1}{2}h^2 a_2^2(f_{xx} + 2k_1 f_{xy} + k_1^2 f_{yy}) + O(h^3).$$

On substituting for k_1 and using (7) we find

$$k_2 = f + ha_2 F + \tfrac{1}{2}h^2 a_2^2 G + O(h^3). \tag{9}$$

Treating k_3 similarly, we obtain

$$\begin{aligned}k_3 = f &+ h\{a_3 f_x + [(a_3 - b_{32})k_1 + b_{32}k_2]f_y\}\\ &+ \tfrac{1}{2}h^2\{a_3^2 f_{xx} + 2a_3[(a_3 - b_{32})k_1 + b_{32}k_2]f_{xy}\\ &+ [(a_3 - b_{32})k_1 + b_{32}k_2]^2 f_{yy}\} + O(h^3).\end{aligned}$$

On substituting for k_1 and k_2 we find, after some manipulation,

$$k_3 = f + ha_3F + h^2(a_2b_{32}Ff_y + \tfrac{1}{2}a_3^2G) + O(h^3). \tag{10}$$

On substituting from (8), (9), and (10) into (5*ii*), the following expansion for $\phi(x, y, h)$, as defined by the method (5), is obtained:

$$\begin{aligned}\phi(x, y, h) = (c_1 + c_2 + c_3)f &+ h(c_2a_2 + c_3a_3)F\\ &+ \tfrac{1}{2}h^2[2c_3a_2b_{32}Ff_y + (c_2a_2^2 + c_3a_3^2)G] + O(h^3).\end{aligned} \tag{11}$$

We now have to match the expansions (6) and (11). Let us see what can be achieved with $R = 1, 2, 3$. (For $R > 3$, (11) is invalid since k_4 will contribute additional terms.)

For $R = 1$, $c_2 = c_3 = 0$ and (11) reduces to

$$\phi(x, y, h) = c_1 f + O(h^3). \tag{12}$$

On setting $c_1 = 1$, (12) differs from the expansion (6) for ϕ_T by a term of order h. Thus the resulting method, which is Euler's rule, has order one.

For $R = 2$, $c_3 = 0$, and (11) reduces to

$$\phi(x, y, h) = (c_1 + c_2)f + hc_2a_2F + \tfrac{1}{2}h^2c_2a_2^2G + O(h^3). \tag{13}$$

To match this expansion with (6) we must satisfy the equations

$$c_1 + c_2 = 1, \qquad c_2a_2 = \tfrac{1}{2}. \tag{14}$$

This is a set of two equations in three unknowns and thus there exists a one-parameter family of solutions. However, it is clear that there is no solution in this family which causes the expansions (6) and (14) to differ by a term of order higher than h^2. Thus there exists an infinite number of two-stage Runge–Kutta methods of order two and none of order more than two. This lack of uniqueness is typical of all Runge–Kutta derivations. Two particular solutions of (14) yield well-known methods.

(i) $c_1 = 0$, $c_2 = 1$, $a_2 = \tfrac{1}{2}$. The resulting method is

$$y_{n+1} - y_n = hf(x_n + \tfrac{1}{2}h, y_n + \tfrac{1}{2}hf(x_n, y_n)). \tag{15}$$

This method, originally due to Runge, is referred to as the *modified Euler method* or the *improved polygon method.*

(ii) $c_1 = \frac{1}{2}, c_2 = \frac{1}{2}, a_2 = 1$. The resulting method,

$$y_{n+1} - y_n = \tfrac{1}{2}h[f(x_n, y_n) + f(x_n + h, y_n + hf(x_n, y_n))] \tag{16}$$

is known as the *improved Euler method.*

For $R = 3$, we can match (11) and (6) up to and including h^2 terms if we satisfy the following set of equations

$$\begin{aligned} c_1 + c_2 + c_3 &= 1, \\ c_2 a_2 + c_3 a_3 &= \tfrac{1}{2}, \\ c_2 a_2^2 + c_3 a_3^2 &= \tfrac{1}{3}, \\ c_3 a_2 b_{32} &= \tfrac{1}{6}. \end{aligned} \tag{17}$$

There are now four equations in six unknowns and there exists a two-parameter family of solutions. Consideration of terms of order h^3, which we have ignored in this derivation, shows that once again no solution of (17) causes the expansions to differ by a term of order higher than h^3. Thus there exists a doubly infinite family of three-stage Runge–Kutta methods of order three, and none of order more than three. Two particular solutions of (17) lead to well known third-order Runge–Kutta methods.

(i) $c_1 = \frac{1}{4}$, $c_2 = 0$, $c_3 = \frac{3}{4}$, $a_2 = \frac{1}{3}$, $a_3 = \frac{2}{3}$, $b_{32} = \frac{2}{3}$. The resulting method may be written

$$\begin{aligned} y_{n+1} - y_n &= \frac{h}{4}(k_1 + 3k_3), \\ k_1 &= f(x_n, y_n), \\ k_2 &= f(x_n + \tfrac{1}{3}h, y_n + \tfrac{1}{3}hk_1), \\ k_3 &= f(x_n + \tfrac{2}{3}h, y_n + \tfrac{2}{3}hk_2). \end{aligned} \tag{18}$$

(Note that we have abused the notation to the extent that hitherto k_r denoted the function $k_r(x, y, h)$, whereas in (18) k_r denotes the function $k_r(x_n, y_n, h)$. No real confusion arises and we shall continue to adopt the convention of (18) when quoting specific Runge–Kutta methods.) The method (18) is known as *Heun's third-order formula.* We observe that the computational advantage in choosing $c_2 = 0$ is somewhat illusory since, although k_2 does not appear in the first equation of (18) it must nevertheless be calculated at each step.

(ii) $c_1 = \frac{1}{6}$, $c_2 = \frac{2}{3}$, $c_3 = \frac{1}{6}$, $a_2 = \frac{1}{2}$, $a_3 = 1$, $b_{32} = 2$. The resulting method,

$$\begin{aligned}
y_{n+1} - y_n &= \frac{h}{6}(k_1 + 4k_2 + k_3),\\
k_1 &= f(x_n, y_n),\\
k_2 &= f(x_n + \tfrac{1}{2}h, y_n + \tfrac{1}{2}hk_1),\\
k_3 &= f(x_n + h, y_n - hk_1 + 2hk_2),
\end{aligned} \tag{19}$$

is *Kutta's third-order rule*. It is the most popular third-order Runge–Kutta method for desk computation (largely because the coefficient $\frac{1}{2}$ is preferable to $\frac{1}{3}$, which appears frequently in (18)).

The derivation of fourth-order Runge–Kutta methods involves tedious manipulation; it transpires that with $R = 4$, fourth order, and no higher, can be attained. In place of (17) a system of 8 equations in 10 unknowns is obtained. (Full details may be found in Ralston.[150]) Two well known fourth-order methods are

$$\begin{aligned}
y_{n+1} - y_n &= \frac{h}{6}(k_1 + 2k_2 + 2k_3 + k_4),\\
k_1 &= f(x_n, y_n),\\
k_2 &= f(x_n + \tfrac{1}{2}h, y_n + \tfrac{1}{2}hk_1),\\
k_3 &= f(x_n + \tfrac{1}{2}h, y_n + \tfrac{1}{2}hk_2),\\
k_4 &= f(x_n + h, y_n + hk_3),
\end{aligned} \tag{20}$$

and

$$\begin{aligned}
y_{n+1} - y_n &= \frac{h}{8}(k_1 + 3k_2 + 3k_3 + k_4),\\
k_1 &= f(x_n, y_n),\\
k_2 &= f(x_n + \tfrac{1}{3}h, y_n + \tfrac{1}{3}hk_1),\\
k_3 &= f(x_n + \tfrac{2}{3}h, y_n - \tfrac{1}{3}hk_1 + hk_2),\\
k_4 &= f(x_n + h, y_n + hk_1 - hk_2 + hk_3).
\end{aligned} \tag{21}$$

The method (20) is undoubtedly the most popular of all Runge–Kutta methods. Indeed it is frequently referred to, somewhat loosely, as 'the fourth-order Runge–Kutta method'. (Occasionally, it is called—even more loosely—'the Runge–Kutta method'.)

Exercises

1. Assuming that the function $f(x, y)$ satisfies the conditions of theorem 1.1, which guarantee the existence of a unique solution of the initial value problem $y' = f(x, y)$, $y(x_0) = y_0$, verify that the function $\phi(x, y, h)$ defined by the general Runge–Kutta method (5) does indeed satisfy conditions (i) and (ii) of theorem 4.1. (Hence, for the general Runge–Kutta method, consistency is necessary and sufficient for convergence.)

2. Interpret geometrically the modified Euler method (15) and the improved Euler method (16).

3. Derive the third-order Runge–Kutta method for which $c_2 = c_3$ and $a_2 = a_3$. (Nyström's third-order method.)

4. The two degrees of freedom in the solution of the system of equations (17) may be removed by making any *two* of the following simplifying assumptions:

$$c_1 = 0, \quad c_2 = 0, \quad c_3 = 0, \quad a_2 = 0, \quad a_3 = 0, \quad b_{32} = 0, \quad a_3 = b_{32}.$$

Consider all possible pairs of these assumptions and show that some lead to contradictions, some lead to methods we have already derived, one leads to an awkward cubic equation for the remaining coefficients, and one leads to a new third-order Runge–Kutta method with simple coefficients.

5. Show that (20), applied to the differential equation $y' = py$, p constant, gives $y_{n+1}/y_n = \exp(ph) + O(ph)^5$.

6. Show that if $f(x, y) \equiv g(x)$, the improved Euler rule (16) reduces to the Trapezoidal rule for quadrature, Kutta's third-order rule (19) and the method (20) reduce to Simpson's rule for quadrature, and method (21) reduces to the three-eighths rule for quadrature. (For this reason (20) and (21) are sometimes called the Kutta–Simpson one-third rule and the Kutta–Simpson three-eighth rule respectively.)

Show also that under the same assumption, Heun's third-order formula (18) reduces to the two-point Radau quadrature formula

$$\int_{-1}^{+1} F(x)\,\mathrm{d}x = \tfrac{1}{2}F(-1) + \tfrac{3}{2}F(\tfrac{1}{3})$$

applied to the integral $\int_{x_n}^{x_n+h} g(x)\,\mathrm{d}x$.

7. Show that the predictor–corrector algorithm consisting of Euler's rule and the Trapezoidal rule applied in *PECE* mode is equivalent to the improved Euler method (16). Find the three-stage Runge–Kutta method which is equivalent to the algorithm consisting of the same predictor–corrector pair applied in $P(EC)^2E$ mode.

4.4 Runge–Kutta methods of order greater than four

We have seen in the previous section that there exists an R-stage method of order R (and none of order greater than R) for $R = 1, 2, 3, 4$. It is tempting, but incorrect, to assume that there exists an R-stage method of order R for general R. The study of higher-order Runge–Kutta methods involves extremely complicated algebra; the best sources of information on this topic are the papers of Butcher.[13,14,16,17,18] In the penultimate of these references, Butcher proves the non-existence of a five-stage method of order five and in the last he proves the following result. Let $p^*(R)$ be the highest order that can be attained by an R-stage method. Then

$$\begin{aligned} &p^*(R) = R, R = 1, 2, 3, 4, \\ &p^*(5) = 4, \\ &p^*(6) = 5, \\ &p^*(7) = 6, \\ &p^*(8) = 6, \\ &p^*(9) = 7, \\ &p^*(R) \leqslant R - 2, R = 10, 11, \ldots. \end{aligned} \tag{22}$$

It is clear from (22) why Runge–Kutta methods of fourth order are the most popular.

We shall quote only two examples of higher order Runge–Kutta methods. These may be of use in finding high-accuracy starting values for linear multistep algorithms. The first example is the *fifth-order Kutta–Nyström method*; it is a six-stage method given by

$$\begin{aligned} y_{n+1} - y_n &= \frac{h}{192}(23k_1 + 125k_3 - 81k_5 + 125k_6), \\ k_1 &= f(x_n, y_n), \\ k_2 &= f(x_n + \tfrac{1}{3}h, y_n + \tfrac{1}{3}hk_1), \\ k_3 &= f(x_n + \tfrac{2}{5}h, y_n + \tfrac{1}{25}h(4k_1 + 6k_2)), \\ k_4 &= f(x_n + h, y_n + \tfrac{1}{4}h(k_1 - 12k_2 + 15k_3)), \\ k_5 &= f(x_n + \tfrac{2}{3}h, y_n + \tfrac{1}{81}h(6k_1 + 90k_2 - 50k_3 + 8k_4)), \\ k_6 &= f(x_n + \tfrac{4}{5}h, y_n + \tfrac{1}{75}h(6k_1 + 36k_2 + 10k_3 + 8k_4)). \end{aligned} \tag{23}$$

Huťa[79] derives the following sixth-order eight-stage method;

$$\begin{aligned} y_{n+1} - y_n = \frac{h}{840}(41k_1 + 216k_3 + 27k_4 + 272k_5 + 27k_6 + 216k_7 \\ + 41k_8), \end{aligned}$$

$$
\begin{aligned}
k_1 &= f(x_n, y_n),\\
k_2 &= f(x_n + \tfrac{1}{9}h, y_n + \tfrac{1}{9}hk_1),\\
k_3 &= f(x_n + \tfrac{1}{6}h, y_n + \tfrac{1}{24}h(k_1 + 3k_2)),\\
k_4 &= f(x_n + \tfrac{1}{3}h, y_n + \tfrac{1}{6}h(k_1 - 3k_2 + 4k_3)),\\
k_5 &= f(x_n + \tfrac{1}{2}h, y_n + \tfrac{1}{8}h(-5k_1 + 27k_2 - 24k_3 + 6k_4)),\\
k_6 &= f(x_n + \tfrac{2}{3}h, y_n + \tfrac{1}{9}h(221k_1 - 981k_2 + 867k_3 - 102k_4 + k_5)),\\
k_7 &= f(x_n + \tfrac{5}{6}h, y_n + \tfrac{1}{48}h(-183k_1 + 678k_2 - 472k_3 - 66k_4\\
&\qquad + 80k_5 + 3k_6)),\\
k_8 &= f(x_n + h, y_n + \tfrac{1}{82}h(716k_1 - 2079k_2 + 1002k_3 + 834k_4\\
&\qquad - 454k_5 - 9k_6 + 72k_7)).
\end{aligned}
\tag{24}
$$

4.5 Error bounds for Runge–Kutta methods

So far we have talked only of the order of an explicit one-step method and have made no mention of its truncation error.

Definition *The* **local truncation error** *at* x_{n+1} *of the general explicit one-step method (1) is defined to be* T_{n+1} *where*

$$T_{n+1} = y(x_{n+1}) - y(x_n) - h\phi(x_n, y(x_n), h) \tag{25}$$

and $y(x)$ *is the theoretical solution of the initial value problem.*

Recall the definition, given in section 2.7, of the local truncation error, T_{n+k}, of a linear multistep method. There it was shown that if we make the localizing assumption that no previous errors have been made (that is, that $y_{n+j} = y(x_{n+j}), j = 0, 1, \dots, k-1$) then the local truncation error at x_{n+k} of an *explicit* linear k-step method satisfies

$$T_{n+k} = y(x_{n+k}) - y_{n+k}.$$

If we make a similar assumption about (1) (namely that $y_n = y(x_n)$) then, from (25) and (1) it follows that

$$T_{n+1} = y(x_{n+1}) - y_{n+1}.$$

Thus the truncation error defined by (25) is local in precisely the sense discussed in section 2.7. The definition of the global truncation error e_{n+1} is exactly as for a linear multistep method; that is $e_{n+1} = y(x_{n+1}) - y_{n+1}$, where now it is no longer assumed that no previous truncation errors have been made.

It was also shown in section 2.7 that, if $y(x)$ is assumed to be sufficiently differentiable, the local truncation error for a linear k-step method of order p can be written in the form

$$T_{n+k} = C_{p+1}h^{p+1}y^{(p+1)}(x_n) + O(h^{p+2}),$$

where C_{p+1} is the error constant and $C_{p+1}h^{p+1}y^{(p+1)}(x_n)$ the principal local truncation error. For the non-linear method (1) the corresponding expression is more complicated, and we must write

$$T_{n+1} = \psi(x_n, y(x_n))h^{p+1} + O(h^{p+2}), \tag{26}$$

where we shall call $\psi(x, y)$ the *principal error function*, and $\psi(x_n, y(x_n))h^{p+1}$ the *principal local truncation error.*

Consider, for example, the general two-evaluation Runge–Kutta method obtained by setting $R = 2$ in (5). By (25), (6), and (13), its local truncation error is

$$\begin{aligned} T_{n+1} = {} & h[f + \tfrac{1}{2}hF + \tfrac{1}{6}h^2(Ff_y + G)]_{\substack{x=x_n \\ y=y(x_n)}} \\ & - h[(c_1 + c_2)f + hc_2a_2F + \tfrac{1}{2}h^2c_2a_2^2G]_{\substack{x=x_n \\ y=y(x_n)}} + O(h^4). \end{aligned}$$

If the order is two, then (14) must hold, and we obtain

$$T_{n+1} = h^3[\tfrac{1}{6}Ff_y + (\tfrac{1}{6} - \tfrac{1}{4}a_2)G]_{\substack{x=x_n \\ y=y(x_n)}} + O(h^4).$$

Thus the principal error function for the general second-order Runge–Kutta method is given by

$$\psi(x, y) = \tfrac{1}{6}Ff_y + (\tfrac{1}{6} - \tfrac{1}{4}a_2)G. \tag{27}$$

Following an argument originally proposed by Lotkin[121] we can find a bound for $\psi(x, y)$, if we assume that the following bounds for f and its partial derivatives hold for $x \in [a, b]$, $y \in (-\infty, \infty)$:

$$|f(x, y)| < Q, \qquad \left|\frac{\partial^{i+j}f(x, y)}{\partial x^i \partial y^j}\right| < P^{i+j}/Q^{j-1}, \qquad i + j \leqslant p, \tag{28}$$

where P and Q are positive constants, and p is the order of the method (in this case, 2). Then

$$\begin{aligned} &|f_y| < P, \\ &|F| = |f_x + ff_y| < PQ + QP = 2PQ, \\ &|G| = |f_{xx} + 2ff_{xy} + f^2f_{yy}| < P^2Q + 2QP^2 + Q^2P^2/Q = 4P^2Q. \end{aligned}$$

Hence, from (27)

$$|\psi(x, y)| < (\tfrac{1}{3} + |\tfrac{2}{3} - a_2|)P^2Q,$$

and we obtain the following bound for the principal local truncation error:

$$|\psi(x_n, y(x_n))h^3| < (\tfrac{1}{3} + |\tfrac{2}{3} - a_2|)h^3P^2Q. \tag{29}$$

It can be shown that the bound we have just found for the principal local truncation error is also a bound for the whole local truncation error T_{n+1}. (See, for example, Henrici,[67] page 75, where, however, the bounds assumed for the partial derivatives of f are not those we have assumed in (28).) This is a consequence of the fact that the Runge–Kutta method is a one-step explicit method. There exists no need for an analysis similar to that given for linear multistep methods in section 3.3, where the difficulty arose essentially from the multistep nature of the method. Thus we may write in place of (29),

$$|T_{n+1}| < (\tfrac{1}{3} + |\tfrac{2}{3} - a_2|)h^3P^2Q. \tag{30}$$

The corresponding bound for the local truncation error of the general third-order Runge–Kutta method obtained by setting $R = 3$ in (5) is shown by Ralston[148] to be as follows, for the case $a_2 \neq 0, a_3 \neq 0, a_2 \neq a_3$:

$$|T_{n+1}| < \{\tfrac{1}{12} + 8|A_1| + |A_2| + |2A_2 + A_3| + |A_2 + A_3| + 2|A_3|\}h^4P^3Q,$$

where

$$\begin{aligned} A_1 &= \tfrac{1}{24} - \tfrac{1}{36}(2a_2 + 2a_3 - 3a_2a_3), \\ A_2 &= \tfrac{1}{24} - \tfrac{1}{12}a_2, \\ A_3 &= \tfrac{1}{8} - \tfrac{1}{6}a_3. \end{aligned} \tag{31}$$

(It is assumed, of course, that the conditions (17) are satisfied and that a_2 and a_3 are the two free parameters.)

In the case of the general fourth-order Runge–Kutta method, the bound for the local truncation error is much more complicated; it may be found in Ralston.[148] For the popular fourth-order method (20), Lotkin,[121] using the above analysis, shows that

$$|T_{n+1}| < \tfrac{73}{720}h^5P^4Q. \tag{32}$$

Alternative bounds for the local truncation error can be found by bounding the partial derivatives of f in a manner other than that of (28). Thus the well-known bound of Bieberbach[5] for the local truncation error of (20) is given by

$$|T_{n+1}| < 6h^5QN(1 + N + N^2 + N^3 + N^4), \tag{33}$$

where, in the neighbourhood $|x - x_0| < A$, $|y - y_0| < B$,

$$|f(x, y)| < Q, \qquad \left|\frac{\partial^{i+j}f(x, y)}{\partial x^i \partial y^j}\right| < N/Q^{j-1},\ i + j \leqslant 4,$$

$$|x - x_0|N < 1 \quad \text{and} \quad AQ < B.$$

Numerical evidence suggests that (32) gives a sharper bound than (33) (see Lotkin[121]).

In section 3.4 we showed that for a linear multistep method the bound for the global truncation error is an order of magnitude greater than the bound for the local truncation error. A similar result holds for the general explicit one step method (1). If the local truncation error T_{n+1} defined by (25) satisfies

$$|T_{n+1}| \leqslant Kh^{p+1}, \tag{34}$$

where K is a constant, then the global truncation error $e_n \equiv y(x_n) - y_n$ satisfies the inequality

$$|e_n| \leqslant \frac{h^pK}{L}[\exp(L(x_n - a)) - 1], \tag{35}$$

where L (>0) is the Lipschitz constant of $f(x, y)$ with respect to y. For a proof of this result, and an extension to take account of round-off error, in the manner of section 3.4, the reader is referred to Henrici[67] (page 73).

Carr[23] gives an alternative bound for the global truncation error of the fourth-order method (20); it applies, however, only to a restricted class of initial value problems. Assume that the local truncation error, T_n, satisfies $|T_n| < E$. Then Carr's theorem states that if $\partial f/\partial y$ is continuous, negative, and bounded from above and below in a region D of the x–y plane by

$$-M_2 < \frac{\partial f}{\partial y} < -M_1 < 0, \tag{36i}$$

then, for all points (x_n, y_n) in a region D^* of the x–y plane, the global truncation error of the method (20) satisfies

$$|e_n| \leqslant 2E/hM_1, \tag{36ii}$$

provided that the steplength is chosen such that

$$h < \min(M_1/M_2^2, 4M_1^3/M_2^4). \tag{36iii}$$

The region D^* is such that if $(x_n, y_n) \in D^*$, then $(x_n, y(x_n)) \in D$. Note that (36), like (35), indicates that the bound for the global error is an order of magnitude greater than that for the local error. Carr's result can also take account of round-off error in the sense that if E bounds not the local truncation error, but the total local error, including round-off, then (36) holds with e_n replaced by $\tilde{e}_n$ (which, in the notation of section 3.4, denotes the total global error). An extension of this result, which applies to a more general class of Runge–Kutta methods, can be found in Galler and Rozenberg.[48]

The error bounds we have discussed in this section may be very hard to apply in practice. The comments we made in section 3.5 on the applicability of error bounds for linear multistep methods are also relevant here. In particular, it is quite impracticable to attempt to form a step-control policy on the basis of (34) and (35). However, there are two good reasons for studying bounds for the local truncation errors of Runge–Kutta methods. Firstly, an important application of Runge–Kutta methods is to provide additional starting values for a multistep predictor–corrector algorithm. In such circumstances, the Runge–Kutta method is applied only for a small number of steps, and a bound on the local error rather than on the global error is adequate. Moreover, it is then necessary to find bounds for the partial derivatives of f only in the immediate vicinity of the initial point. If, in our choice of steplength for the Runge–Kutta starting method (which need not be the same as the steplength of the predictor–corrector algorithm) we are guided by an error bound which turns out to be too conservative, then at least the resulting computational inefficiency is restricted to the starting process; we have the consolation of knowing that in the remainder of the numerical solution starting errors will not dominate.

The second point concerns the choice for the free parameters in Runge–Kutta methods. In the classical methods discussed in section 4.3, these were chosen in order to give simple coefficients. Ralston[148] has investigated the possibility of choosing these coefficients to minimize the bound for the local truncation error. We shall discuss the resulting methods in section 4.8.

Example 1 *It is proposed to use the fourth-order Runge–Kutta method (20) to obtain starting values for the initial value problem*

$$y' = \exp(10(x - y)), \qquad y(0) = 0{\cdot}1.$$

The error in these values is to be less than 10^{-8}. *Use Lotkin's bound for the local truncation error to suggest a suitable value for the steplength.*

$$f(x, y) = \exp(10(x - y)), \qquad \frac{\partial^{i+j} f(x, y)}{\partial x^i \, \partial y^j} = (-1)^j 10^{i+j} \exp(10(x - y)).$$

Assume that $|f| < Q$ in the immediate vicinity of the origin.

Then,

$$|f_x|, |f_y| < 10Q,$$
$$|f_{xx}|, |f_{xy}|, |f_{yy}| < 10^2 Q,$$
$$|f_{xxx}|, |f_{xxy}|, |f_{xyy}|, |f_{yyy}| < 10^3 Q,$$
$$|f_{xxxx}|, |f_{xxxy}|, |f_{xxyy}|, |f_{xyyy}|, |f_{yyyy}| < 10^4 Q.$$

But, by (28) we must find P and Q such that

$$|f| < Q,$$
$$|f_x| < QP, |f_y| < P,$$
$$|f_{xx}| < QP^2, |f_{xy}| < P^2, |f_{yy}| < P^2/Q,$$
$$|f_{xxx}| < QP^3, |f_{xxy}| < P^3, |f_{xyy}| < P^3/Q, |f_{yyy}| < P^3/Q^2,$$
$$|f_{xxxx}| < QP^4, |f_{xxxy}| < P^4, |f_{xxyy}| < P^4/Q, |f_{xyyy}| < P^4/Q^2, |f_{yyyy}| < P^4/Q^3.$$

It is clear that these bounds are satisfied if we choose $P = 10$, $0 < Q \leqslant 1$. Now $f(x_0, y_0) = f(0, 0{\cdot}1) = \exp(-1) = 0{\cdot}368$. We make the assumption (strictly unjustifiable, since we do not yet know how the solution varies in the immediate vicinity of the initial point) that $|f|$ is bounded by 0·368 in the vicinity of the origin. Thus we choose $P = 10$, $Q = 0{\cdot}368$. By Lotkin's bound (32)

$$|T_{n+1}| < \tfrac{73}{720} 10^4 \,.\, 0{\cdot}368 h^5.$$

Thus for $|T_{n+1}| < 10^{-8}$, we require

$$h^5 < \frac{72}{73} \frac{10^{-10}}{3{\cdot}68}$$

or, approximately, $h < 0{\cdot}8 \times 10^{-2}$.

Example 2 Solve the initial value problem of example 1 in the interval $0 \leqslant x \leqslant 1$, *using the fourth-order Runge–Kutta method (20) with steplength* 0·01. *Compare the actual errors with the global error bounds given by (35) and (36), using (32) to bound the local error.*

The theoretical solution of the initial value problem is

$$y(x) = \tfrac{1}{10} \ln(\exp(10x) + \exp(1) - 1).$$

The numerical solution obtained at every tenth step is displayed in table 13.

Our first observation is that the application, in example 1, of Lotkin's bound to suggest a suitable steplength worked well, since with a value of h close to that suggested in example 1 we have, in fact, succeeded in achieving eight-decimal accuracy (apart from round-off) throughout the interval [0, 1].

From the numerical solution, we may deduce that for $x \in [0, 1]$,

$$-0{\cdot}1 \leqslant x - y < 0$$

Table 13

x	Numerical solution	Theoretical solution	Error-bound (35)
0	0·100,000,00	0·100,000,00	
0·1	0·148,988,02	0·148,988,01	$1{\cdot}8 \times 10^{-6}$
0·2	0·220,908,06	0·220,908,05	$6{\cdot}5 \times 10^{-6}$
0·3	0·308,208,52	0·308,208,51	$1{\cdot}9 \times 10^{-5}$
0·4	0·403,098,64	0·403,098,64	$5{\cdot}4 \times 10^{-5}$
0·5	0·501,151,12	0·501,151,12	$1{\cdot}5 \times 10^{-4}$
0·6	0·600,425,02	0·600,425,02	$4{\cdot}1 \times 10^{-4}$
0·7	0·700,156,57	0·700,156,56	$1{\cdot}1 \times 10^{-3}$
0·8	0·800,057,63	0·800,057,63	$2{\cdot}9 \times 10^{-3}$
0·9	0·900,021,20	0·900,021,20	$8{\cdot}2 \times 10^{-3}$
1·0	1·000,007,80	1·000,007,80	$2{\cdot}2 \times 10^{-2}$

and hence that

$$\exp(-1) \leqslant f(x, y) < 1. \tag{37}$$

Thus, using the results of example 1, we may set $P = 10$, $Q = 1$ in (32) to obtain

$$|T_{n+1}| < \tfrac{73}{720} 10^4 h^5 = 1014h^5,$$

and can therefore take $K = 1014$ in (34). In order to calculate the Lipschitz constant L, we observe that

$$\frac{\partial f}{\partial y} = -10 \exp(10(x - y)) = -10f(x, y).$$

From (37) it follows that for $x \in [0, 1]$,

$$-10 \exp(-1) \geqslant \frac{\partial f}{\partial y} > -10 \tag{38}$$

and we may take $L = 10$. Then (35) yields the bound

$$|e_n| \leqslant 1014 \times 10^{-9} [\exp(10x_n) - 1].$$

The values of this bound for $x_n = 0\ (0{\cdot}1)\ 1{\cdot}0$ are displayed in table 13.
From (38) we see that the bound given by (36) can be applied with

$$M_1 = 10 \exp(-1), \qquad M_2 = 10,$$

provided that

$$h < \min[\tfrac{1}{10} \exp(-1), \tfrac{2}{5} \exp(-3)] = 0{\cdot}02.$$

Our choice of 0·01 for h satisfies this requirement and thus (36*ii*) holds with $E = Kh^{p+1} = 1014 \times 10^{-10}$. We thus find the following bound for e_n:

$$|e_n| < \frac{2028 \times 10^{-8}}{10 \exp(-1)} = 5{\cdot}5 \times 10^{-6}.$$

Thus both bounds for the global truncation error turn out to be very conservative.

Exercises

8. Compute Bieberbach's bound (33) for the local truncation error for the method and problem of example 1, and compare it with Lotkin's bound.

9. Repeat the analysis of examples 1 and 2 for the initial value problem

$$y' = -10(y-1)^2, \qquad y(0) = 2,$$

whose theoretical solution is $y(x) = 1 + 1/(1 + 10x)$.

10. Use (31) to compare bounds for the local truncation errors of the third-order methods (18) and (19).

4.6 Error estimates for Runge–Kutta methods

In the previous section we have seen that bounds for the local truncation error do not form a suitable basis for monitoring local truncation error with a view to constructing a step-control policy similar to that developed for predictor–corrector methods. What is needed, in place of a bound, is a readily computable *estimate* of the local truncation error, similar to that we obtained by Milne's device for predictor–corrector pairs. Several such estimates exist and we shall discuss a number of them briefly in this section. As we shall see, none of them is as satisfactory as Milne's device.

The most commonly used estimate arises from an application of the process of the *deferred approach to the limit*, alternatively called *Richardson extrapolation* (Richardson[151]). Under the usual localizing assumption that no previous errors have been made, we have seen that we may write, using (26)

$$y(x_{n+1}) - y_{n+1} = T_{n+1} = \psi(x_n, y(x_n))h^{p+1} + O(h^{p+2}), \tag{39}$$

where p is the order of the Runge–Kutta method. Now let us compute y^*_{n+1}, a second approximation to $y(x_{n+1})$, obtained by applying the same method at x_{n-1} with steplength $2h$. Under the same localizing assumption, it follows that

$$\begin{aligned} y(x_{n+1}) - y^*_{n+1} &= \psi(x_{n-1}, y(x_{n-1}))(2h)^{p+1} + O(h^{p+2}) \\ &= \psi(x_n, y(x_n))(2h)^{p+1} + O(h^{p+2}), \end{aligned} \tag{40}$$

on expanding $\psi(x_{n-1}, y(x_{n-1}))$ about $(x_n, y(x_n))$. On subtracting (39) from (40) we obtain

$$y_{n+1} - y^*_{n+1} = (2^{p+1} - 1)\psi(x_n, y(x_n))h^{p+1} + O(h^{p+2}).$$

Thus the principal local truncation error, which is taken as an estimate for the local truncation error, may be written as

$$\psi(x_n, y(x_n))h^{p+1} = (y_{n+1} - y^*_{n+1})/(2^{p+1} - 1). \tag{41}$$

Thus to apply Richardson extrapolation we compute over two successive steps using steplength h, and then recompute over the double step using steplength $2h$. The difference between the values for y so obtained, divided by 31 in the case of a fourth-order method, is then an estimate of the local truncation error. This estimate is usually quite adequate for step-control purposes, but it involves a considerable increase in computational effort; thus to obtain such an estimate at every second step will call for an increase of roughly 50% in computational effort. This compares badly with Milne's device for predictor–corrector methods, where the estimate was obtained for negligible additional computational effort. Richardson extrapolation could obviously be applied to linear multistep methods and, indeed, has wide applications throughout numerical analysis. (See chapter 6.)

Estimates for the local truncation error which do not involve additional evaluations of the function $f(x, y)$ have been considered by Kuntzmann.[97] These estimates, however, involve data calculated at a number of previous steps and essentially estimate the average local truncation error over these steps. In a situation where the local truncation error is changing rapidly —and this is the important case from the point of view of step-control policy—such estimates can be misleading. One such estimate, quoted by Scraton,[161] and which applies for the fourth-order Runge–Kutta method (20) is

$$30T_{n+3} = 10y_n + 9y_{n+1} - 18y_{n+2} - y_{n+3} + 3h[f_n + 6f_{n+1} + 3f_{n+2}], \tag{42}$$

where $f_{n+j} = f(x_{n+j}, y_{n+j})$, $j = 0, 1, 2$; note that these evaluations of f will already have been made in applying (20).

The idea of deriving a special Runge–Kutta method which admits an easily calculated error estimate which does not depend on quantities calculated at previous steps was first proposed by Merson.[127] Merson's method is

$$\begin{aligned}
y_{n+1} - y_n &= \frac{h}{6}(k_1 + 4k_4 + k_5),\\
k_1 &= f(x_n, y_n),\\
k_2 &= f(x_n + \tfrac{1}{3}h, y_n + \tfrac{1}{3}hk_1),\\
k_3 &= f(x_n + \tfrac{1}{3}h, y_n + \tfrac{1}{6}hk_1 + \tfrac{1}{6}hk_2),\\
k_4 &= f(x_n + \tfrac{1}{2}h, y_n + \tfrac{1}{8}hk_1 + \tfrac{3}{8}hk_3),\\
k_5 &= f(x_n + h, y_n + \tfrac{1}{2}hk_1 - \tfrac{3}{2}hk_3 + 2hk_4).
\end{aligned} \tag{43}$$

The method has order four and an estimate of the local truncation error is given by

$$30T_{n+1} = h(-2k_1 + 9k_3 - 8k_4 + k_5). \tag{44}$$

This method has been widely used for non-linear problems, although, as pointed out by Scraton,[161] the error estimate is valid only when the differential equation is linear in both x and y, that is of the form

$$y' = ax + by + c.$$

When (43) is applied to a non-linear differential equation, the error estimate (44) frequently grossly overestimates the local truncation error and occasionally (England[41]) underestimates it.

A fourth-order method which admits an error estimate which is valid for a non-linear differential equation is derived by Scraton.[161] It is

$$\begin{aligned}
y_{n+1} + y_n &= h[\tfrac{17}{162}k_1 + \tfrac{81}{170}k_3 + \tfrac{32}{135}k_4 + \tfrac{250}{1377}k_5],\\
k_1 &= f(x_n, y_n),\\
k_2 &= f(x_n + \tfrac{2}{9}h, y_n + \tfrac{2}{9}hk_1),\\
k_3 &= f(x_n + \tfrac{1}{3}h, y_n + \tfrac{1}{12}hk_1 + \tfrac{1}{4}hk_2),\\
k_4 &= f\left(x_n + \frac{3}{4}h, y_n + \frac{3h}{128}(23k_1 - 81k_2 + 90k_3)\right),\\
k_5 &= f\left(x_n + \frac{9}{10}h, y_n + \frac{9h}{10{,}000}\right.\\
&\qquad \left.\times(-345k_1 + 2025k_2 - 1224k_3 + 544k_4)\right).
\end{aligned} \tag{45}$$

The estimate for the local truncation error of (45) is given by

$$T_{n+1} = hqr/s,$$

where

$$\begin{aligned}
q &= -\tfrac{1}{18}k_1 + \tfrac{27}{170}k_3 - \tfrac{4}{15}k_4 + \tfrac{25}{153}k_5,\\
r &= \tfrac{19}{24}k_1 - \tfrac{27}{8}k_2 + \tfrac{57}{20}k_3 - \tfrac{4}{15}k_4,\\
s &= k_4 - k_1.
\end{aligned} \tag{46}$$

Note that the methods of both Merson and Scraton do not require additional function evaluations in order to compute the error estimate. However, when we observe that both methods have fourth order and require five function evaluations per step, whereas we know that there exist fourth-order Runge–Kutta methods which require only four evaluations per step, we see that additional function evaluations are in effect required if we demand an error estimate.

Scraton's estimate, although more realistic than Merson's when applied to a general non-linear differential equation, has the disadvantage that it is not linear in the k_r. As a result it is applicable only to a single differential equation and does not extend to a system of equations. In order to find a method which admits an error estimate which is linear in the k_r, and thus holds for a general non-linear differential equation or system of equations, it is necessary to make further sacrifices in the form of additional function evaluations. Thus England[41] gives the following fourth-order six-stage method:

$$
\begin{aligned}
y_{n+1} - y_n &= \frac{h}{6}(k_1 + 4k_3 + k_4), \\
k_1 &= f(x_n, y_n), \\
k_2 &= f(x_n + \tfrac{1}{2}h, y_n + \tfrac{1}{2}hk_1), \\
k_3 &= f(x_n + \tfrac{1}{2}h, y_n + \tfrac{1}{4}hk_1 + \tfrac{1}{4}hk_2), \\
k_4 &= f(x_n + h, y_n - hk_2 + 2hk_3), \\
k_5 &= f\left(x_n + \frac{2}{3}h, y_n + \frac{h}{27}(7k_1 + 10k_2 + k_4)\right), \\
k_6 &= f\left(x_n + \frac{1}{5}h, y_n + \frac{h}{625} \right. \\
&\qquad \left. \times (28k_1 - 125k_2 + 546k_3 + 54k_4 - 378k_5)\right).
\end{aligned} \tag{47}
$$

The associated estimate for the local truncation error is

$$
T_{n+1} = \frac{h}{336}(-42k_1 - 224k_3 - 21k_4 + 162k_5 + 125k_6). \tag{48}
$$

(Note that if (47) is used without the estimate (48) it is essentially a four-stage method.) Further Runge–Kutta processes which admit error estimates have been derived by Shintani.[165,166,167]

We can conclude that all the estimates for the local truncation error of Runge–Kutta methods we have discussed either average the error over a number of steps or require, in one way or another, additional function evaluations. None is as easily obtained as in the case of linear multistep methods used in predictor–corrector modes.

Example 3 *For the initial value problem*

$$y' = 1 + y^2, \qquad y(0) = 1,$$

whose theoretical solution is $y(x) = \tan(x + \pi/4)$, *compute the numerical solution for three steps, using the fourth-order Runge–Kutta method* (20), *with a steplength of* 0·1. *Recompute for one step using a steplength of* 0·2. *Compare the actual error with the estimate for the local truncation error given by* (i) *formula* (42) *and* (ii) *Richardson extrapolation.*

The numerical results are shown in table 14. For this example, the errors rise sharply as the solution moves away from the initial point, and it is not surprising that formula (42), which averages the error over three steps, gives a poor estimate.

Table 14

h	x	Theoretical solution	Numerical solution	Actual error
0·1	0·1	1·223,048,88	1·223,048,91	-3×10^{-8}
	0·2	1·508,497,65	1·508,496,16	$+149 \times 10^{-8}$
	0·3	1·895,765,12	1·895,754,15	$+1097 \times 10^{-8}$
0·2	0·2	1·508,497,65	1·508,456,13	

It in fact estimates the local truncation error to be -2376×10^{-8}. Richardson extrapolation, on the other hand, gives a good estimate. Applying (41) we find the estimate

$$T_{n+1} = (1{\cdot}508{,}496{,}16 - 1{\cdot}508{,}456{,}13)/31 = 129 \times 10^{-8},$$

which is an adequate estimate of the actual error at $x = 0{\cdot}2$. (Strictly, the actual error at 0·2 is a global error but, in view of the smallness of the local error at 0·1, it is largely comprised of the local error at 0·2.)

Example 4 *For the problem of example 3, compute, for one step of length* 0·1, *numerical solutions using the methods of Merson* (43), *Scraton* (45), *and England* (47). *Compare the actual (local) truncation errors with the appropriate error estimate given respectively by* (44), (46), *and* (48). *Finally, assuming exact values for* y *at* $x = -0{\cdot}3, -0{\cdot}2, -0{\cdot}1,$ *and* 0, *compute the numerical solution for* y *at* $x = 0{\cdot}1$ *using Hamming's predictor–corrector pair (example 11 of chapter 3, page 93) in PECE mode. Compare the actual (local) truncation error with the error estimate afforded by Milne's device.*

Actual and estimated errors at $x = 0{\cdot}1$ are shown in table 15. Note that in this instance all errors are local. Due to the non-linearity of the differential equation, the error estimated by Merson's process exaggerates the actual error by a factor of 10.

Table 15

	Actual error	Estimated error
Merson	180×10^{-8}	1788×10^{-8}
Scraton	-38×10^{-8}	-20×10^{-8}
England	203×10^{-8}	165×10^{-8}
Hamming (*PECE*)	-3520×10^{-8}	-6504×10^{-8}

(In general, when $f(x, y)$ is not linear in both x and y, the quantity on the right-hand side of (44) is $O(h^4)$ not $O(h^5)$; see exercise 12 below.) The error estimates by the other three processes are adequate for the purposes of step-control.

Exercises

11. For each of the four methods considered in example 4, continue the solution for one more step and use Richardson extrapolation to estimate the error at $x = 0{\cdot}2$. Compare with the actual error and with the estimate at $x = 0{\cdot}2$ obtained as in example 4.

12. If $f(x, y) \equiv g(x)$, the exact solution of the initial value problem $y' = f(x, y)$, $y(x_0) = y_0$ after one step is given by

$$y(x_1) - y_0 = \int_{x_0}^{x_0+h} g(x)\,\mathrm{d}x.$$

Observing that (43) and (47) both then reduce to a statement of Simpson's rule for quadrature, write down the exact principal local truncation error for these methods. Show that the error estimated by (48) is exactly this expression but that that estimated by (44) is $O(h^4)$ unless $y(x)$ is a polynomial of degree at most three.

4.7 Weak stability theory for Runge–Kutta methods

It is possible to develop a theory of weak stability for Runge–Kutta methods along exactly the same lines as we pursued in section 3.6 for linear multistep methods. The manipulation is, however, rather tedious and we can shorten the argument considerably by drawing upon our experience of the analysis of section 3.6. There it was shown that the linearized equation satisfied by the total *global* error $\tilde{e}_n = y(x_n) - \tilde{y}_n$ (which includes round-off as well as truncation error) generated by the linear multistep method

$$\sum_{j=0}^{k} \alpha_j y_{n+j} = h \sum_{j=0}^{k} \beta_j f_{n+j} \tag{49}$$

is of the form

$$\sum_{j=0}^{k} (\alpha_j - h\lambda\beta_j)\tilde{e}_{n+j} = \phi,$$

provided that we make the assumptions

$$\partial f/\partial y = \lambda, \quad \text{constant.} \tag{50i}$$

and,

$$\text{local error} = \text{constant.} \tag{50ii}$$

Subsequently, we made no use of the function ϕ and, in effect, defined intervals of absolute and relative stability in terms of the behaviour of the solution of the equation

$$\sum_{j=0}^{k} (\alpha_j - h\lambda\beta_j)\tilde{e}_{n+j} = 0. \tag{51}$$

Let us, instead, apply (49) directly to the test equation

$$y' = \lambda y,$$

for which the assumption (50i) is obviously valid. We obtain

$$\sum_{j=0}^{k} (\alpha_j - h\lambda\beta_j)y_{n+j} = 0, \tag{52}$$

which is exactly the same equation as (51) with simply a change of argument. Thus, for example, we obtain exactly the same results as in section 3.6 if we define the interval of absolute stability to be that interval of the $h\lambda$ line for which all solutions of the difference equation, obtained by applying the method (49) to the test equation $y' = \lambda y$ tend to zero as n tends to infinity. To interpret the result in terms of the general equation $y' = f(x, y)$ subject to the unavoidable assumptions (50) we simply take λ to be an estimate of $\partial f/\partial y$. In this approach we are linearizing the original differential equation, as opposed to linearizing the error equation.

Let us now apply the Runge–Kutta method (5) with $R = 3$ to the test equation $y' = \lambda y$

$$\begin{aligned}
k_1 &= f(x, y) = \lambda y,\\
k_2 &= f(x + ha_2, y + ha_2 k_1) = \lambda(y + ha_2\lambda y) = \lambda y(1 + a_2 h\lambda),\\
k_3 &= f(x + ha_3, y + h(a_3 - b_{32})k_1 + hb_{32}k_2)\\
&= \lambda[y + h\lambda y(a_3 - b_{32}) + h\lambda y b_{32}(1 + a_2 h\lambda)]\\
&= \lambda y(1 + a_3 h\lambda + a_2 b_{32} h^2\lambda^2).
\end{aligned}$$

Hence

$$\phi(x, y, h) = c_1 k_1 + c_2 k_2 + c_3 k_3$$
$$= \lambda y[(c_1 + c_2 + c_3) + (c_2 a_2 + c_3 a_3)h\lambda + c_3 a_2 b_{32} h^2 \lambda^2],$$

and from (5*i*) we obtain the difference equation

$$y_{n+1} - y_n = h\lambda[(c_1 + c_2 + c_3) + (c_2 a_2 + c_3 a_3)h\lambda + c_3 a_2 b_{32} h^2 \lambda^2] y_n.$$

Introducing the previous notation $\bar{h} = h\lambda$, we obtain

$$y_{n+1}/y_n = 1 + (c_1 + c_2 + c_3)\bar{h} + (c_2 a_2 + c_3 a_3)\bar{h}^2 + c_3 a_2 b_{32} \bar{h}^3.$$

The general solution of this equation is

$$y_n = d_1 r_1^n,$$

where d_1 is an arbitrary constant, and

$$r_1 = 1 + (c_1 + c_2 + c_3)\bar{h} + (c_2 a_2 + c_3 a_3)\bar{h}^2 + c_3 a_2 b_{32} \bar{h}^3. \qquad (53)$$

We can then define the three-stage Runge–Kutta method to be *absolutely stable on the interval* (α, β) if r_1, given by (53) satisfies $|r_1| < 1$ whenever $\bar{h} \in (\alpha, \beta)$.

On comparing this definition of absolute stability with that given in section 3.6 for a linear multistep method, we observe that for the Runge–Kutta method there is only one 'root' to be controlled, as compared with k such roots for the linear k-step method. It follows that relative stability, as defined in section 3.6, is not applicable to Runge–Kutta methods (although we could frame a definition similar to some of the alternatives mentioned in section 3.6).

If the Runge–Kutta method under discussion is consistent, then $c_1 + c_2 + c_3 = 1$, and from (53) we may write

$$r_1 = 1 + \bar{h} + O(\bar{h}^2).$$

Hence for sufficiently small positive $\bar{h}$, $r_1 > 1$, and we may conclude, as in the case of linear multistep methods, that the interval of absolute stability has the form $(\alpha, 0)$. Moreover, if the three-stage method has order three, then equations (17) hold, and (53) yields

$$r_1 = 1 + \bar{h} + \tfrac{1}{2}\bar{h}^2 + \tfrac{1}{6}\bar{h}^3. \qquad (54)$$

A plot of this function against $\bar{h}$ reveals that $|r_1| < 1$ whenever $\bar{h} \in (-2{\cdot}51, 0)$. Recalling that (17) does not specify the coefficients of the method uniquely, we may conclude that *all three-stage Runge–Kutta methods of order three have the same interval of absolute stability, namely* $(-2{\cdot}51, 0)$.

We can obtain a generalization of this result from the following alternative approach. In section 4.3 we saw that if the general R-stage Runge–Kutta method (5) has order p, then $\phi(x, y, h)$, defined by (5*ii*), differs from $\phi_T(x, y, h)$, defined by (4), by terms of order h^p. Thus for a method of order p, by (5*i*)

$$y_{n+1} - y_n = h\phi_T(x_n, y_n, h) + O(h^{p+1})$$
$$= hf(x_n, y_n) + \frac{h^2}{2!}f^{(1)}(x_n, y_n) + \dots$$
$$+ \frac{h^p}{p!}f^{(p-1)}(x_n, y_n) + O(h^{p+1}).$$

For the test equation $y' = \lambda y$, $f(x_n, y_n) = \lambda y_n$, $f^{(q)}(x_n, y_n) = \lambda^{q+1}y_n$, $q = 1, 2, \dots, p-1$. Hence

$$y_{n+1} - y_n = \left(h\lambda + \frac{1}{2!}h^2\lambda^2 + \dots + \frac{1}{p!}h^p\lambda^p\right)y_n + O(h^{p+1})$$

or

$$y_{n+1}/y_n = r_1 = 1 + \bar{h} + \frac{1}{2!}\bar{h}^2 + \dots + \frac{1}{p!}\bar{h}^p + O(\bar{h}^{p+1}). \quad (55)\dagger$$

On the other hand, it is clear from a generalization of the analysis leading to (53) that for an R-stage method, r_1 is a polynomial of degree R in $\bar{h}$. In section 4.4 we saw if an R-stage method has order p, then $R \geqslant p$, and R can equal p only for $p = 1, 2, 3, 4$. Hence for a p-stage method of order p, ($p \leqslant 4$), we have

$$r_1 = 1 + \bar{h} + \frac{1}{2}\bar{h}^2 + \dots + \frac{1}{p!}h^p \quad (56)$$

irrespective of the values given to the parameters left free after satisfying the order requirements. It follows that, for a given p, $p = 1, 2, 3, 4$, *all p-stage Runge–Kutta methods of order p have the same interval of absolute stability.* These intervals are given in table 16, where R_p denotes *any* p-stage Runge–Kutta method of order p, $p = 1, 2, 3, 4$.

If the R-stage method has order $p < R$ (and this will always be the case for $p > 4$) then r_1 takes the form

$$r_1 = 1 + \bar{h} + \frac{1}{2}\bar{h}^2 + \dots + \frac{1}{p!}\bar{h}^p + \sum_{q=p+1}^{R} \gamma_q\bar{h}^q, \quad (57)$$

† Note that (55) may be written in the form $r_1 = \exp(\bar{h}) + O(\bar{h}^{p+1})$; compare this with equation (43) of section 3.6 (page 66).

Table 16

Method	r_1	Interval of absolute stability
R_1	$1 + \bar{h}$	$(-2, 0)$
R_2	$1 + \bar{h} + \frac{1}{2}\bar{h}^2$	$(-2, 0)$
R_3	$1 + \bar{h} + \frac{1}{2}\bar{h}^2 + \frac{1}{6}\bar{h}^3$	$(-2{\cdot}51, 0)$
R_4	$1 + \bar{h} + \frac{1}{2}\bar{h}^2 + \frac{1}{6}\bar{h}^3 + \frac{1}{24}\bar{h}^4$	$(-2{\cdot}78, 0)$

where the coefficients γ_q are functions of the coefficients of the Runge–Kutta method, but are not determined by the order requirement. In such a case the interval will depend on the particular choice for the free parameters. If, for example, for $R = 3$ we consider methods of order two, then only the first two of the equations (17) need be satisfied and we obtain from (53)

$$r_1 = 1 + \bar{h} + \tfrac{1}{2}\bar{h}^2 + \gamma_3\bar{h}^3,$$

where $\gamma_3 = c_3 a_2 b_{32}$. If $\gamma_3 = 0$, the interval of absolute stability is clearly $(-2, 0)$, whereas for $\gamma_3 = \frac{1}{6}$ it is, from table 16, $(-2{\cdot}51, 0)$. (Note that $\gamma_3 = \frac{1}{6}$ is not a sufficient condition for the order of the method to be three, since the third of the equations (17) is not necessarily satisfied.) For $\gamma_3 = \frac{1}{12}$, the interval of absolute stability becomes $(-4{\cdot}52, 0)$.

Example 5 Illustrate the effect of weak instability by using the fourth-order Runge–Kutta method (20) to compute numerical solutions for the following initial value problems:

(*i*) $y' = -20y$, $y(0) = 1$; *use steplengths* 0·1 *and* 0·2.
(*ii*) $y' = -5xy^2 + 5/x - 1/x^2$, $y(1) = 1$; *use steplengths* 0·2, 0·3, *and* 0·4.

(i) For this problem, $\partial f/\partial y = -20$ and hence $\bar{h} = -20h$. Since the interval of absolute stability for the method (20) is, by table 16, $(-2{\cdot}78, 0)$, it follows that $\bar{h}$ lies inside the interval of absolute stability when $h = 0{\cdot}1$ but outside it when $h = 0{\cdot}2$. Actual global errors for these two cases are quoted in table 17 and clearly demonstrate the effect.

Table 17

x	Global error, $h = 0{\cdot}1$	Global error, $h = 0{\cdot}2$
0·0	0	0
0·2	−0·092,795	+4·98
0·4	−0·012,010	25·0
0·6	−0·001,366	125·0
0·8	−0·000,152	625·0
1·0	−0·000,017	3125·0

(ii) For this problem $\partial f/\partial y = -10xy = -10$ on the theoretical solution $y = 1/x$. ($\partial f/\partial y$ is not constant on neighbouring integral curves, however.) Taking $\lambda = -10$, it follows that $\bar{h}$ lies inside the interval of absolute stability when $h = 0{\cdot}2$ but outside it when $h = 0{\cdot}3$ or $h = 0{\cdot}4$. Actual global errors are quoted in table 18. Absolute

Table 18

x	Global error, $h = 0{\cdot}2$	Global error, $h = 0{\cdot}3$	Global error, $h = 0{\cdot}4$
1·0	0	0	0
2·2	−0·001,373	−0·041,828	+0·164,223
3·4	−0·000,321	−0·026,870	$>10^{10}$
4·6	−0·000,121	−0·019,615	
5·8	−0·000,058	−0·015,407	
7·0	−0·000,033	−0·012,675	
⋮	⋮	⋮	
25·0	−0·000,001	−0·003,452	

stability for $h = 0{\cdot}2$ and absolute instability for $h = 0{\cdot}4$ are clearly demonstrated by the numerical results.

4.8 Runge–Kutta methods with special properties

We have already seen that when a Runge–Kutta method of desired order is derived, there are, in general, a number of free parameters which cannot be used to increase the order further. In section 4.3, where we derived classical Runge–Kutta methods, these parameters were chosen in such a way that the resulting methods had simple coefficients, convenient for desk computation. In this section, we consider various other ways in which the free parameters may be specified to advantage. We have already seen instances of this in section 4.6, where we considered special Runge–Kutta methods which admitted certain estimates of the local truncation error.

Perhaps the most important task to which free parameters can be applied is the reduction of the local truncation error. Since this error depends in a complicated manner on the function $f(x, y)$, we cannot hope to specify parameter values which minimize the local truncation error, but we can specify them so as to minimize a bound for the local truncation error. This idea was proposed by Ralston[148] who derived Runge–Kutta methods which have minimum bounds of the type due to Lotkin,[121] discussed in section 4.5. Thus, for the third-order three-stage Runge–Kutta method, Lotkin's bound for the local truncation error, given by (31) is seen to be a function of the free parameters a_2, a_3. This bound takes its minimum value when $a_2 = \frac{1}{2}$ and $a_3 = \frac{3}{4}$, yielding the method

$$
\begin{aligned}
y_{n+1} - y_n &= \frac{h}{9}(2k_1 + 3k_2 + 4k_3),\\
k_1 &= f(x_n, y_n),\\
k_2 &= f(x_n + \tfrac{1}{2}h, y_n + \tfrac{1}{2}hk_1),\\
k_3 &= f(x_n + \tfrac{3}{4}h, y_n + \tfrac{3}{4}hk_2).
\end{aligned}
\tag{58}
$$

The bound for the local truncation error for this method is given by

$$|T_{n+1}| < \tfrac{1}{9}h^4P^3Q,$$

in the notation of section 4.5. Four-stage Runge–Kutta methods of order four also have two free parameters, which may be taken to be a_2 and a_3. Ralston shows that the bound for the local truncation error is then minimized when

$$a_2 = 0{\cdot}4, \qquad a_3 = 0{\cdot}455{,}737{,}25, \tag{59}$$

yielding the method

$$
\begin{aligned}
y_{n+1} - y_n &= h(0{\cdot}174{,}760{,}28k_1 - 0{\cdot}551{,}480{,}66k_2 + 1{\cdot}205{,}535{,}60k_3\\
&\qquad + 0{\cdot}171{,}184{,}78k_4),\\
k_1 &= f(x_n, y_n),\\
k_2 &= f(x_n + 0{\cdot}4h, y_n + 0{\cdot}4hk_1),\\
k_3 &= f(x_n + 0{\cdot}455{,}737{,}25h, y_n + 0{\cdot}296{,}977{,}61hk_1\\
&\qquad + 0{\cdot}158{,}759{,}64hk_2),\\
k_4 &= f(x_n + h, y_n + 0{\cdot}218{,}100{,}40hk_1 - 3{\cdot}050{,}965{,}16hk_2\\
&\qquad + 3{\cdot}832{,}864{,}76hk_3).
\end{aligned}
\tag{60}
$$

The local truncation error for this method satisfies

$$|T_{n+1}| < 0{\cdot}0546h^5P^4Q.$$

Comparing this with (32) we see that the error bound for (60) is approximately half of that for the classical fourth-order method (20).

There are other ways in which we may attempt to use the free parameters in order to improve local accuracy. Thus King[85] derives Runge–Kutta methods for which the bound on the local truncation error is minimized subject to a constraint which has the effect of causing the order of the method to increase in the case when $f(x, y)$ is independent of y. An example is the following fourth-order method for which the local truncation error satisfies

$$|T_{n+1}| < 0{\cdot}0944h^5P^4Q$$

in general, but satisfies

$$|T_{n+1}| < 1{\cdot}389 \times 10^{-5}h^6P^5Q \tag{61}$$

when f is independent of y:

$$\begin{aligned} y_{n+1} - y_n &= h(0{\cdot}376{,}403{,}062{,}7k_2 + 0{\cdot}512{,}485{,}826{,}2k_3 \\ &\quad + 0{\cdot}111{,}111{,}111{,}1k_4), \\ k_1 &= f(x_n, y_n), \\ k_2 &= f(x_n + 0{\cdot}155{,}051{,}025{,}7h,\ y_n + 0{\cdot}155{,}051{,}025{,}7hk_1), \\ k_3 &= f(x_n + 0{\cdot}644{,}948{,}974{,}3h,\ y_n - 0{\cdot}831{,}918{,}358{,}8hk_1 \\ &\quad + 1{\cdot}476{,}867{,}333hk_2), \\ k_4 &= f(x_n + h,\ y_n + 3{\cdot}311{,}862{,}178hk_1 - 3{\cdot}949{,}489{,}743hk_2 \\ &\quad + 1{\cdot}637{,}627{,}564hk_3). \end{aligned} \tag{62}$$

Such methods can be expected to lead to some increase in local accuracy for those initial value problems for which $f(x, y)$ is not strongly dependent on y. Moreover (62) could do double duty in a programme library either as an initial value problem algorithm or as a quadrature algorithm, for which it is well suited by virtue of the small error bound (61). (Runge–Kutta methods of order five and six, which reduce to quadrature formulae of order six and seven respectively when f is independent of y, may be found in Luther[122,123] and Luther and Konen.[124])

It is noticeable that none of the above attempts to reduce local truncation error by minimizing bounds results in a spectacular reduction of the bound. Hull and Johnston[74] investigate fourth-order four-stage Runge–Kutta methods which minimize a variety of different estimates for the local truncation error. In all cases they find that the optimum lies near $a_2 = 0{\cdot}35$ and $a_3 = 0{\cdot}45$ (cf. (59)) and that the value of the estimate near the optimum is not particularly sensitive to changes in the parameter values. For a variety of numerical examples, however, the *actual* local truncation error, as opposed to estimates and bounds, does turn out to be sensitive to changes in the parameters.

Exercises

13. (Ralston.[148]) Find the second-order two-stage Runge–Kutta method with minimum bound for the local truncation error.

14. (King.[85]) Show that the method

$$\begin{aligned} y_{n+1} - y_n &= \frac{h}{10}(k_1 + 5k_2 + 4k_3), \\ k_1 &= f(x_n, y_n), \\ k_2 &= f(x_n + \tfrac{1}{3}h, y_n + \tfrac{1}{3}hk_1), \\ k_3 &= f(x_n + \tfrac{5}{6}h, y_n - \tfrac{5}{12}hk_1 + \tfrac{5}{4}hk_2), \end{aligned}$$

has order three. Find a bound for the local truncation error and compare it with that of (58). Show that in the case where $f(x, y)$ is independent of y, the method reduces to a quadrature formula of order four.

Another area where we can look for some advantage from a judicious choice for the free parameters concerns the weak stability characteristics of Runge–Kutta methods. It follows from the discussion of section 4.7 that for R-stage methods, where $R \leqslant 4$, no increase in the interval of absolute stability can be gained by altering the values of the parameters left free after the maximum attainable order has been achieved. This is not so for $R > 4$, and Lawson[107] chooses free parameters in a fifth-order six-stage method so as to increase the interval of absolute stability. The resulting method is

$$\begin{aligned} y_{n+1} - y_n &= \frac{h}{90}(7k_1 + 32k_3 + 12k_4 + 32k_5 + 7k_6), \\ k_1 &= f(x_n, y_n), \\ k_2 &= f(x_n + \tfrac{1}{2}h, y_n + \tfrac{1}{2}hk_1), \\ k_3 &= f(x_n + \tfrac{1}{4}h, y_n + \tfrac{1}{16}h(3k_1 + k_2)), \\ k_4 &= f(x_n + \tfrac{1}{2}h, y_n + \tfrac{1}{2}hk_3), \\ k_5 &= f(x_n + \tfrac{3}{4}h, y_n + \tfrac{3}{16}h(-k_2 + 2k_3 + 3k_4)), \\ k_6 &= f(x_n + h, y_n + \tfrac{1}{7}h(k_1 + 4k_2 + 6k_3 - 12k_4 + 8k_5)). \end{aligned} \tag{63}$$

This method has an interval of absolute stability of approximately $(-5{\cdot}7, 0)$, which is rather more than twice the length of the interval for a fourth-order four-stage method. Hence, for problems for which considerations of stability, rather than of local accuracy, limit the size of the steplength, this method could be used with a steplength twice that permitted by a fourth-order method, resulting in a smaller total number of function evaluations over a given interval of integration. A sixth-order seven-stage method with an interval of absolute stability of approximately $(-6{\cdot}2, 0)$ may be found in Lawson.[108]

Finally, free parameters may also be specified in such a way as to reduce storage requirements in the computer. Such requirements are negligible for a single differential equation but can be significant when a large system of (N) simultaneous equations is being integrated. Gill[57] derives a fourth-order four-stage Runge–Kutta method which requires only $3N + A$ storage locations, where A is independent of N, whereas the general four-stage Runge–Kutta method requires $4N + A$. Conte and Reeves[29] effect a similar economy for third-order methods and the idea has been further developed by Blum[8] and Fyfe.[47] However, for modern computers, storage difficulties seldom arise with Runge–Kutta processes, and methods of the Gill type are now less important than they once were. In passing, we note that Gill's method also has an interesting property of reducing the accumulation of round-off error. For a discussion of this aspect, the reader is referred to the papers of Gill[57] and Blum.[8]

4.9 Comparison with predictor–corrector methods

We have now reached the stage where we have examined in some detail various properties of the two best-known classes of methods for the numerical solution of initial value problems, namely predictor–corrector and Runge–Kutta. It is natural to ask 'Which is the better class?' In an attempt to make this question more precise, let us consider the comparison with respect to the following specific properties:

(i) local accuracy, of which order is a crude estimate,
(ii) weak stability characteristics,
(iii) computational cost, measured by the number of function evaluations per step, and
(iv) programming ease.

These properties are, as we have seen, interrelated and, moreover, the form of the interrelation for Runge–Kutta methods is quite different from that for predictor–corrector methods. Thus, for example, we have seen that the maximum order of a Runge–Kutta method depends directly on the number of function evaluations per step, while a predictor–corrector pair of *any* order will require precisely two function evaluations per step if it is used in *PECE* mode. Thus substantial difficulties are encountered in any attempt to quantify the comparison between the two classes of method.

Following the discussion of section 3.13, only *PECE* and $P(EC)^2$ modes, both of which require two evaluations per step, will be considered. In the remainder of this section the phrase 'predictor–corrector method' will imply that the algorithm will be employed in one of these two modes.

Let us first compare the local accuracy of methods which have the same order. We assume that the predictor and corrector both have order p; it follows from section 3.10 that the principal local truncation error of a predictor–corrector method is

$$C_{p+1}h^{p+1}y^{(p+1)}(x_n), \tag{64}$$

where C_{p+1} is the error constant of the corrector. The principal local truncation error of a Runge–Kutta method of order p is, by (26), of the form

$$\psi(x_n, y(x_n))h^{p+1} \tag{65}$$

and a direct comparison of principal local truncation errors for an arbitrary initial value problem is not possible. Computational experience indicates that for *fourth-order* methods, the local accuracy of Runge–Kutta methods is frequently superior to that of predictor–corrector methods. However, the balance can change if we make the comparison between methods of the same order, subject to the constraint that both methods involve the same number of function evaluations over a given interval. For $p \leqslant 4$, there exist Runge–Kutta methods of order p which call for p function evaluations per step, and no Runge–Kutta methods of order p which call for less. Since all predictor–corrector methods call for two function evaluations per step, we may use, in the predictor–corrector method, a step-length $2/p$ times that used in the Runge–Kutta; both methods will then use the same number of function evaluations over a given interval. Thus the principal local truncation error (64) of the predictor–corrector method is effectively replaced by

$$C_{p+1}(2h/p)^{p+1}y^{(p+1)}(x_n). \tag{66}$$

On this basis, the predictor–corrector method is frequently the more accurate. This effect is illustrated in the following example.

Example 6 For the initial value problem $y' = \lambda y$, $y(0) = 1$, let the principal local truncation errors at the first step given by (64), (65), and (66) be denoted by T_{PC}, T_{RK}, and T^*_{PC} respectively, where the Runge–Kutta method is any fourth-order four-stage method and the predictor–corrector method is a fourth-order method for which the corrector is a zero-stable three-step method. Show that for **any** *such predictor–corrector method*

$$|T_{RK}| > |T^*_{PC}|,$$

while for some such predictor–corrector methods

$$|T_{PC}| > |T_{RK}|.$$

From section 4.7 it follows that y_1 calculated by any fourth-order four-stage Runge–Kutta method, is given by

$$y_1 = 1 + h\lambda + \tfrac{1}{2}h^2\lambda^2 + \tfrac{1}{6}h^3\lambda^3 + \tfrac{1}{24}h^4\lambda^4.$$

Since the theoretical solution is $\exp(\lambda x_1) = \exp(\lambda h)$, it follows that

$$T_{RK} = h^5\lambda^5/120.$$

Since $y^{(5)}(x_1) = \lambda^5 \exp(\lambda x_1) = \lambda^5(1 + O(h))$, the errors given by (64) and (66) at the first step are

$$T_{PC} = C_5 h^5 \lambda^5, \qquad T^*_{PC} = C_5 h^5\lambda^5/32.$$

Consider the general fourth-order three-step corrector quoted in section 2.10. The error constant is $C_5 = -(19 + 11a + 19b)/720$, where the parameters a and b must be chosen so that the condition of zero-stability is satisfied. Using the Routh–Hurwitz criterion, we find that this is so if and only if

$$1 + a + b > 0, \qquad 1 - b > 0, \qquad 1 - a + b > 0.$$

A simple calculation shows that the maximum value of $|19 + 11a + 19b|/720$, subject to the above constraints, is less than $\frac{1}{12}$. Hence, for all zero-stable three-step correctors of order four

$$|T^*_{PC}| < \tfrac{1}{12}h^5|\lambda^5|/32 < |T_{RK}|.$$

On the other hand, for say $a = b = 0$,

$$|T_{PC}| = \tfrac{19}{720}h^5|\lambda^5| > |T_{RK}|.$$

Example 7 Compare the **global** *accuracy of fourth-order Runge–Kutta and PECE algorithms by computing numerical solutions to the initial value problem $y' = -y$, $y(0) = 1$, in the range $0 \leqslant x \leqslant 10$ using the following algorithms:*

(*1*) RK; $h = h_0$,
(*2*) ABM; $h = h_0$,
(*3*) CK; $h = h_0$,
(*4*) ABM; $h = \frac{1}{2}h_0$,
(*5*) CK; $h = \frac{1}{2}h_0$.

Here RK *refers to the fourth-order Runge–Kutta method (20),* ABM *to the fourth-order Adams–Bashforth–Moulton pair (equation (77) of chapter 3) in PECE mode, and* CK *to the Crane–Klopfenstein pair (equations (78), (77ii) of chapter 3) in PECE mode. Choose $h_0 = 0{\cdot}25$.*

The global errors resulting from these computations are displayed in table 19. The Runge–Kutta method is seen to be more accurate than either of the two *PECE* algorithms, if all three methods use the same steplength. However, if the two *PECE* algorithms are applied with half the steplength employed in the Runge–Kutta method—in which case all three methods require the same total number of function evaluations over the interval [0, 10]—then the *PECE* algorithms emerge as the more accurate.

Table 19
Errors $\times 10^8$

x	RK; $h = h_0$	ABM; $h = h_0$	CK; $h = h_0$	ABM; $h = \frac{1}{2}h_0$	CK; $h = \frac{1}{2}h_0$
0	0	0	0	0	0
1·0	−1476	3297	3635	289	310
2·0	−1086	5912	6646	276	295
3·0	−599	3910	4348	164	176
4·0	−294	2077	2303	84	89
5·0	−135	999	1106	39	42
6·0	−60	453	502	17	19
7·0	−26	198	219	7	8
8·0	−11	84	93	3	3
9·0	−4	35	39	1	1
10·0	−2	15	16	1	1

Exercise

15. Compare, in the manner of example 6, the principal local truncation errors (64), (65), and (66) for the problem $y' = \lambda y$, $y(0) = 1$, using Runge–Kutta and predictor–corrector methods of order three, and of order two.

Finally, if we attempt to compare local accuracy on the basis of keeping the total number of function evaluations for a given interval constant, paying no regard to order (which in practical terms appears to be an appropriate viewpoint), the comparison becomes meaningless, since we can arbitrarily increase the order of the predictor–corrector method without changing the number of function evaluations. Practical limits to such an increase would be set by weak stability considerations and by the demand for an excessive number of starting values.

The manner in which the size of the interval of absolute stability varies with order is, for Runge–Kutta methods, quite different from that for linear multistep methods. An increase in order is accompanied by a small increase in the size of the interval in the case of Runge–Kutta methods of order up to four, and by a substantial decrease in the case of linear multistep methods. (Compare table 16 of section 4.7 with table 8 of section 3.8.)

It is also noticeable from tables 16 and 8 that the stability characteristics of Runge–Kutta methods (which are explicit) are much superior to those of the explicit, Adams–Bashforth methods. This is the main reason why an explicit Runge–Kutta method makes a viable algorithm whereas an explicit linear multistep method on its own does not. Turning now to predictor–corrector pairs, we see that, whereas changing the number of function evaluations per step for a Runge–Kutta method (i.e. changing the attainable order) has remarkably little effect on the size of the stability interval, the analysis of section 3.11 showed that, for a predictor–corrector algorithm, changing the number of function evaluations per step (i.e. changing the mode) could produce substantial alterations in the size of the interval.

We now list intervals of absolute stability for the general fourth-order four-stage Runge–Kutta method, for the fourth-order Adams–Bashforth–Moulton pair and for the Crane–Klopfenstein algorithm, which was designed to have as large an interval as possible (see section 3.13):

Runge–Kutta:	$(-2{\cdot}78, 0)$,
Adams–Bashforth–Moulton ($PECE$):	$(-1{\cdot}25, 0)$,
Adams–Bashforth–Moulton ($P(EC)^2$):	$(-0{\cdot}90, 0)$,
Crane–Klopfenstein ($PECE$):	$(-2{\cdot}48, 0)$.

If we now make the comparison on the basis of the same number of function evaluations over a given interval we can, as we have seen, apply the predictor–corrector methods with half the steplength of the Runge–Kutta method. The intervals of absolute stability of the predictor–corrector methods are then effectively doubled. The Crane–Klopfenstein algorithm is then much superior to the Runge–Kutta, while the Adams–Bashforth–Moulton $PECE$ algorithm and the Runge–Kutta method are comparable. On this basis, the Runge–Kutta methods designed by Lawson to have extended intervals of absolute stability also compare badly with the Crane–Klopfenstein algorithm. Lawson's fifth-order method (63), for example, has an interval $(-5{\cdot}7, 0)$ but requires six function evaluations per step. Thus, for equivalent computational effort, we may multiply the interval of the Crane–Klopfenstein algorithm by 3, obtaining $(-7{\cdot}44, 0)$; (but recall that the Crane–Klopfenstein algorithm has only order four).

Lastly, Runge–Kutta methods do not leave much room for manoeuvre if the initial value problem poses serious difficulties of weak stability. Thus Runge–Kutta methods prove inadequate for the stiff systems we discuss in chapter 8. In contrast, if we are prepared to correct to convergence with a predictor–corrector pair for which the corrector is the

Trapezoidal rule, we can extend the interval of absolute stability to $-\infty$.

Comparison on the basis of ease of programming can be made at two distinct levels. If we merely wish to write a programme which will perform the operations of the algorithm, then there is no doubt that Runge–Kutta methods are much easier to implement than predictor–corrector methods. The former need no special starting procedures and, if a change of steplength is required at some stage, this can be achieved at negligible effort; these are the most telling advantages of Runge–Kutta methods. If, however, we wish to write a programme which incorporates automatic steplength selection, based on the sort of policy we discussed in chapter 3, then we need to have available estimates of the local truncation error and of $\lambda \approx \partial f/\partial y$. If a predictor–corrector method is the basis of the algorithm, then both these estimates can be easily obtained. If a Runge–Kutta method is the basis, however, calculation of effective estimates is more complicated, and substantially raises the overall computational cost.

The conclusions of this section may be summarized by restricting the comparison to methods which have the same order and which require the same number of function evaluations over a given interval. Then on the question of local accuracy, we can draw no firm quantitative conclusions but the evidence suggests that predictor–corrector methods are preferable. From the point of view of weak stability characteristics, predictor–corrector methods are clearly to be preferred. Runge–Kutta methods are easily programmed for a computation with fixed steplength, but if a step-control policy is to be incorporated, the ease with which predictor–corrector methods allow us to monitor local truncation error and estimates for $\partial f/\partial y$ is a strong argument in favour of basing library routines on predictor–corrector methods. Runge–Kutta methods are then appropriate for providing starting values and effecting changes of steplength.

The discussions of this section illustrate some of the difficulties encountered in attempting to compare different numerical algorithms for initial value problems. For an interesting theoretical approach to problems of comparison, the reader is referred to papers by Hull[71] and Stewart.[175] Comparative studies based on computational results have been produced by Crane and Fox,[31] Clark,[27] and Hull, Enright, Fellen, and Sedgwick.[73]

4.10 Implicit Runge–Kutta methods

We conclude this chapter with a discussion of two recent developments in the field of Runge–Kutta methods, each of which goes some way towards overcoming the deficiencies of conventional Runge–Kutta methods which we have discussed in the preceding section.

All the methods discussed so far in this chapter have been explicit. As pointed out by Butcher,[13] it is also possible to consider implicit Runge–Kutta methods. The general R-stage implicit Runge–Kutta method is defined by

$$y_{n+1} - y_n = h\phi(x_n, y_n, h), \tag{67i}$$

$$\left.\begin{aligned} \phi(x, y, h) &= \sum_{r=1}^{R} c_r k_r, \\ k_r &= f\left(x + ha_r, y + h\sum_{s=1}^{R} b_{rs}k_s\right), \qquad r = 1, 2, \ldots, R, \end{aligned}\right\} \tag{67ii}$$

$$a_r = \sum_{s=1}^{R} b_{rs}, \qquad r = 1, 2, \ldots, R. \tag{67iii}$$

The functions k_r are no longer defined explicitly but by a set of R implicit equations, in general non-linear; thus the apparent explicitness of (67*i*) is somewhat misleading. Order is defined precisely as for the general explicit one-step method; in particular, order p will be achieved if we can choose values for the constants c_r, a_r, b_{rs} such that the expansion of the function $\phi(x, y, h)$ defined by (67*ii*) in powers of h differs from the expansion for $\phi_T(x, y, h)$, given by (4), only in the pth and higher powers of h.

The derivation of such methods is rather complicated; a general treatment is given by Butcher.[15] We consider here only the two-stage method obtained by setting $R = 2$ in (67). Even with this restriction, orders of up to four are possible, so that we must have available an expansion for $\phi_T(x, y, h)$ which specifies one more power of h than does (6). The necessary expansion is

$$\begin{aligned} \phi_T(x, y, h) = f &+ \tfrac{1}{2}hF + \tfrac{1}{6}h^2(Ff_y + G) \\ &+ \tfrac{1}{24}h^3[(3f_{xy} + 3ff_{yy} + f_y^2)F + Gf_y + H] + O(h^4), \end{aligned} \tag{68}$$

where, extending the notation of (7),

$$\begin{aligned} F &= f_x + ff_y, \\ G &= f_{xx} + 2ff_{xy} + f^2f_{yy}, \\ H &= f_{xxx} + 3ff_{xxy} + 3f^2f_{xyy} + f^3f_{yyy}. \end{aligned}$$

From (67*ii*), with $R = 2$, we have for $r = 1, 2$,

$$k_r = f(x + ha_r, y + b_{r1}hk_1 + b_{r2}hk_2).$$

Expanding as a Taylor series about (x, y) we obtain

$$\begin{aligned} k_r = f &+ h[a_r f_x + (b_{r1}k_1 + b_{r2}k_2)f_y] \\ &+ \tfrac{1}{2}h^2[a_r^2 f_{xx} + 2a_r(b_{r1}k_1 + b_{r2}k_2)f_{xy} + (b_{r1}k_1 + b_{r2}k_2)^2 f_{yy}] \\ &+ \tfrac{1}{6}h^3[a_r^3 f_{xxx} + 3a_r^2(b_{r1}k_1 + b_{r2}k_2)f_{xxy} + 3a_r(b_{r1}k_1 + b_{r2}k_2)^2 f_{xyy} \\ &+ (b_{r1}k_1 + b_{r2}k_2)^3 f_{yyy}] + O(h^4), \qquad r = 1, 2. \end{aligned} \tag{69}$$

Since these two equations are implicit, we can no longer proceed by successive substitution as in section 4.3. Let us assume, instead, that the solutions for k_1 and k_2 may be expressed in the form

$$k_r = A_r + hB_r + h^2C_r + h^3D_r + O(h^4), \qquad r = 1, 2. \tag{70}$$

Substituting for k_r by (70) in (69) we obtain

$$\begin{aligned} &A_r + hB_r + h^2C_r + h^3D_r \\ &\quad = f + h\{a_r f_x + [b_{r1}(A_1 + hB_1 + h^2C_1) \\ &\qquad + b_{r2}(A_2 + hB_2 + h^2C_2)]f_y\} + \tfrac{1}{2}h^2\{a_r^2 f_{xx} + 2a_r[b_{r1}(A_1 + hB_1) \\ &\qquad + b_{r2}(A_2 + hB_2)]f_{xy} + [b_{r1}(A_1 + hB_1) + b_{r2}(A_2 + hB_2)]^2 f_{yy}\} \\ &\qquad + \tfrac{1}{6}h^3\{a_r^3 f_{xxx} + 3a_r^2(b_{r1}A_1 + b_{r2}A_2)f_{xxy} \\ &\qquad + 3a_r(b_{r1}A_1 + b_{r2}A_2)^2 f_{xyy} + (b_{r1}A_1 + b_{r2}A_2)^3 f_{yyy}\} \\ &\qquad + O(h^4), \qquad r = 1, 2. \end{aligned}$$

On equating powers of h we obtain

$$\begin{aligned} A_r &= f, \\ B_r &= a_r f_x + (b_{r1}A_1 + b_{r2}A_2)f_y, \\ C_r &= (b_{r1}B_1 + b_{r2}B_2)f_y + \tfrac{1}{2}a_r^2 f_{xx} + a_r(b_{r1}A_1 + b_{r2}A_2)f_{xy} \\ &\quad + \tfrac{1}{2}(b_{r1}A_1 + b_{r2}A_2)^2 f_{yy}, \\ D_r &= (b_{r1}C_1 + b_{r2}C_2)f_y + a_r(b_{r1}B_1 + b_{r2}B_2)f_{xy} \\ &\quad + (b_{r1}A_1 + b_{r2}A_2)(b_{r1}B_1 + b_{r2}B_2)f_{yy} + \tfrac{1}{6}a_r^3 f_{xxx} \\ &\quad + \tfrac{1}{2}a_r^2(b_{r1}A_1 + b_{r2}A_2)f_{xxy} + \tfrac{1}{2}a_r(b_{r1}A_1 + b_{r2}A_2)^2 f_{xyy} \\ &\quad + \tfrac{1}{6}(b_{r1}A_1 + b_{r2}A_2)^3 f_{yyy}, \qquad r = 1, 2. \end{aligned}$$

This set of equations is seen to be explicit and can be solved by successive substitution. Making use of (67*iii*), we obtain the following solution

$$\begin{aligned}
A_r &= f, \\
B_r &= a_r F, \\
C_r &= (b_{r1}a_1 + b_{r2}a_2)Ff_y + \tfrac{1}{2}a_r^2 G, \\
D_r &= [b_{r1}(b_{11}a_1 + b_{12}a_2) + b_{r2}(b_{21}a_1 + b_{22}a_2)]Ff_y^2 \\
&\quad + a_r(b_{r1}a_1 + b_{r2}a_2)F(f_{xy} + ff_{yy}) + \tfrac{1}{2}(b_{r1}a_1^2 + b_{r2}a_2^2)Gf_y \\
&\quad + \tfrac{1}{6}a_r^3 H, \qquad r = 1, 2.
\end{aligned} \tag{71}$$

Using (70), the expansion for $\phi(x, y, h)$ defined by (67*ii*) may be written

$$\phi(x, y, h) = c_1A_1 + c_2A_2 + h(c_1B_1 + c_2B_2) + h^2(c_1C_1 + c_2C_2) + h^3(c_1D_1 + c_2D_2) + O(h^4),$$

where the coefficients $A_r, B_r, C_r, D_r, r = 1, 2$, are given by (71). Comparing this with the expansion (68) for $\phi_T(x, y, h)$, we see that the two-stage implicit Runge–Kutta method will have order one if

$$c_1 + c_2 = 1, \tag{72i}$$

order two if, in addition,

$$c_1a_1 + c_2a_2 = \tfrac{1}{2}, \tag{72ii}$$

order three if, in addition,

$$\begin{aligned}
c_1(b_{11}a_1 + b_{12}a_2) + c_2(b_{21}a_1 + b_{22}a_2) &= \tfrac{1}{6}, \\
c_1a_1^2 + c_2a_2^2 &= \tfrac{1}{3},
\end{aligned} \tag{72iii}$$

and order four if, in addition,

$$\begin{aligned}
(c_1b_{11} + c_2b_{21})(b_{11}a_1 + b_{12}a_2) + (c_1b_{12} + c_2b_{22})(b_{21}a_1 + b_{22}a_2) &= \tfrac{1}{24}, \\
c_1a_1(b_{11}a_1 + b_{12}a_2) + c_2a_2(b_{21}a_1 + b_{22}a_2) &= \tfrac{1}{8}, \\
c_1(b_{11}a_1^2 + b_{12}a_2^2) + c_2(b_{21}a_1^2 + b_{22}a_2^2) &= \tfrac{1}{12}, \\
c_1a_1^3 + c_2a_2^3 &= \tfrac{1}{4}.
\end{aligned} \tag{72iv}$$

Moreover, from (67*iii*) we must have

$$\begin{aligned}
a_1 &= b_{11} + b_{12}, \\
a_2 &= b_{21} + b_{22}.
\end{aligned}$$

In view of this last requirement, there are in effect only six undetermined coefficients, namely $c_1, c_2, a_1, a_2, b_{12}, b_{21}$. It is of interest to note that this is precisely the number of undetermined coefficients we had at our disposal when we derived, in section 4.3, the general three-stage explicit Runge–Kutta method. There, however, the *form* of the expansion for $\phi(x, y, h)$ precluded any possibility of attaining an order greater than three. In the present case, the form of the expansion for $\phi(x, y, h)$ holds out a possibility of attaining order four. At first sight this hope appears over-ambitious, since we would need to satisfy the eight conditions (72) with only six coefficients at our disposal. However, if we solve (72*i*), (72*ii*) and the last equations of (72*iii*) and of (72*iv*), we find

$$c_1 = c_2 = \tfrac{1}{2}, \qquad a_1 = \frac{1}{2} \pm \frac{\sqrt{3}}{6}, \qquad a_2 = \frac{1}{2} \mp \frac{\sqrt{3}}{6}, \tag{73}$$

where superior alternative signs are to be taken together. On substituting these values into the rest of the equations, we find that *all* of the remaining equations (72) are satisfied if

$$b_{11} = b_{22} = \tfrac{1}{4}, \qquad b_{12} = a_1 - \tfrac{1}{4}, \qquad b_{21} = a_2 - \tfrac{1}{4}.$$

It follows from symmetry that the two solutions indicated by the alternative signs, in (73) lead to the same method. Thus there exists a unique two-stage implicit Runge–Kutta method of order four, defined by

$$y_{n+1} - y_n = \frac{h}{2}(k_1 + k_2),$$

$$k_1 = f\left(x_n + \left(\frac{1}{2} + \frac{\sqrt{3}}{6}\right)h, \quad y_n + \frac{1}{4}hk_1 + \left(\frac{1}{4} + \frac{\sqrt{3}}{6}\right)hk_2\right), \tag{74}$$

$$k_2 = f\left(x_n + \left(\frac{1}{2} - \frac{\sqrt{3}}{6}\right)h, \quad y_n + \left(\frac{1}{4} - \frac{\sqrt{3}}{6}\right)hk_1 + \frac{1}{4}hk_2\right),$$

a method originally proposed by Hammer and Hollingsworth.[64]

In contrast with the results of section 4.4 on attainable order for an R-stage explicit Runge–Kutta, the result we have derived above extends without modification to general R; for any $R \geqslant 2$, there exists an R-stage implicit Runge–Kutta method of order $2R$ (see Butcher[13]). Implicit Runge–Kutta methods with $R > 2$ may be found in Butcher.[15] This considerable increase in attainable order for a given number of stages does not necessarily make implicit Runge–Kutta methods more computationally efficient than explicit ones. In order to apply an R-stage implicit Runge–Kutta method, at each stage it is necessary to solve, by some iterative process, the system of R implicit non-linear equations which define the k_r. Butcher[15] proves that the iteration

$$k_r^{[t+1]} = f\left(x + ha_r, y + h\sum_{s=1}^{r-1} b_{rs}k_s^{[t+1]} + h\sum_{s=r}^{R} b_{rs}k_s^{[t]}\right),$$
$$t = 0, 1, 2, \ldots, \qquad r = 1, 2, \ldots, R, \quad (75)$$

will converge provided that

$$h < 1/[L(u + v)], \quad (76)$$

where L is the Lipschitz constant of f with respect to y, and

$$u = \max\{|b_{11}| + |b_{12}| + \ldots + |b_{1R}|, |b_{22}| + |b_{23}| + \ldots + |b_{2R}|, \ldots, |b_{RR}|\},$$
$$v = \max\{|b_{21}|, |b_{31}| + |b_{32}|, \ldots, |b_{R1}| + |b_{R2}| + \ldots + |b_{RR-1}|\}.$$

Besides placing the additional constraint (76) on the steplength, using the iteration (75) means that it is no longer true that an R-stage method calls for precisely R function evaluations per step; the total number of function evaluations per step depends on the particular problem. The prospect of solving at each step a system of non-linear equations is a considerable computational deterrent. A somewhat less formidable problem arises in the case when $b_{rs} = 0$ for $r < s$; the resulting methods are called by Butcher 'semi-explicit'. (For explicit methods $b_{rs} = 0$ for $r \leqslant s$.) That an R-stage semi-explicit method can attain higher order than an R-stage explicit method is demonstrated by the following fourth-order three-stage method quoted by Butcher.[15]

$$y_{n+1} - y_n = \frac{h}{6}(k_1 + 4k_2 + k_3),$$
$$k_1 = f(x_n, y_n), \quad (77)$$
$$k_2 = f(x_n + \tfrac{1}{2}h, y_n + \tfrac{1}{4}hk_1 + \tfrac{1}{4}hk_2),$$
$$k_3 = f(x_n + h, y_n + hk_2).$$

The main interest of implicit Runge–Kutta methods lies in their weak stability characteristics, which are much superior to those of explicit methods. Thus, for example, if we repeat the analysis of section 4.7 for the method (74) we obtain

$$\frac{y_{n+1}}{y_n} = r_1 = \frac{1 + \frac{1}{2}\bar{h} + \frac{1}{12}\bar{h}^2}{1 - \frac{1}{2}\bar{h} + \frac{1}{12}\bar{h}^2}. \tag{78}$$

This is a fourth-order *rational* approximation (it is in fact the (2, 2) Padé approximation—see section 8.7) to exp $(\bar{h})$, whereas a fourth-order explicit Runge–Kutta method produced a fourth-order *polynomial* approximation to the exponential. It follows from (78) that the interval of absolute stability for the method (74) is $(-\infty, 0)$; the Trapezoidal rule has the same interval, but its order is only half that of (74). Note, however, that if the iteration scheme (75) is to be employed, (76), as well as the size of the local truncation error, limits the choice for $\bar{h}$. (The Trapezoidal rule is similarly limited by equation (55) of chapter 3.)

Our conclusion is that implicit Runge–Kutta methods can offer substantially improved weak stability characteristics but at such a computational cost that they are only to be contemplated when stability considerations are paramount; the stiff systems we shall discuss in chapter 8 provide such a situation.

Exercises

16. Prove formally that the alternative solutions indicated in (73) both lead to the same method (74).

17. Show that when $f(x, y) \equiv g(x)$, the implicit method (74) reduces to a quadrature formula which is equivalent to the two-point Legendre–Gauss quadrature formula

$$\int_{-1}^{+1} F(x)\,\mathrm{d}x = F(-1/\sqrt{3}) + F(1/\sqrt{3}).$$

18. In addition to (74), Hammer and Hollingsworth[64] propose the method

$$y_{n+\frac{2}{3}} = y_n + \frac{h}{3}(y'_{n+\frac{2}{3}} + y'_n),$$

$$y_{n+1} = y_n + \frac{h}{4}(3y'_{n+\frac{2}{3}} + y'_n),$$

where $y'_n = f(x_n, y_n)$, $y'_{n+\frac{2}{3}} = f(x_{n+\frac{2}{3}}, y_{n+\frac{2}{3}})$. Write this method in the form (67), and use (72) to show that it is of third order.

19. Prove that the semi-explicit method (77) has order four and find its interval of absolute stability.

4.11 Block methods

The novel property of the methods we shall discuss in this section is that of simultaneously producing approximations to the solution of the initial value problem at a block of points $x_{n+1}, x_{n+2}, \ldots, x_{n+N}$. Although these methods will be formulated in terms of linear multistep methods, we shall see that they are equivalent to certain Runge–Kutta methods and preserve the traditional Runge–Kutta advantages of being self-starting and of permitting easy change of steplength. Their advantage over conventional Runge–Kutta methods lies in the fact that they are less expensive in terms of function evaluations for given order.

Block methods appear to have been first proposed by Milne[131] who advocated their use only as a means of obtaining starting values for predictor–corrector algorithms; Sarafyan[158] also considers them in this light. Rosser,[155] however, develops Milne's proposals into algorithms for general use. Following Rosser, we develop a fourth-order explicit block method with $N = 2$. The method will thus simultaneously generate approximations y_{n+1} and y_{n+2} to $y(x_{n+1})$ and $y(x_{n+2})$ respectively. Let $Y_{mr}, m = 1, 2$, denote an approximation to $y(x_{n+m})$ which is locally in error by a term of order h^{r+2}. By this we mean that, under the usual localizing assumption that $y_n = y(x_n)$, $y(x_{n+m}) - Y_{mr} = O(h^{r+2})$, $m = 1, 2$. The approximations Y_{mr}, $r = 0, 1, 2, \ldots$, are built up by using the following linear multistep methods.

$$
\begin{aligned}
y_{n+1} - y_n &= \frac{h}{2}(f_{n+1} + f_n); &&\text{order 2,}\\
y_{n+2} - y_n &= 2hf_{n+1}; &&\text{order 2,}\\
y_{n+1} - y_n &= \frac{h}{12}(-f_{n+2} + 8f_{n+1} + 5f_n); &&\text{order 3,}\\
y_{n+2} - y_n &= \frac{h}{3}(f_{n+2} + 4f_{n+1} + f_n); &&\text{order 4.}
\end{aligned}
\tag{79}
$$

(Note that the third of (79) is not a linear multistep method of the class discussed in chapters 2 and 3, since it does not satisfy the condition $\alpha_k = +1$.) Let

$$
Y_{0r} = y_n, \qquad Y'_{0r} = f(x_n, y_n), \qquad r = 0, 1, 2, \ldots .
$$

The algorithm uses (79) in the following sequence of steps:

$$
\left.
\begin{aligned}
Y_{10} &= y_n + hY'_{00},\\
Y'_{10} &= f(x_{n+1}, Y_{10}),\\
Y_{11} &= y_n + \frac{h}{2}(Y'_{10} + Y'_{00}),\\
Y'_{11} &= f(x_{n+1}, Y_{11}),\\
Y_{21} &= y_n + 2hY'_{11},\\
Y'_{21} &= f(x_{n+2}, Y_{21}),\\
Y_{12} &= y_n + \frac{h}{12}(-Y'_{21} + 8Y'_{11} + 5Y'_{01}),\\
Y'_{12} &= f(x_{n+1}, Y_{12}),\\
Y_{22} &= y_n + \frac{h}{3}(Y'_{21} + 4Y'_{12} + Y'_{02}),\\
Y'_{22} &= f(x_{n+2}, Y_{22}),\\
Y_{23} &= y_n + \frac{h}{3}(Y'_{22} + 4Y'_{12} + Y'_{02}).
\end{aligned}
\right\} \tag{80}
$$

A simple calculation establishes that the approximations defined by (80) are indeed of the order indicated by the notation. Thus, if $y_n = y(x_n)$,

$$
\begin{aligned}
Y_{10} &= y_n + hf_n = y(x_{n+1}) + O(h^2),\\
Y_{11} &= y_n + \frac{h}{2}[f(x_n, y_n) + f(x_{n+1}, y(x_{n+1}) + O(h^2))]\\
&= y_n + \frac{h}{2}[f(x_n, y_n) + f(x_{n+1}, y(x_{n+1}))] + O(h^3)\\
&= y(x_{n+1}) + O(h^3),
\end{aligned}
$$

and so on. One step of the algorithm is completed by setting

$$y_{n+2} = Y_{23}, \qquad y_{n+1} = Y_{12}.$$

Thus the local error in y_{n+2} is $O(h^5)$ while that in y_{n+1} is $O(h^4)$. This lower accuracy at internal points of the block is a feature of block methods. However, in order to apply the algorithm at the next block to obtain values y_{n+3}, y_{n+4}, the only necessary starting value is y_{n+2}, and the loss of accuracy in y_{n+1} does not affect subsequent points. (Contrast with conventional linear multistep methods.) Thus the algorithm remains effectively of order four.

Let us now consider the number of function evaluations demanded by this algorithm. Examination of (80) reveals that in order to proceed to the next block using as starting value y_{n+2}, which has local error of order h^5, it would be adequate to have an estimate for $f(x_{n+2}, y_{n+2})$ which is locally in error by $O(h^4)$. Such an estimate is already available in the shape of Y'_{22} defined by the penultimate equation of (80). Thus it is unnecessary to make a function evaluation at the initial part of the new block. Thus, at all blocks except the first, we may dispense with the first function evaluation. On this understanding, (80) calls for five function evaluations for the two-step block and is considerably more economical than a conventional fourth-order Runge–Kutta method, which would require eight function evaluations for two steps.

Let us now consider (80) from a different notational viewpoint. Let

$$k_1 = f(x_n, y_n), \qquad k_2 = Y'_{10}, \qquad k_3 = Y'_{11}, \qquad k_4 = Y'_{21},$$
$$k_5 = Y'_{12}, \qquad k_6 = Y'_{22}.$$

Then (80) may be written in the form

$$\begin{aligned}
y_{n+2} - y_n &= \frac{h}{3}(k_1 + 4k_5 + k_6),\\
k_1 &= f(x_n, y_n),\\
k_2 &= f(x_n + h, y_n + hk_1),\\
k_3 &= f\left(x_n + h, y_n + \frac{h}{2}k_1 + \frac{h}{2}k_2\right),\\
k_4 &= f(x_n + 2h, y_n + 2hk_3),\\
k_5 &= f\left(x_n + h, y_n + \frac{h}{12}(5k_1 + 8k_3 - k_4)\right),\\
k_6 &= f\left(x_n + 2h, y_n + \frac{h}{3}(k_1 + k_4 + 4k_5)\right).
\end{aligned} \tag{81}$$

To reduce (81) to conventional Runge–Kutta form, define the new steplength $\tilde{h} = 2h$, let $\tilde{x}_n = x_0 + n\tilde{h}\,(= x_{2n})$, and let $\tilde{y}_n$ be an approximation to $y(\tilde{x}_n)$. Then (81) becomes

$$\tilde{y}_{n+1} - \tilde{y}_n = \frac{\tilde{h}}{6}(k_1 + 4k_5 + k_6),$$

$$\begin{aligned}
k_1 &= f(\tilde{x}_n, \tilde{y}_n), \\
k_2 &= f(\tilde{x}_n + \tfrac{1}{2}\tilde{h}, \tilde{y}_n + \tfrac{1}{2}\tilde{h}k_1), \\
k_3 &= f(\tilde{x}_n + \tfrac{1}{2}\tilde{h}, \tilde{y}_n + \tfrac{1}{4}\tilde{h}k_1 + \tfrac{1}{4}\tilde{h}k_2), \\
k_4 &= f(\tilde{x}_n + \tilde{h}, \tilde{y}_n + \tilde{h}k_3), \\
k_5 &= f\left(\tilde{x}_n + \frac{1}{2}\tilde{h}, \tilde{y}_n + \frac{\tilde{h}}{24}(5k_1 + 8k_3 - k_4)\right), \\
k_6 &= f\left(\tilde{x}_n + \tilde{h}, \tilde{y}_n + \frac{\tilde{h}}{6}(k_1 + k_4 + 4k_5)\right).
\end{aligned} \tag{82}$$

Thus the algorithm (80) is equivalent to the six-stage explicit Runge–Kutta method (82), which can be shown to have order four. Its special interest lies in the fact that $\tilde{y}_n + (\tilde{h}/24)(5k_1 + 8k_3 - k_4)$, which has to be calculated in order to evaluate k_5, is a third-order approximation to $\tilde{y}_{n+\frac{1}{2}}$, and k_6 is a third-order approximation to $f(\tilde{x}_{n+1}, \tilde{y}_{n+1})$.

Rosser's paper contains a large array of linear multistep methods of the type (79) which permit us to formulate many algorithms of the type (80). In addition it includes discussion of other computational devices leading to further economies, and also considers the possibility of formulating algorithms which include evaluations of $\partial f/\partial y$. This last device results in substantially improved weak stability characteristics. (It follows from (82) that the weak stability properties of methods of the type (80) are similar to those of explicit Runge–Kutta methods; that is, the expression for r_1 is polynomial in $\bar{h}$.) One might point out, however, that if conventional linear multistep methods are modified to include evaluations of $\partial f/\partial y$ they too enjoy substantially improved weak stability properties (Rahme,[146] Lambert[100]). Rosser also derives, from a heuristic argument, the following error-control policy for the algorithm (80). The steplength is increased, decreased, or left unchanged according to whether Δ is much less than, much greater than, or approximately equal to an agreed tolerance, where

$$\Delta = h|Y'_{22} - 2Y'_{12} + Y'_{02}|. \tag{83}$$

However, the right-hand side of (83) is, in general, of order h^3 while the local truncation error is of order h^5. Thus (83) does not constitute an estimate of the principal local truncation error and we conclude that (80), like all Runge–Kutta processes, does not admit a local error estimate comparable, in accuracy and ease of computation, with that given by Milne's device for a predictor–corrector algorithm. In terms of function evaluations per step, (80), as we have seen, is roughly comparable with a fourth-order *PECE* algorithm (but the latter would produce fourth-order results at *all* the points of the discretization). In any event, Rosser shows that the economies afforded by the block approach can also be obtained in formulations which are basically predictor–corrector rather than Runge–Kutta in character. Thus the balance of competitiveness between Runge–Kutta and predictor–corrector methods, discussed in section 4.9, remains substantially unaltered.

Block algorithms of the type (80) are clearly explicit; implicit block methods are also possible. An early example due to Clippinger and Dimsdale and quoted by Kunz[98] (page 206) has recently been analysed by Shampine and Watts.[162] The algorithm is

$$\begin{aligned} y_{n+1} - \tfrac{1}{2}y_{n+2} &= \tfrac{1}{2}y_n + \frac{h}{4}(f_n - f_{n+2}), \\ y_{n+2} &= y_n + \frac{h}{3}(f_n + 4f_{n+1} + f_{n+2}). \end{aligned} \tag{84}$$

Butcher[15] points out that this is equivalent to the implicit three-stage Runge–Kutta method

$$\begin{aligned} y_{n+1} - y_n &= \frac{h}{6}(k_1 + 4k_2 + k_3), \\ k_1 &= f(x_n, y_n), \\ k_2 &= f\left(x_n + \frac{1}{2}h, y_n + \frac{h}{24}(5k_1 + 8k_2 - k_3)\right), \\ k_3 &= f\left(x_n + h, y_n + \frac{h}{6}(k_1 + 4k_2 + k_3)\right), \end{aligned} \tag{85}$$

where the steplength in (85) is twice that in (84). In the light of our investigation of implicit Runge–Kutta methods, it is not too surprising that Shampine and Watts find that the algorithm (84) has an interval of absolute stability of $(-\infty, 0)$. The algorithm is fourth-order in y_{n+2} and third-order in y_{n+1}; once again the lower accuracy at internal points of the block does not affect the accuracy at subsequent blocks. To obtain the advantages of a semi-infinite interval of absolute stability it is, of course, necessary to iterate the algorithm (84) to convergence. Shampine and Watts also consider the possibility of using (84) with a suitable block predictor in $PECE$ and $P(EC)^2$ modes. Disappointingly, the intervals of absolute stability then shrink to $(-0{\cdot}439, 0)$ and $(-0{\cdot}410, 0)$ respectively; these compare poorly with the stability intervals for conventional predictor–corrector and Runge–Kutta methods.

Exercises

20. (Rosser.[155]) Show that the algorithm obtained from the first three equations of (80) by setting $y_{n+1} = Y_{11}$ is equivalent to the improved Euler method (16).

21. By considering the terms in f_x, f_{xx}, f_{xxx}, etc., in the Taylor expansion for Δ given by (83), show that, in general, Δ is of order h^3.

22. Write down algorithm (85) with h replaced by $2h$ and y_{n+1} replaced by y_{n+2}. Deduce the equivalence of (84) and (85).

23. (Shampine and Watts.[162]) Let $\mathbf{y}_m = (y_{n+1}, y_{n+2})^T$, $\mathbf{F}(\mathbf{y}_m) = (f_{n+1}, f_{n+2})^T$. Show that (84) may be written in the form

$$\mathbf{y}_m = hB\mathbf{F}(\mathbf{y}_m) + \mathbf{e}y_n + h\mathbf{d}f_n,$$

where the elements of the matrix B and the vectors $\mathbf{e}$ and $\mathbf{d}$ are numerical coefficients. Hence write down an iterative form of the algorithm.

Find the elements of B, $\mathbf{e}$, and $\mathbf{d}$. Why must $\mathbf{e}$ have unit elements for all such block implicit methods?

5

Hybrid methods

5.1 Hybrid formulae

Writing in 1955, Kopal[89] (page 218) makes the following comment: '... extrapolation and substitution methods ... can be regarded as two extreme ways for a construction of numerical solutions of ordinary differential equations—leaving a vast no-man's-land in between, the exploration of which has barely as yet begun.' In this context 'extrapolation methods' means methods of linear multistep type and 'substitution methods' means methods of Runge–Kutta type. From our discussion in section 4.9 on the relative merits of linear multistep and Runge–Kutta methods, it emerged that the former class of methods, though generally the more efficient in terms of accuracy and weak stability properties for a given number of function evaluations per step, suffered the disadvantage of requiring additional starting values and special procedures for changing steplength. These difficulties would be reduced, without sacrifice, if we could lower the stepnumber of a linear multistep method without reducing its order. The difficulty here lies in satisfying the essential condition of zero-stability. Thus we saw in section 2.9 that an implicit linear k-step method is capable of achieving order $2k$, but if the condition of zero-stability is to be satisfied then the maximum attainable order is only $k + 2$ when k is even and $k + 1$ when k is odd. This 'zero-stability barrier' was circumvented by the introduction, in 1964–5, of modified linear multistep formulae which incorporate a function evaluation at an off-step point. Such formulae, simultaneously proposed by Gragg and Stetter,[61] Butcher,[19] and Gear[50] were christened 'hybrid' by the last author—an apt name since, whilst retaining certain linear multistep characteristics, hybrid methods share with Runge–Kutta methods the property of utilizing data at points other than the step points $\{x_n | x_n = a + nh\}$. Thus we may regard the introduction of hybrid formulae as an important step into the no-man's-land described by Kopal.

We define a *k-step hybrid formula* to be a formula of the type

$$\sum_{j=0}^{k} \alpha_j y_{n+j} = h \sum_{j=0}^{k} \beta_j f_{n+j} + h\beta_\nu f_{n+\nu}, \tag{1}$$

where $\alpha_k = +1$, α_0 and β_0 are not both zero, $\nu \notin \{0, 1, \ldots, k\}$, and, consistently with our previous usage, $f_{n+\nu} = f(x_{n+\nu}, y_{n+\nu})$. In order to implement such a formula, even when it is explicit (that is, has $\beta_k = 0$) a special predictor to estimate $y_{n+\nu}$ is necessary. Thus a hybrid formula, unlike a linear multistep method, cannot be regarded as a *method* in its own right. We postpone until section 5.2 a discussion of suitable predictors for use with (1).

We say that the formula (1) is *zero-stable* if no root of the polynomial $\rho(\zeta) = \sum_{j=0}^{k} \alpha_j \zeta^j$ has modulus greater than one and if every root with modulus one is simple. Analogously to the definitions given in chapter 2 for linear multistep methods, we say that (1) has order p and error constant C_{p+1} if $C_0 = C_1 = \ldots = C_p = 0$, $C_{p+1} \neq 0$, where, for a suitably differentiable test function $y(x)$,

$$\begin{aligned}\mathscr{L}[y(x); h] &\equiv \sum_{j=0}^{k} [\alpha_j y(x + jh) - h\beta_j y'(x + jh)] - h\beta_\nu y'(x + \nu h) \\ &= C_0 y(x) + C_1 h y^{(1)}(x) + \ldots + C_q h^q y^{(q)}(x) + \ldots.\end{aligned}$$

The principal local truncation error at x_{n+k} is then defined to be $C_{p+1} h^{p+1} y^{(p+1)}(x_n)$. Order and error constant are most conveniently established by defining, in the manner of equation (23) of chapter 2, the constants D_q as follows:

$$\begin{aligned}D_0 &= \alpha_0 + \alpha_1 + \alpha_2 + \ldots + \alpha_k, \\ D_1 &= -t\alpha_0 + (1 - t)\alpha_1 + (2 - t)\alpha_2 + \ldots + (k - t)\alpha_k \\ &\quad - (\beta_0 + \beta_1 + \beta_2 + \ldots + \beta_k + \beta_\nu), \\ D_q &= \frac{1}{q!}[(-t)^q\alpha_0 + (1 - t)^q\alpha_1 + (2 - t)^q\alpha_2 + \ldots + (k - t)^q\alpha_k] \\ &\quad - \frac{1}{(q - 1)!}[(-t)^{q-1}\beta_0 + (1 - t)^{q-1}\beta_1 + (2 - t)^{q-1}\beta_2 \\ &\quad + \ldots + (k - t)^{q-1}\beta_k + (\nu - t)^{q-1}\beta_\nu], \qquad q = 2, 3, \ldots.\end{aligned} \tag{2}$$

The order of (1) is p if and only if, for any t, $D_0 = D_1 = \ldots = D_p = 0$, $D_{p+1} \neq 0$; C_{p+1} is then equal to D_{p+1}.

Let us consider the case of the implicit two-step formula of class (1); it is

$$y_{n+2} + \alpha_1 y_{n+1} + \alpha_0 y_n = h(\beta_2 f_{n+2} + \beta_1 f_{n+1} + \beta_0 f_n + \beta_\nu f_{n+\nu}),$$
$$\nu \notin \{0, 1, 2\}. \quad (3)$$

We start by choosing

$$\alpha_1 = -(1 + a); \qquad \alpha_0 = a, \qquad -1 \leqslant a < 1. \quad (4)$$

This choice ensures that the method has order at least zero and is zero-stable. What order can we hope to attain? From (2) with $t = 1$, we obtain

$$\begin{aligned}
D_0 &= 0,\\
D_1 &= -a + 1 - (\beta_0 + \beta_1 + \beta_2 + \beta_\nu),\\
D_2 &= \frac{1}{2!}(a + 1) - [-\beta_0 + \beta_2 + (\nu - 1)\beta_\nu],\\
D_3 &= \frac{1}{3!}(-a + 1) - \frac{1}{2!}[\beta_0 + \beta_2 + (\nu - 1)^2\beta_\nu],\\
D_4 &= \frac{1}{4!}(a + 1) - \frac{1}{3!}[-\beta_0 + \beta_2 + (\nu - 1)^3\beta_\nu],\\
D_5 &= \frac{1}{5!}(-a + 1) - \frac{1}{4!}[\beta_0 + \beta_2 + (\nu - 1)^4\beta_\nu],\\
D_6 &= \frac{1}{6!}(a + 1) - \frac{1}{5!}[-\beta_0 + \beta_2 + (\nu - 1)^5\beta_\nu].
\end{aligned} \quad (5)$$

Excluding a, which we intend to retain for the moment, there are five undetermined parameters $\beta_0, \beta_1, \beta_2, \beta_\nu$, and ν. We might thus expect to attain order five by solving the set of five equations $D_1 = D_2 = \ldots = D_5 = 0$. Note, however, that these equations, besides being non-linear in ν, are subject to the constraint $\nu \notin \{0, 1, 2\}$, and we cannot easily say whether a solution exists. Gragg and Stetter[61] have proved the following result. With the formula (1) associate the polynomials $\rho(\zeta) = \sum_{j=0}^{k} \alpha_j \zeta^j$, $\sigma(\zeta) = \sum_{j=0}^{k'} \beta_j \zeta^j$, where $k' \leqslant k$, ρ satisfies the condition of zero-stability, and $\rho(1) = 0$. (The only cases of interest are $k' = k$ corresponding to (1) being implicit and $k' = k - 1$ corresponding to it being explicit.) Then for all but a finite number of inadmissible such polynomials ρ, there exist uniquely a polynomial σ, a constant β_ν, and a real number $\nu \notin \{0, 1, \ldots, k'\}$ such that the order of (1) is at least $k' + 3$. (The constraints which preclude certain polynomials ρ do not reduce the number of free parameters: for full details see Gragg and Stetter's paper.) A corresponding result for conventional linear multistep methods (see Henrici,[67] page 226) guarantees, for any ρ subject to $\rho(1) = 0$, the unique existence of a σ of degree k' such that the order of the associated linear multistep method is at least $k' + 1$. Thus Gragg and Stetter's result shows that, with certain exceptions, we can utilize *both* of the new parameters v and β_v we have introduced to raise the order of (1) to two above that attainable by a linear multistep method having the same left-hand side and the same value for k'; note that this is done without jeopardizing control over zero-stability. Returning to our specific example (3), it is convenient to set $v - 1 = s$, where $s \notin \{-1, 0, 1\}$, in (5). Solving $D_1 = D_2 = \ldots = D_5 = 0$, we obtain

$$\begin{aligned}
\beta_2 &= \tfrac{1}{12}(5 + a) + \tfrac{1}{8}(1 + a)/(s - 1),\\
\beta_1 &= \tfrac{2}{3}(1 - a) - \tfrac{1}{4}(1 + a)/s,\\
\beta_0 &= -\tfrac{1}{12}(5a + 1) + \tfrac{1}{8}(1 + a)/(s + 1),\\
\beta_v &= -\tfrac{1}{4}(1 + a)/s(s^2 - 1),\\
s &= \tfrac{8}{15}(1 - a)/(1 + a).
\end{aligned} \tag{6}$$

There are certain special cases which must be excluded from this solution. Firstly, if $a = -1$, s is infinite. Secondly, from the last of the equations (6) it follows that the zero-stability condition $-1 \leqslant a < 1$, assumed in (4) is satisfied if and only if $s > 0$. However, we still have to satisfy the requirement $s \notin \{-1, 0, 1\}$ and hence the case $s = 1$, or, equivalently, $a = -\frac{7}{23}$, must be excluded. Hence we must add to (6) the conditions

$$-1 < a < 1, \qquad a \neq -\tfrac{7}{23}. \tag{7}$$

(It is shown by Gragg and Stetter that when $a = -1$ or $-\frac{7}{23}$, the polynomial ρ falls into their class of inadmissible polynomials.) We have thus shown that the method (3) with coefficients given by (4), (6), and (7) has order five and is zero-stable—a result consistent with Gragg and Stetter's general theorem. From (5) and (6) the error constant is seen to be

$$D_6 = \frac{1}{2.6!}(1 + a)(3s^2 - 1), \tag{8}$$

where s is given in terms of a by the last of (6). We now ask whether there remains an allowable choice for a which causes the order of (3) to rise to six. From (8), this will be the case if $a = -1$, $s = +1/\sqrt{3}$, or $s = -1/\sqrt{3}$. The first of these possibilities is excluded by (7), and the last is excluded by the condition of zero-stability which, as we have seen, is equivalent to $s > 0$. The choice $s = +1/\sqrt{3}$ is, however, allowable, and gives the following values for the parameters in (3), which now has order six:

$$
\begin{aligned}
\alpha_2 &= 1, & \beta_2 &= (\sqrt{3} + 1)/[(8 + 5\sqrt{3})(3 - \sqrt{3})],\\
\alpha_1 &= -16/(8 + 5\sqrt{3}), & \beta_1 &= 8\sqrt{3}/[3(8 + 5\sqrt{3})],\\
\alpha_0 &= (8 - 5\sqrt{3})/(8 + 5\sqrt{3}), & \beta_0 &= (\sqrt{3} - 1)/[(8 + 5\sqrt{3})(3 + \sqrt{3})],\\
v &= 1 + 1/\sqrt{3}, & \beta_v &= 6\sqrt{3}/(8 + 0\sqrt{3}),
\end{aligned}
\tag{9}
$$

$$\text{Error constant} = D_7 = -8\sqrt{3}/[(8 + 5\sqrt{3}).9.7!] \approx -0{\cdot}000{,}018{,}3.$$

The spurious root of ρ for this formula is $(8 - 5\sqrt{3})/(8 + 5\sqrt{3}) \approx -0{\cdot}040$, corroborating that the formula is indeed zero-stable. We have thus succeeded in deriving a zero-stable implicit two-step hybrid formula of order six. Since, with only seven parameters at our disposal at the outset, we could not hope for order greater than six, we have in fact achieved zero-stability with no sacrifice in order.

This result may not appear to be too surprising for the case $k = k' = 2$ when we recall that there exists a conventional linear two-step method (Simpson's rule) which has maximal order $2k$ and is zero-stable. When $k > 2$, however, a linear multistep method, explicit or implicit, must sacrifice attainable order if it is to be zero-stable (see section 2.9). On the other hand, Gragg and Stetter[61] find that this is not the case for hybrid formulae of class (1). They derive zero-stable hybrid formulae of maximal order $k + k' + 2$ for $k \leqslant 4$, $k' = k, k - 1$. Kohfeld and Thompson[87] take the investigation further and derive maximal order k-step hybrid formulae for $k \leqslant 10$, $k' = k, k - 1$. They find that such formulae are zero-stable if and only if $k \leqslant 6$ when $k = k'$ or $k \leqslant 4$ when $k = k' - 1$. Kohfeld and Thompson's paper contains numerical coefficients and error constants for a large number of maximal hybrid formulae and is a useful reference for those wishing to construct high accuracy algorithms based on such formulae.

Butcher[19] and Gear[50] adopt a somewhat different viewpoint, and specify the off-step point x_{n+v} in advance. In general, this results in their maximal order zero-stable formulae being, for given k, k', one order less than those of Gragg and Stetter. There are, however, practical advantages in this approach. Thus if v is taken to be $k - \frac{1}{2}$ and the values for y at appropriate off-step points are stored, it is clearly possible to halve the steplength at some stage of the computation without calculating any new starting values. Butcher derives zero-stable implicit k-step hybrid formulae of order $2k + 1$ with $v = k - \frac{1}{2}$ for $k = 1, 2, 3$. For $k > 3$, the choice $v = k - \frac{1}{2}$ leads to zero-instability, but he finds zero-stable formulae with $v = k - \frac{1}{3}$ for $k = 4, 5, 6$. As with Kohfeld and Thompson's paper, the complete range of zero-stable hybrid formulae of the Butcher–Gear type is too lengthy to be quoted here, and the interested reader is again referred to the original papers. Finally, we can obtain one of Butcher's formulae from equations (3), (4), and (6) by setting $s = \frac{1}{2}$; it is

$$y_{n+2} - \frac{1}{31}(32y_{n+1} - y_n) = \frac{h}{93}(15f_{n+2} + 12f_{n+1} - f_n + 64f_{n+\frac{3}{2}}), \quad (10)$$

a zero-stable formula of order five, whose error constant is $D_6 = \frac{1}{5580}$.

Exercises

1. (Gragg and Stetter.[61]) Find the unique values of β_0, β_v, and v for which the formula $y_{n+1} - y_n = h(\beta_0 f_n + \beta_v f_{n+v})$ has order three. (Compare with a method of Hammer and Hollingsworth[64] quoted in exercise 18 of chapter 4, page 155.)

2. (Gragg and Stetter.[61]) Find the implicit one-step hybrid formula of maximal order, and interpret it in terms of conventional linear multistep methods.

3. Derive the explicit two-step formula corresponding to the implicit formula defined by (3), (4), (6), and (7). In the explicit case are there any restrictions analogous to (7)? Find also the explicit formula of maximal order corresponding to the implicit formula defined by (9).

4. Gear[50] quotes the following hybrid formula:

$$y_{n+2}^p = 2y_{n+1} - y_n + \frac{h}{3}(4f_{n+\frac{3}{2}} - 3f_{n+1} - f_n),$$

$$y_{n+2} = y_{n+2}^p - 6\alpha(y_{n+1} - y_n) + \alpha h(f_{n+2} - 4f_{n+\frac{3}{2}} + 7f_{n+1} + 2f_n).$$

Find the range of α for which this formula is zero-stable. Find also its order and error constant. Show that when α is chosen so that the error constant vanishes the formula is identical with (10).

5.2 Hybrid predictor–corrector methods

The value of f_{n+v}, which must be computed before any of the hybrid formulae discussed in the previous section can be implemented, may be obtained by the use of a special predictor of the form

$$y_{n+v} + \sum_{j=0}^{k-1} \bar{\alpha}_j y_{n+j} = h \sum_{j=0}^{k-1} \bar{\beta}_j f_{n+j}, \quad (11)$$

followed by the evaluation $f_{n+v} = f(x_{n+v}, y_{n+v})$. The order of the predictor (11) is defined, in the usual way, to be p if, for a suitably differentiable test function $y(x)$,

$$y(x + vh) + \sum_{j=0}^{k-1} [\bar{\alpha}_j y(x + jh) - h\bar{\beta}_j y'(x + jh)] \equiv \bar{\mathscr{L}}[y(x); h]$$

$$= \bar{C}_0 y(x) + \bar{C}_1 h y'(x) + \ldots + \bar{C}_q h^q y^{(q)}(x) + \ldots,$$

where $\bar{C}_0 = \bar{C}_1 = \ldots = \bar{C}_p = 0, \bar{C}_{p+1} \neq 0$; the error constant is then $\bar{C}_{p+1}$ and the principal local truncation error $\bar{C}_{p+1}h^{p+1}y^{(p+1)}(x_n)$. Again, it is convenient to define order through the constants $\bar{D}_q$ defined as follows:

$$\begin{aligned}
\bar{D}_0 &= \bar{\alpha}_0 + \bar{\alpha}_1 + \ldots + \bar{\alpha}_{k-1} + 1,\\
\bar{D}_1 &= -t\bar{\alpha}_0 + (1 - t)\bar{\alpha}_1 + \ldots + (k - 1 - t)\bar{\alpha}_{k-1} + v - t\\
&\quad -(\bar{\beta}_0 + \bar{\beta}_1 + \ldots + \bar{\beta}_{k-1}),\\
\bar{D}_q &= \frac{1}{q!}[(-t)^q\bar{\alpha}_0 + (1 - t)^q\bar{\alpha}_1 + \ldots + (k - 1 - t)^q\bar{\alpha}_{k-1} + (v - t)^q]\\
&\quad - \frac{1}{(q - 1)!}[(-t)^{q-1}\bar{\beta}_0 + (1 - t)^{q-1}\bar{\beta}_1 + \ldots\\
&\quad + (k - 1 - t)^{q-1}\bar{\beta}_{k-1}], \qquad q = 2, 3, \ldots.
\end{aligned} \tag{12}$$

The order of (11) is then p if and only if, for any t, $\bar{D}_0 = \bar{D}_1 = \ldots = \bar{D}_p = 0$, $\bar{D}_{p+1} \neq 0$; $\bar{D}_{p+1}$ then equals $\bar{C}_{p+1}$. Note that (11) is of quite a different form from (1); thus, for example, we cannot talk of the zero-stability of (11), since the expression $\sum_{j=0}^{k-1} \bar{\alpha}_j \zeta^j + \zeta^v$ is no longer a polynomial.

If the hybrid formula (1) is explicit (that is, has $\beta_k = 0$), then we may clearly form an algorithm on the basis of (1) and (11) alone. If, however, (1) is implicit, then we need in addition to (11) a conventional predictor to calculate a value $y_{n+k}^{[0]}$. Thus, in formulating mnemonics to extend those used in chapter 3 to describe conventional predictor–corrector modes, it is necessary to distinguish between explicit and implicit formulae of class (1). We shall indicate by P_{H} an application of an explicit hybrid formula of class (1), by C_{H} an application of an implicit formula of the same class, and by P_v an application of a formula of class (11). P_k indicates an application of a conventional explicit linear multistep method involving no data at off-step points, and E an evaluation of f in terms of known arguments. Thus, in any algorithm, P_{H} or C_{H} must be preceded by P_vE. Thus, the use of (11) together with an explicit formula of class (1) may be described as a

$P_\nu E\, P_{\text{H}} E$ mode: if an implicit formula of class (1) is used, then modes $P_\nu E\, P_k E\, C_{\text{H}} E$ and $P_\nu E\, P_k E\, C_{\text{H}}$ are, for example, possible. These modes are formally defined below. As in the case of conventional predictor–corrector algorithms, we frequently have to use predictors whose stepnumbers exceed that of the hybrid formula P_{H} or C_{H}. We take account of this by removing the constraint that α_0 and β_0 do not both vanish, and replace it by the restriction that not all of $\alpha_0, \beta_0, \bar{\alpha}_0, \bar{\beta}_0, \alpha_0^*$ and β_0^* vanish. (As in chapter 3, $\alpha_j^*, \beta_j^*, j = 0, 1, 2, \ldots, k, (\beta_k^* = 0)$ are the coefficients of a conventional linear multistep predictor.)

$$
\begin{aligned}
P_\nu E\, P_{\text{H}} E: \quad & P_\nu: \quad y_{n+\nu} + \sum_{j=0}^{k-1} \bar{\alpha}_j y_{n+j} = h \sum_{j=0}^{k-1} \bar{\beta}_j f_{n+j}, \\
& E: \quad f_{n+\nu} = f(x_{n+\nu}, y_{n+\nu}), \\
& P_{\text{H}}: \quad \sum_{j=0}^{k} \alpha_j y_{n+j} = h \sum_{j=0}^{k-1} \beta_j f_{n+j} + h\beta_\nu f_{n+\nu}, \\
& E: \quad f_{n+k} = f(x_{n+k}, y_{n+k}).
\end{aligned} \tag{13}
$$

$$
\begin{aligned}
P_\nu E\, P_k E\, C_{\text{H}} E: \quad & P_\nu: \quad y_{n+\nu} + \sum_{j=0}^{k-1} \bar{\alpha}_j y_{n+j}^{[1]} = h \sum_{j=0}^{k-1} \bar{\beta}_j f_{n+j}^{[1]}, \\
& E: \quad f_{n+\nu} = f(x_{n+\nu}, y_{n+\nu}), \\
& P_k: \quad y_{n+k}^{[0]} + \sum_{j=0}^{k-1} \alpha_j^* y_{n+j}^{[1]} = h \sum_{j=0}^{k-1} \beta_j^* f_{n+j}^{[1]}, \\
& E: \quad f_{n+k}^{[0]} = f(x_{n+k}, y_{n+k}^{[0]}), \\
& C_{\text{H}}: \quad \sum_{j=0}^{k} \alpha_j y_{n+j}^{[1]} = h\beta_k f_{n+k}^{[0]} + h \sum_{j=0}^{k-1} \beta_j f_{n+j}^{[1]} + h\beta_\nu f_{n+\nu}, \\
& E: \quad f_{n+k}^{[1]} = f(x_{n+k}, y_{n+k}^{[1]}).
\end{aligned} \tag{14}
$$

$$
\begin{aligned}
P_\nu E\, P_k E\, C_{\text{H}}: \quad & P_\nu: \quad y_{n+\nu} + \sum_{j=0}^{k-1} \bar{\alpha}_j y_{n+j}^{[1]} = h \sum_{j=0}^{k-1} \bar{\beta}_j f_{n+j}^{[0]}, \\
& E: \quad f_{n+\nu} = f(x_{n+\nu}, y_{n+\nu}), \\
& P_k: \quad y_{n+k}^{[0]} + \sum_{j=0}^{k-1} \alpha_j^* y_{n+j}^{[1]} = h \sum_{j=0}^{k-1} \beta_j^* f_{n+j}^{[0]}, \\
& E: \quad f_{n+k}^{[0]} = f(x_{n+k}, y_{n+k}^{[0]}), \\
& C_{\text{H}}: \quad \sum_{j=0}^{k} \alpha_j y_{n+j}^{[1]} = h \sum_{j=0}^{k} \beta_j f_{n+j}^{[0]} + h\beta_\nu f_{n+\nu}.
\end{aligned} \tag{15}
$$

All of the algorithms discussed by Gragg and Stetter[61] and Gear[50] fall into one or other of these three modes. Butcher,[19] however, introduces a new possibility which turns out to have certain advantages. Consider, for example, the mode defined by (14). After the stage $P_\nu E$ has been concluded, it would be possible to replace the conventional predictor P_k by one which involved, on its right-hand side, the value $f_{n+\nu}$, which is available. This is tantamount to replacing P_k by an explicit hybrid formula of class (1), that is, by P_{H}. We thus obtain the mode $P_\nu E\, P_{\mathrm{H}} E\, C_{\mathrm{H}} E$; its formal definition is given by (14) with the P_k stage replaced by P_{H}, where

$$P_{\mathrm{H}}:\quad y_{n+k}^{[0]} + \sum_{j=0}^{k-1} \alpha_j^* y_{n+j}^{[1]} = h \sum_{j=0}^{k-1} \beta_j^* f_{n+j}^{[1]} + h\beta_\nu^* f_{n+\nu}. \tag{16}$$

Clearly, we can similarly construct the mode $P_\nu E\, P_{\mathrm{H}} E\, C_{\mathrm{H}}$.

Exercise

5. The introduction of (16) into (14) allows the predictor for y_{n+k} to make use of the information generated by the previously applied predictor for $y_{n+\nu}$. In this sense, the rôles of the predictors for y_{n+k} and $y_{n+\nu}$ may be interchanged by introducing an *implicit* formula (C_ν) obtained by adding the term $h\bar{\beta}_k f_{n+k}$ to the right-hand side of (11). Thus, formally define two new modes, $P_k E\, C_\nu E\, C_{\mathrm{H}} E$ and $P_k E\, C_\nu E\, C_{\mathrm{H}}$.

5.3 The local truncation error of hybrid methods

We denote by T_{n+k}, $\bar{T}_{n+\nu}$, and T_{n+k}^* the local truncation errors of, respectively, the hybrid formula of class (1), the predictor of class (11) used to estimate $y_{n+\nu}$, and the predictor used to estimate y_{n+k}. These local truncation errors are defined in a manner analogous to that of section 2.7. Thus, if $y(x)$ represents the theoretical solution of the given initial value problem, then

$$T_{n+k} = \sum_{j=0}^{k} \alpha_j y(x_n + jh) - h \sum_{j=0}^{k'} \beta_j y'(x_n + jh) - h\beta_\nu y'(x_n + \nu h), \tag{17i}$$

where $k' = k - 1$ if the hybrid formula is explicit, and $k' = k$ if it is implicit. Similarly,

$$\bar{T}_{n+\nu} = y(x_n + \nu h) + \sum_{j=0}^{k-1} \bar{\alpha}_j y(x_n + jh) - h \sum_{j=0}^{k-1} \bar{\beta}_j y'(x_n + jh). \tag{17ii}$$

For either of the modes $P_\nu E\, P_k E\, C_{\mathrm{H}} E$, $P_\nu E\, P_k E\, C_{\mathrm{H}}$,

$$T_{n+k}^* = y(x_{n+k}) + \sum_{j=0}^{k-1} \alpha_j^* y(x_n + jh) - h \sum_{j=0}^{k-1} \beta_j^* y'(x_n + jh), \tag{17iii}$$

while, for either of the modes $P_\nu E\, P_H E\, C_H E$, $P_\nu E\, P_H E\, C_H$,

$$T^*_{n+k} = y(x_{n+k}) + \sum_{j=0}^{k-1} \alpha^*_j y(x_n + jh) - h \sum_{j=0}^{k-1} \beta^*_j y'(x_n + jh) - h\beta^*_\nu y'(x_{n+\nu}) \tag{17iv}$$

(see equation (16)). We now establish the local truncation errors of the various algorithms described in the preceding section in terms of T_{n+k}, $\bar{T}_{n+\nu}$, and T^*_{n+k}.

Considering the first stage of the $P_\nu E\, P_H E$ mode defined by (13), we may write

$$y(x_n + \nu h) + \sum_{j=0}^{k-1} \bar{\alpha}_j y(x_n + jh) = h \sum_{j=0}^{k-1} \bar{\beta}_j f(x_n + jh, y(x_n + jh)) + \bar{T}_{n+\nu},$$

and

$$y_{n+\nu} + \sum_{j=0}^{k-1} \bar{\alpha}_j y_{n+j} = h \sum_{j=0}^{k-1} \bar{\beta}_j f(x_{n+j}, y_{n+j}).$$

On subtracting, and making the usual localizing assumption (see section 2.7) that $y_{n+j} = y(x_{n+j})$, $j = 0, 1, \ldots, (k-1)$, we obtain

$$y(x_{n+\nu}) - y_{n+\nu} = \bar{T}_{n+\nu}.$$

Similarly, corresponding to the third stage of (13), we may write

$$\sum_{j=0}^{k} \alpha_j y(x_n + jh) = h \sum_{j=0}^{k-1} \beta_j f(x_n + jh, y(x_n + jh)) + h\beta_\nu f(x_n + \nu h, y(x_n + \nu h)) + T_{n+k},$$

and

$$\sum_{j=0}^{k} \alpha_j y_{n+j} = h \sum_{j=0}^{k-1} \beta_j f(x_{n+j}, y_{n+j}) + h\beta_\nu f(x_{n+\nu}, y_{n+\nu}),$$

giving, on subtraction,

$$\begin{aligned} y(x_{n+k}) - y_{n+k} &= h\beta_\nu[f(x_{n+\nu}, y(x_{n+\nu})) - f(x_{n+\nu}, y_{n+\nu})] + T_{n+k} \\ &= h\beta_\nu[y(x_{n+\nu}) - y_{n+\nu}] \frac{\partial f(x_{n+\nu}, \eta_{n+\nu})}{\partial y} + T_{n+k}, \end{aligned}$$

for some $\eta_{n+\nu}$ in the interval whose endpoints are $y_{n+\nu}$ and $y(x_{n+\nu})$. If we make the further assumption that, locally, we may take $\partial f/\partial y = \lambda$, constant, we obtain

$$y(x_{n+k}) - y_{n+k} = T_{n+k} + h\lambda\beta_\nu \bar{T}_{n+\nu},$$

an expression for the local truncation error of the algorithm (13). In an analogous manner, we can establish the local truncation errors of the algorithms defined by (14), (15), and (16). The results are as follows, where T_{n+k}, $\bar{T}_{n+v}$, and T^*_{n+k} are given by (17):

Mode	*Local truncation error*	
$P_v E\, P_H E$	$T_{n+k} + h\lambda\beta_v\bar{T}_{n+v}$,	(18*i*)
$P_v E\, P_k E\, C_H E$ / $P_v E\, P_k E\, C_H$	$T_{n+k} + h\lambda(\beta_v\bar{T}_{n+v} + \beta_k T^*_{n+k})$,	(18*ii*)
$P_v E\, P_H E\, C_H E$ / $P_v E\, P_H E\, C_H$	$T_{n+k} + h\lambda(\beta_v\bar{T}_{n+v} + \beta_k T^*_{n+k}) + h^2\lambda^2\beta_k\beta^*_v\bar{T}_{n+v}$.	(18*iii*)

If we now assume that the theoretical solution $y(x)$ of the initial value problem is sufficiently differentiable, and let the hybrid formula (P_H or C_H), the predictor for y_{n+v} (P_v), and the predictor for y_{n+k} (P_k or P_H) have orders $p, \bar{p}, p^*$, and error constants $C_{p+1}, \bar{C}_{\bar{p}+1}, C^*_{p^*+1}$, respectively, then

$$
\begin{aligned}
T_{n+k} &= h^{p+1}C_{p+1}y^{(p+1)}(x_n) + O(h^{p+2}),\\
\bar{T}_{n+v} &= h^{\bar{p}+1}\bar{C}_{\bar{p}+1}y^{(\bar{p}+1)}(x_n) + O(h^{\bar{p}+2}),\\
T^*_{n+k} &= h^{p^*+1}C^*_{p^*+1}y^{(p^*+1)}(x_n) + O(h^{p^*+2}).
\end{aligned}
\tag{19}
$$

We have already discussed in section 5.1 the order, p, that can be attained by a zero-stable hybrid formula of given stepnumber k. It is clear from (11) that $\bar{p} = 2k - 1$ is attainable. (Formulae of class (11) with $\bar{p} = 2k - 1$ are identical with Hermite interpolation formulae; see, for example, Kopal.[89]) The maximal value for p^* is clearly $2k - 1$ for $P_v E\, P_k E\, C_H(E)$ modes, and $2k$ for $P_v E\, P_H E\, C_H(E)$ modes. The maximal attainable orders of the various formulae appearing in the modes listed in (18) may be summarized as follows, where it is assumed that each formula has stepnumber k:

Explicit hybrid:	P_H (Gragg and Stetter, $k \leqslant 4$):	$p = 2k + 1$,	
Implicit hybrid:	C_H (Gragg and Stetter, $k \leqslant 6$):	$p = 2k + 2$,	
	C_H (Butcher, Gear, $k \leqslant 6$):	$p = 2k + 1$,	(20)
Predictor for y_{n+v}:	P_v:	$\bar{p} = 2k - 1$,	
Predictor for y_{n+k}:	P_k:	$p^* = 2k - 1$,	
	P_H:	$p^* = 2k$.	

In choosing predictors for use in hybrid algorithms, there are several factors to be considered. Firstly, if one of the modes $P_vE\ P_HE, P_vE\ P_kE\ C_H(E)$ is chosen, where the hybrid formula P_H or C_H is of Gragg–Stetter type and has maximal attainable order, then it follows from (18), (19), and (20) that the order of the local truncation error of the algorithm cannot equal that of the hybrid formula alone, unless the stepnumber of the predictor(s) exceeds that of the hybrid formula. In a sense, this tends to defeat the purpose of hybrid formulae, which were motivated by the desire to increase the order without increasing the stepnumber. The same statement would hold for hybrid correctors of the Butcher–Gear type if used in $P_vE\ P_kE\ C_H(E)$ mode. If, however, $P_vE\ P_HE\ C_H(E)$ modes are employed, then it is possible to find algorithms whose order and stepnumber are those of the hybrid corrector. From (20), the maximal attainable order of the predictor P_H is seen to be $2k$. Let us, however, retain a free parameter in the coefficients of P_H, thus reducing its order to $2k - 1$. Then, from (18), (19), and (20), the local truncation error of the algorithm is

$$h^{2k+2}C_{2k+2}y^{(2k+2)}(x_n) + h^{2k+1}\lambda(\beta_v\bar{C}_{2k} + \beta_kC^*_{2k})y^{(2k)}(x_n) + O(h^{2k+2}),$$

where we assume that the corrector C_H has order $2k + 1$. If we use the free parameter in the coefficients defining P_H in such a way that the equation

$$\beta_v\bar{C}_{2k} + \beta_kC^*_{2k} = 0 \tag{21}$$

is satisfied, then the algorithm has local truncation error $y(x_{n+k}) - y^{[1]}_{n+k} = O(h^{2k+2})$, that is, is of order $2k + 1$. The condition (21) essentially cancels out the principal parts of the local truncation errors of the predictors P_v and P_H. We shall call algorithms which employ this device 'error-balanced'.

Consider, for example, the zero-stable two-step hybrid corrector (10) which has order 5. The unique two-step third-order predictor P_v of class (11) which estimates $y_{n+\frac{3}{2}}$ is

$$y_{n+\frac{3}{2}} - y_n = \frac{h}{8}(9f_{n+1} + 3f_n), \tag{22}$$

and has error constant $\bar{C}_4 = \frac{3}{128}$. The two-step hybrid predictor of class (1) used to estimate y_{n+k} has the form

$$y_{n+2} + \alpha^*_1y_{n+1} + \alpha^*_0y_n = h(\beta^*_1f_{n+1} + \beta^*_0f_n + \beta^*_{\frac{3}{2}}f_{n+\frac{3}{2}}).$$

Retaining $\beta^*_1 = \beta$ as parameter, we find from (2) that if we set

$$\alpha^*_0 = -(1 + 6\beta)/5, \qquad \alpha^*_1 = (-4 + 6\beta)/5,$$
$$\beta^*_0 = (2 + 7\beta)/15, \qquad \beta^*_{\frac{3}{2}} = (16 - 4\beta)/15,$$

then $p^* = 3$ and $C_4^* = (1 + \beta)/30$. Hence, condition (21) is satisfied if

$$\beta_{\frac{3}{2}}\bar{C}_4 + \beta_2 C_4^* = 0,$$

that is, if

$$\tfrac{64}{93} \cdot \tfrac{3}{128} + \tfrac{15}{93} \cdot \tfrac{1}{30}(1 + \beta) = 0,$$

or $\beta = -4$. With this value, the predictor P_H is

$$y_{n+2} - \tfrac{1}{5}(28y_{n+1} - 23y_n) = \frac{2h}{15}(-30f_{n+1} - 13f_n + 16f_{n+\frac{1}{2}}). \qquad (23)$$

The algorithm consisting of (10) as C_H, (22) as P_v and (23) as P_H combined in $P_vE\ P_HE\ C_H(E)$ mode is a two-step algorithm of order five. This algorithm is one of those derived by Butcher;[19] similar algorithms with stepnumber k and order $2k + 1$, $k = 3, 4, 5, 6$, may be found in Butcher's paper.

Exercises

6. Verify the local truncation errors given by (18*ii*) and (18*iii*).

7. Find an expression similar to those of (18) for the local truncation errors of the modes $P_kE\ C_vE\ C_H(E)$ defined in exercise 5 (page 170). Deduce that there exist error-balanced algorithms based on these modes with stepnumber k and order $2k + 1$.

8. Derive the two-step $P_kE\ C_vE\ C_HE$ algorithm of order five which uses formula (10) as C_H.

9. (Butcher.[19]) Derive the one-step error-balanced $P_vE\ P_HE\ C_HE$ algorithm of order three which uses as C_H the formula

$$y_{n+1} - y_n = \frac{h}{6}(f_{n+1} + f_n + 4f_{n+\frac{1}{2}}),$$

derived in exercise 2. Show that this algorithm is identical with Kutta's third-order rule (equation (19) of chapter 4).

So far, we have been satisfied if the order of the hybrid predictor–corrector algorithm equals that of the hybrid formula alone. This point of view ignores an advantage possessed by hybrid formulae—particularly those of Gragg and Stetter type—namely that they have remarkably small error constants. Consider, for example, an algorithm in $P_vE\ P_kE\ C_H(E)$ mode using, as C_H, the following one-step implicit hybrid formula of Gragg and Stetter:

$$C_H\colon\ y_{n+1} - y_n = \frac{h}{6}(f_{n+1} + f_n + 4f_{n+\frac{1}{2}}). \qquad (24)$$

This is, of course, Simpson's rule, compressed to one step (see exercise 9). Its order is thus four, but its error constant is $C_5 = -\frac{1}{2880} \approx -0{\cdot}000347$ (that is, $1/2^5$ times the error constant of Simpson's rule, since, in effect, h has been replaced by $h/2$). It is clear from (20) that there do not exist one-step predictors P_v and P_k of order three, which, by (18), would be necessary if the order of the algorithm is to be four. Thus we are forced to increase the stepnumber to two, by replacing (24) by the equivalent form

$$C_{\mathrm{H}}:\quad y_{n+2} - y_{n+1} = \frac{h}{6}(f_{n+2} + f_{n+1} + 4f_{n+\frac{3}{2}}). \tag{25i}$$

The unique two-step third-order predictors P_v and P_k are then as follows:

$$P_v:\quad y_{n+\frac{3}{2}} - y_n = \frac{3h}{8}(3f_{n+1} + f_n);\qquad \bar{p} = 3,\quad \bar{C}_4 = \tfrac{3}{128} \approx 0{\cdot}0234, \tag{25ii}$$

$$P_k:\quad y_{n+2} + 4y_{n+1} - 5y_n = h(4f_{n+1} + 2f_n);\qquad p^* = 3,\quad C_4^* = \tfrac{1}{6} \approx 0{\cdot}167. \tag{25iii}$$

Thus, by (18) and (19), the local truncation error of an algorithm in $P_vE\,P_kE\,C_{\mathrm{H}}(E)$ mode based on (25) is

$$\begin{aligned} &-0{\cdot}000{,}347h^5y^{(5)}(x_n) + h\lambda[\tfrac{2}{3}(0{\cdot}0234) + \tfrac{1}{6}(0{\cdot}167)]h^4y^{(4)}(x_n) + O(h^6)\\ &\qquad = h^5[-0{\cdot}000{,}347y^{(5)}(x_n) + 0{\cdot}0434\lambda y^{(4)}(x_n)] + O(h^6). \end{aligned}$$

Considering, for example, the equation $y' = \lambda y$, for which $y^{(s)}(x_n) = \lambda^s y(x_n)$, $s = 0, 1, 2, \ldots$, we see that the advantage of the small error constant in (24) is being lost. In view of this, we might replace (25*ii*) and (25*iii*) by fourth-order predictors. This calls for yet another increase in stepnumber. Since three-step predictors can, by (20), attain order five, we can arbitrarily choose to set $\bar{\alpha}_0 = 0 = \alpha_0^*$. Thus (25) is replaced by

$$C_{\mathrm{H}}:\quad y_{n+3} - y_{n+2} = \frac{h}{6}(f_{n+3} + f_{n+2} + 4f_{n+\frac{5}{2}}), \tag{26i}$$

$$P_v:\quad y_{n+\frac{5}{2}} + \tfrac{9}{16}y_{n+2} - \tfrac{25}{16}y_{n+1} = \frac{h}{64}(87f_{n+2} + 48f_{n+1} - 3f_n);$$

$$\bar{p} = 4,\quad \bar{C}_5 = \tfrac{51}{3840} \approx 0{\cdot}0133, \tag{26ii}$$

$$P_k:\quad y_{n+3} + 8y_{n+2} - 9y_{n+1} = \frac{h}{3}(17f_{n+2} + 14f_{n+1} - f_n);$$

$$p^* = 4,\quad C_5^* = \tfrac{1}{9} \approx 0{\cdot}111. \tag{26iii}$$

The local truncation error of a $P_\nu E\ P_k E\ C_H(E)$ algorithm based on (26) is

$$(-0{\cdot}000{,}347 + 0{\cdot}0274h\lambda)h^5y^{(5)}(x_n) + O(h^6). \tag{27}$$

Thus, even when the predictors have the same order as the corrector, the truncation errors of the predictors will dominate unless $|h\lambda|$ is small—in this case, of the order of 0·01. Kohfeld and Thompson[87] show that this difficulty gets more severe as the order of the Gragg and Stetter formula is increased. Thus they show, for example, that if sixth-order predictors are incorporated with the sixth-order hybrid corrector (9) in a $P_\nu E\ P_k E\ C_H(E)$ mode, then the local truncation error is

$$(-0{\cdot}000{,}018{,}3 + 0{\cdot}006{,}56h\lambda)h^7y(x_n) + O(h^8),$$

and we now require $|h\lambda|$ to be of the order of 0·003 if the truncation errors of the predictors are not to dominate that of the corrector.

Kohfeld and Thompson propose the following way out of this difficulty. We shall replace the conventional predictor P_k by an explicit hybrid formula, P_H, of class (1) which, unlike the Butcher algorithms already discussed, uses a *new* off-step point, $x_{n+\mu}$, chosen to maximize the order. A new predictor, P_μ, of class (11) is required to predict $y_{n+\mu}$, and the predictor, P_ν, for $y_{n+\nu}$ is modified to make use of the value $f_{n+\mu}$ which is then available. Thus a Kohfeld–Thompson algorithm may be described as a $P_\mu E\ P_\nu E\ P_H E\ C_H(E)$ mode. In the case when the main hybrid formula, C_H, is (26*i*), these predictors take the following form:

$$P_\mu:\quad y_{n+\mu} + \bar{\bar{\alpha}}_2 y_{n+2} + \bar{\bar{\alpha}}_1 y_{n+1} + \bar{\bar{\alpha}}_0 y_n = h(\bar{\bar{\beta}}_2 f_{n+2} + \bar{\bar{\beta}}_1 f_{n+1} + \bar{\bar{\beta}}_0 f_n), \tag{28i}$$

$$P_\nu:\quad y_{n+\frac{5}{2}} + \bar{\alpha}_2 y_{n+2} + \bar{\alpha}_1 y_{n+1} + \bar{\alpha}_0 y_n = h(\bar{\beta}_2 f_{n+2} + \bar{\beta}_1 f_{n+1} + \bar{\beta}_\mu f_{n+\mu}), \tag{28ii}$$

$$P_H:\quad y_{n+3} + \alpha_2^* y_{n+2} = h(\beta_2^* f_{n+2} + \beta_1^* f_{n+1} + \beta_\mu^* f_{n+\mu}). \tag{28iii}$$

We extend our previous notation by letting the order, error constant, and local truncation error of P_μ be denoted by $\bar{\bar{p}}$, $\bar{\bar{C}}_{\bar{\bar{p}}+1}$, and $\bar{\bar{T}}_{n+\mu}$ respectively. The local truncation error of the algorithm is found, in the same way as before, to be

$$T_{n+k} + h\lambda(\beta_k T^*_{n+k} + \beta_\nu \bar{T}_{n+\nu}) + h^2\lambda^2(\beta_k\beta_\mu^* + \beta_\nu\bar{\beta}_\mu)\bar{\bar{T}}_{n+\mu}. \tag{29}$$

The predictor (28*iii*) attains maximal order when

$$\mu = \tfrac{27}{10},\quad \alpha_2^* = -1,\quad \beta_2^* = \tfrac{221}{714},\quad \beta_1^* = -\tfrac{7}{714},\quad \beta_\mu^* = \tfrac{500}{714}, \tag{30}$$

giving $p^* = 4$, $C_5^* = \frac{1}{576} \approx 0{\cdot}001{,}736$. (The formula so defined is one derived by Gragg and Stetter.[61] Note that it does not attain the maximal order indicated by (20); this is a consequence of the fact that an order has been sacrificed to set $\alpha_1^* = 0$, that is, to make the formula of Adams type.) Clearly, with μ so specified, it is possible for P_ν, given by (28*ii*), to attain order five. Instead, a parameter is retained by choosing

$$\begin{aligned}
&\bar{\alpha}_2 = \tfrac{45}{64} - \tfrac{357}{25}\beta, \qquad \bar{\alpha}_1 = -\tfrac{25}{16} + \tfrac{3213}{250}\beta, \qquad \bar{\alpha}_0 = -\tfrac{9}{64} + \tfrac{357}{250}\beta,\\
&\bar{\beta}_2 = \tfrac{45}{32} - \tfrac{4063}{500}\beta, \qquad \bar{\beta}_1 = \tfrac{15}{16} - \tfrac{4291}{500}\beta, \qquad \bar{\beta}_\mu = \beta,
\end{aligned} \tag{31}$$

whence

$$\bar{p} = 4, \qquad \bar{C}_5 = (\tfrac{45}{32} - \tfrac{33677}{2000}\beta)/120.$$

The device of error-balancing can now be employed to set to zero the leading term in $\beta_k T^*_{n+k} + \beta_\nu \bar{T}_{n+\nu}$. Thus β is chosen so that $\frac{1}{6}C_5^* + \frac{2}{3}\bar{C}_5 = 0$, that is,

$$\beta = 10{,}000/115{,}464. \tag{32}$$

With this value, (31) now specifies P_ν uniquely. Thus, in (29),

$$T_{n+k} = C_5 h^5 y^{(5)}(x_n) + O(h^6)$$

(where we recall that C_5 is small) and

$$h\lambda(\beta_k T^*_{n+k} + \beta_\nu \bar{T}_{n+\nu}) = O(h^7).$$

The coefficients in P_μ are chosen so that it has order five, with the result that the last term in (29) is now $O(h^8)$. (Observe that P_μ is the only formula in the algorithm which does not use a value of f at an off-step point; our experience suggests that it will not have a particularly small error constant.) The exact values of the coefficients in P_μ, when $\mu = 2{\cdot}7$, are as follows:

$$\begin{aligned}
\bar{\bar{\alpha}}_2 &= 5{\cdot}793{,}727{,}5, & \bar{\bar{\beta}}_2 &= 3{\cdot}686{,}917{,}5,\\
\bar{\bar{\alpha}}_1 &= -3{\cdot}5721, & \bar{\bar{\beta}}_1 &= 6{\cdot}072{,}57,\\
\bar{\bar{\alpha}}_0 &= -3{\cdot}221{,}627{,}5, & \bar{\bar{\beta}}_0 &= 0{\cdot}955{,}867{,}5,
\end{aligned} \tag{33}$$

whence order $\bar{\bar{p}} = 5$ and error constant $\bar{\bar{C}}_6 = 0{\cdot}0142$.

The $P_\mu E\, P_\nu E\, P_H E\, C_H(E)$ modes defined by (28), (30), (31), (32), (33), and (26*i*) now constitute hybrid algorithms in which all predictors are sufficiently accurate to enable full advantage to be taken of the remarkably small error constant of the corrector C_H. In addition to the above algorithm, similar algorithms based on correctors like (24) but with stepnumbers 2, 3, 4, 5, and 6 may be found in Kohfeld and Thompson,[87] the coefficients being specified to up to fifteen decimal places. (Note that if these algorithms are to succeed in extracting the potentially high accuracy of the hybrid correctors, the coefficients must themselves be computed to high accuracy.)

We observe that the Kohfeld–Thompson algorithms require four or three function evaluations per step, depending on whether or not the final function evaluation is made; this contrasts with conventional predictor–corrector algorithms, which require, respectively, two or one evaluations. However, computational experience suggests that the accuracy of the Kohfeld–Thompson algorithms is such that they are in effect less costly in computing effort for given accuracy than conventional predictor–corrector algorithms.

Finally, an extension of the analysis of section 3.10 shows that an error estimate by Milne's device can be obtained for a hybrid algorithm (other than $P_vE\,P_HE$ of course) provided that the order of the predictor P_k (or P_H) for y_{n+k} equals that of the hybrid corrector C_H. Thus Kohfeld–Thompson algorithms permit error estimation, while error-balanced algorithms of the Butcher type do not.

We conclude from this section that hybrid formulae can be used to advantage in two complementary ways. Error-balanced algorithms of the Butcher type achieve high order with no increase in stepnumber over that of the hybrid formula; they thus call for a minimal amount of complication in the way of starting values and step-changing procedures, but do not permit direct error estimation. Kohfeld–Thompson algorithms, on the other hand, have a larger stepnumber than the hybrid formula, and, moreover, call for three or four evaluations per step; they do, however, permit direct error estimation, and are capable of extremely high accuracy.

Example 1 *For a* **linear** *initial value problem (i.e. one for which $f(x, y) = g(x)y$) it is, of course, possible to solve for y_{n+k} the equation given by an implicit hybrid corrector C_H without recourse to a predictor P_k (or P_H). Thus, by comparing the numerical solution y_{n+k} with the theoretical solution $y(x_{n+k})$ under the assumption that all previous numerical values for y (including those at off-step, points) coincide with the theoretical solution, it is possible to demonstrate the high accuracy of the C_H formula after one step. (a) Do this in the case when C_H is the fourth-order formula (24) and the initial value problem is $y' = -y$, $y(0) = 1$, taking the steplength to be 0·2. Illustrate the loss of accuracy due to predictors by performing one step, for the same problem, using the fourth-order algorithms (b) (25), (c) (26), and (d) the Kohfeld–Thompson algorithm defined by (28), (30), (31), (32), (33), and (26i); take all necessary starting values from the theoretical solution of the initial value problem. (e) Compare the above results with those obtained from the fifth-order Butcher algorithm defined by (22), (23), and (10).*

Since algorithms (b) and (c) have stepnumber 3, the earliest step at which a comparison can be made is that which yields y_3, where $x_j = 0{\cdot}2j$, $j = 0, 1, \ldots.$

(a) For the given problem, (24) gives

$$y_3 - y_2 = \frac{0{\cdot}2}{6}(-y_3 - y_2 - 4y_{\frac{5}{2}}),$$

or

$$3{\cdot}1y_3 = 2{\cdot}9y_2 - 0{\cdot}4y_{\frac{5}{2}}.$$

Assuming that $y_2 \equiv y(x_2) = \exp(-0{\cdot}4) = 0{\cdot}670{,}320{,}046$, and

$$y_{\frac{5}{2}} \equiv y(x_{\frac{5}{2}}) = \exp(-0{\cdot}5) = 0{\cdot}606{,}530{,}660,$$

we find that $y_3 = 0{\cdot}548{,}811{,}571$, whereas the theoretical solution is

$$y(x_3) = \exp(-0{\cdot}6) = 0{\cdot}548{,}811{,}636.$$

For algorithms (b), (c), and (d) the necessary starting values are taken to be

$$y_0 = \exp(0) = 1, \qquad y_1 = \exp(-0{\cdot}2) = 0{\cdot}818{,}730{,}753,$$
$$y_2 = \exp(-0{\cdot}4) = 0{\cdot}670{,}320{,}046.$$

(b) Setting $n = 1$ in (25), we obtain

P_v: $y_{\frac{5}{2}} - y_1 = (0{\cdot}6/8)(-3y_2 - y_1)$, whence, using the above starting values,

$$y_{\frac{5}{2}} = 0{\cdot}606{,}503{,}936.$$

P_k: $y_3^{[0]} + 4y_2 - 5y_1 = 0{\cdot}2(-4y_2 - 2y_1)$, giving

$$y_3^{[0]} = 0{\cdot}548{,}625{,}243.$$

C_H: $y_3^{[1]} - y_2 = (0{\cdot}2/6)(-y_3^{[0]} - y_2 - 4y_{\frac{5}{2}})$, giving

$$y_3^{[1]} = 0{\cdot}548{,}821{,}345.$$

(c) Setting $n = 0$ in (26) and proceeding in a similar manner, we obtain

$$y_{\frac{5}{2}} = 0{\cdot}606{,}533{,}900 \qquad y_3^{[0]} = 0{\cdot}548{,}838{,}321, \qquad y_3^{[1]} = 0{\cdot}548{,}810{,}247.$$

(d) Setting $n = 0$ in (28) and using the coefficients defined by the appropriate equations, we obtain

$$y_{2\cdot7} = 0{\cdot}582{,}747{,}535, \qquad y_{\frac{5}{2}} = 0{\cdot}606{,}530{,}329, \qquad y_3^{[0]} = 0{\cdot}548{,}812{,}096,$$
$$y_3^{[1]} = 0{\cdot}548{,}811{,}597.$$

(e) Setting $n = 1$ in (22), (23), and (10), we obtain

$$y_{\frac{5}{2}} = 0{\cdot}606{,}503{,}936, \qquad y_3^{[0]} = 0{\cdot}548{,}938{,}479, \qquad y_3^{[1]} = 0{\cdot}548{,}811{,}230.$$

These results are summarized in table 20, which displays the errors after one step of each algorithm. The stepnumber is quoted, as is the number of function evaluations which would be necessary if the algorithm were applied to a general non-linear problem. (It is assumed that in each case a final evaluation would be made.)

It is clear from table 20 that, due to the inadequate accuracy of the predictors, algorithms (25) and (26) fail to exploit the potential accuracy of the C_H formula. The Kohfeld–Thompson algorithm, on the other hand, is more accurate than the C_H formula. The fifth-order Butcher algorithm turns out to be less accurate than the fourth-order Kohfeld–Thompson algorithm—but it requires one less starting value

Table 20

	Method	Order	Error × 10^9	Stepnumber	Number of evaluations
(a)	C_H	4	+65	1	—
(b)	(25)	4	−9709	2	3
(c)	(26)	4	+1389	3	3
(d)	Kohfeld–Thompson	4	+39	3	4
(e)	Butcher	5	+406	2	3

and one less evaluation per step. We conclude from this example that, due to the strong dependence of the accuracy on the error constants, it is not possible to make a general comparison between hybrid algorithms and conventional predictor–corrector and Runge–Kutta algorithms, in the spirit of section 4.9.

Finally, the mechanism of error-balancing can clearly be seen at work in the results of (e). For,

$$y(x_{\frac{5}{2}}) - y_{\frac{5}{2}} = \exp(-0{\cdot}5) - y_{\frac{5}{2}} \approx 3 \times 10^{-5} \qquad (= \bar{T}_{n+\nu}),$$
$$y(x_3) - y_3^{[0]} = \exp(-0{\cdot}6) - y_3^{[0]} \approx -1 \times 10^{-4} \qquad (= T^*_{n+k}),$$

and (18*iii*) would lead us to expect that the predictors would contribute an error of order 10^{-5} to the local truncation error of the algorithm; due to error-balancing, the local truncation error is, from table 20, approximately 4×10^{-7}.

Exercises

10. Repeat example 1 when the linear initial value problem is replaced by the non-linear problem $y' = 4xy^{\frac{1}{2}}$, $y(0) = 1$, whose theoretical solution is $y(x) = (1 + x^2)^2$. (Note that the equation for y_{n+k} given by C_H is now quadratic, and for the purposes of (a), can be solved by the usual formula for the roots of a quadratic.)

11. Find the error constant of the third-order one-step explicit hybrid formula

$$P_H\colon \quad y_{n+1} - y_n = \frac{h}{4}(3f_{n+\frac{2}{3}} + f_n)$$

(see exercise 1, page 167). Find the unique third-order two-step predictor P_ν which could be used in conjunction with the above hybrid formula to form a $P_\nu E\, P_H E$ algorithm. Examine the local truncation error of this algorithm and find the approximate range of $h\lambda$ for which the local truncation error of P_ν will not dominate this error.

12. For the problem of example 1, compare, after one step, the accuracy of the $P_\nu E\, P_H E$ algorithm derived in exercise 11 with that of the P_H formula alone. Repeat for the problem of exercise 10.

5.4 Weak stability of hybrid methods

It is straightforward to find, in a manner analogous to that of section 3.11, stability polynomials for the various hybrid predictor–corrector modes we have described. Thus, for the mode $P_\nu E\, P_{\mathrm{H}}E$ defined by (13), if we introduce the notation

$$\begin{aligned}
\bar{\rho}(\zeta) &\equiv \sum_{j=0}^{k-1} \bar{\alpha}_j \zeta^j, \qquad & \bar{\sigma}(\zeta) &\equiv \sum_{j=0}^{k-1} \bar{\beta}_j \zeta^j,\\
\rho(\zeta) &\equiv \sum_{j=0}^{k} \alpha_j \zeta^j, \qquad & \sigma(\zeta) &\equiv \sum_{j=0}^{k-1} \beta_j \zeta^j,
\end{aligned} \tag{34}$$

the stability polynomial is found to be

$$\pi(r, \bar{h}) = \rho(r) - \bar{h}\sigma(r) + \bar{h}\beta_\nu[\bar{\rho}(r) - \bar{h}\bar{\sigma}(r)]. \tag{35}$$

Stability polynomials for the other modes we have discussed can all be deduced from those for the most complicated modes considered, namely the $P_\mu E\, P_\nu E\, P_{\mathrm{H}}E\, C_{\mathrm{H}}(E)$ modes of Kohfeld and Thompson, which may be formally defined as follows:

$$\begin{aligned}
P_\mu&: \quad y_{n+\mu} + \sum_{j=0}^{k-1} \bar{\bar{\alpha}}_j y_{n+j}^{[1]} = h \sum_{j=0}^{k-1} \bar{\bar{\beta}}_j f_{n+j}^{[t]},\\
P_\nu&: \quad y_{n+\nu} + \sum_{j=0}^{k-1} \bar{\alpha}_j y_{n+j}^{[1]} = h \sum_{j=0}^{k-1} \bar{\beta}_j f_{n+j}^{[t]} + h\bar{\beta}_\mu f_{n+\mu},\\
P_{\mathrm{H}}&: \quad y_{n+k}^{[0]} + \sum_{j=0}^{k-1} \alpha_j^* y_{n+j}^{[1]} = h \sum_{j=0}^{k-1} \beta_j^* f_{n+j}^{[t]} + h\beta_\mu^* f_{n+\mu},\\
C_{\mathrm{H}}&: \quad \sum_{j=0}^{k} \alpha_j y_{n+j}^{[1]} = h\beta_k f_{n+k}^{[0]} + h \sum_{j=0}^{k-1} \beta_j f_{n+j}^{[t]} + h\beta_\nu f_{n+\nu},
\end{aligned} \tag{36}$$

where $t = 1$ when the mode includes the final evaluation (E) and $t = 0$ when it does not. We define the following polynomials:

$$\begin{aligned}
\bar{\bar{\rho}}(\zeta) &\equiv \sum_{j=0}^{k-1} \bar{\bar{\alpha}}_j \zeta^j, \qquad & \bar{\bar{\sigma}}(\zeta) &\equiv \sum_{j=0}^{k-1} \bar{\bar{\beta}}_j \zeta^j,\\
\bar{\rho}(\zeta) &\equiv \sum_{j=0}^{k-1} \bar{\alpha}_j \zeta^j, \qquad & \bar{\sigma}(\zeta) &\equiv \sum_{j=0}^{k-1} \bar{\beta}_j \zeta^j,\\
\rho^*(\zeta) &\equiv \sum_{j=0}^{k} \alpha_j^* \zeta^j, \qquad & \sigma^*(\zeta) &\equiv \sum_{j=0}^{k-1} \beta_j^* \zeta^j,\\
\rho(\zeta) &\equiv \sum_{j=0}^{k} \alpha_j \zeta^j, \qquad & \sigma(\zeta) &\equiv \sum_{j=0}^{k} \beta_j \zeta^j.
\end{aligned} \tag{37}$$

(Note that ρ and σ are associated with P_H in (34), but with C_H in (37).) Then, for the mode $P_\mu E\ P_\nu E\ P_H E\ C_H E$, the stability polynomial is

$$\pi(r, \bar{h}) = \rho - \bar{h}\sigma + \bar{h}\beta_k(\rho^* - \bar{h}\sigma^*) + \bar{h}\beta_\nu(\bar{\rho} - \bar{h}\bar{\sigma}) + \bar{h}^2(\beta_k\beta_\mu^* + \beta_\nu\bar{\beta}_\mu)(\bar{\bar{\rho}} - \bar{h}\bar{\bar{\sigma}}) \quad (38i)$$

while, for the $P_\mu E\ P_\nu E\ P_H E\ C_H$ mode, it is

$$\begin{aligned}\pi(r, \bar{h}) = {} & r^k(\rho - \bar{h}\sigma) + \bar{h}(\rho^*\sigma - \rho\sigma^*) + \bar{h}\beta_\nu[r^k(\bar{\rho} - \bar{h}\bar{\sigma}) + \bar{h}(\rho^*\bar{\sigma} - \bar{\rho}\sigma^*)] \\ & + \bar{h}^2\beta_\nu\bar{\beta}_\mu[r^k(\bar{\bar{\rho}} - \bar{h}\bar{\bar{\sigma}}) + \bar{h}(\rho^*\bar{\bar{\sigma}} - \bar{\bar{\rho}}\sigma^*)] \\ & + \bar{h}^2\beta_\mu^*[(\bar{\bar{\rho}}\sigma - \rho\bar{\bar{\sigma}}) + \bar{h}\beta_\nu(\bar{\rho}\bar{\bar{\sigma}} - \bar{\bar{\rho}}\bar{\sigma})],\end{aligned} \quad (38ii)$$

where we have written ρ for $\rho(r)$, etc., throughout. On comparing (36) with (14), (15), and (16), it is evident that we recover the modes $P_\nu E\ P_H E\ C_H(E)$ from (36) by setting

$$\beta_\mu^* = \beta_\nu^*, \qquad \bar{\beta}_\mu = 0, \qquad \bar{\bar{\rho}}(\zeta) \equiv \bar{\rho}(\zeta), \qquad \bar{\bar{\sigma}}(\zeta) \equiv \bar{\sigma}(\zeta), \quad (39i)$$

since the P_μ and P_ν steps are then identical. If we further set

$$\beta_\nu^* = 0 \quad (39ii)$$

we recover the modes $P_\nu E\ P_k E\ C_H(E)$. By substituting (39) into (38), stability polynomials for all the modes we have discussed can be obtained.

Intervals of absolute and relative stability are defined as in section 3.6, and can be established by any of the methods discussed in sections 3.7 and 3.11. Due to the increased complexity of the stability polynomials, it is doubtful whether methods other than the root locus and boundary locus methods are practicable, except in the simplest modes. (See exercise 15 below.) The literature appears to contain virtually no quantitative results on stability intervals for hybrid methods, and it is incumbent on the user to compute them himself—but see example 2.

Example 2 Show that the $P_\nu E\ P_H E$ algorithm for which P_ν is the fourth-order predictor

$$y_{n+2\cdot7} - 2\cdot267{,}025y_{n+2} - 3\cdot5721y_{n+1} + 4\cdot839{,}125y_n = h(f_{n+2} - 4\cdot6751f_{n+1} - 1\cdot731{,}05f_n)$$

and P_H is the hybrid formula (see (30))

$$y_{n+3} - y_{n+2} = \frac{h}{714}(221f_{n+2} - 7f_{n+1} + 500f_{n+2\cdot7})$$

is absolutely stable for $\bar{h} = 0\cdot2$, and absolutely unstable for $\bar{h} = 2\cdot0$. Illustrate by using the algorithm with $h = 0\cdot01$ and $0\cdot1$ to solve the initial value problem $y' = -10(y-1)^2$, $y(0) = 2$.

Application of (35) yields the stability polynomial

$$\pi(r, \bar{h}) = r^3 + A(\bar{h})r^2 + B(\bar{h})r + C(\bar{h}),$$

where

$$A(\bar{h}) = -1 - 1{\cdot}897\bar{h} - 0{\cdot}700\bar{h}^2,$$
$$B(\bar{h}) = -2{\cdot}492\bar{h} + 3{\cdot}274\bar{h}^2,$$

and

$$C(\bar{h}) = 3{\cdot}389\bar{h} + 1{\cdot}212\bar{h}^2.$$

We find that

$$\begin{aligned}\pi(r, 0{\cdot}2) &= r^3 - 0{\cdot}649r^2 + 0{\cdot}629r - 0{\cdot}630\\ &= (r - 0{\cdot}82)(r^2 + 0{\cdot}171r + 0{\cdot}769).\end{aligned}$$

Thus $\pi(r, 0{\cdot}2)$ has a real root at 0·82, and a pair of complex roots on the circle of radius $\sqrt{0{\cdot}769}$. All of these roots lie within the unit circle, and the algorithm is thus absolutely stable when $\bar{h} = 0{\cdot}2$. Also,

$$\begin{aligned}\pi(r, 2{\cdot}0) &= r^3 - 0{\cdot}006r^2 + 18{\cdot}080r - 1{\cdot}930.\\ &= (r - 0{\cdot}107)(r^2 + 0{\cdot}101r + 18{\cdot}091),\end{aligned}$$

so that $\pi(r, 2{\cdot}0)$ has a real root at 0·107 and a pair of complex roots on the circle of radius $\sqrt{18{\cdot}091}$. The algorithm is thus clearly absolutely unstable when $\bar{h} = 2{\cdot}0$.

For the given initial value problem, $\partial f/\partial y$ is initially -20, and thus, in the immediate neighbourhood of the origin, we estimate $\bar{h}$ to be 0·2, 2·0 when $h = 0{\cdot}01, 0{\cdot}1$ respectively. The errors in the numerical solutions given by the algorithm with these two steplengths are shown in table 21. Starting values are taken from the theoretical solution $y(x) = 1 + 1/(1 + 10x)$.

Table 21

x	Errors in numerical solutions	
	$h = 0{\cdot}01$	$h = 0{\cdot}1$
0·03	567×10^{-8}	—
0·04	723×10^{-8}	—
0·05	491×10^{-8}	—
0·10	379×10^{-8}	—
0·20	186×10^{-8}	—
0·30	106×10^{-8}	0·118
0·40	68×10^{-8}	0·072
0·50	48×10^{-8}	0·287
0·60	34×10^{-8}	0·361
0·70	27×10^{-8}	1·929
0·80	21×10^{-8}	40·131
0·90	17×10^{-8}	2×10^6

Note that for both solutions the local truncation errors are initially large. (At the origin, $y^{(5)}(x)$ is very large indeed.) As x increases, the local truncation errors grow smaller, and the initial errors are damped out when $h = 0{\cdot}01$, but not when $h = 0{\cdot}1$, corroborating our finding that the former corresponds to $\bar{h}$ being inside, and the latter to it being outside, an interval of absolute stability.

Exercises

13. Show that, if we permit $\nu = k$, the k-step $P_\nu E\, P_H E$ mode becomes identical with a conventional k-step predictor–corrector $PECE$ mode. The stability polynomial of the latter is, by section 3.11, $\pi(r, \bar{h}) = R(r) - \bar{h}S(r) + \bar{h}\beta_k[R^*(r) - \bar{h}S^*(r)]$, where the predictor has characteristic polynomials R^*, S^*, and the corrector R, S. On comparing this with (35), we note that ρ and R both have degree k but that σ has degree $k - 1$ while S has degree k. Explain why the stability polynomials are nevertheless identical. Why is it not possible to define a hybrid $P_\nu E\, P_H$ mode which would coincide with a conventional PEC mode when $\nu = k$?

14. Use (38) to find the stability polynomial of the hybrid algorithm derived in exercise 9 (page 174), and demonstrate that it is identical with the stability polynomial of Kutta's third-order rule, as derived in section 4.7.

15. The third-order predictor (P_ν)

$$y_{n+1{\cdot}7} + 1{\cdot}156y_{n+1} - 2{\cdot}156y_n = h(2{\cdot}023f_{n+1} + 0{\cdot}833f_n)$$

is used with the explicit hybrid formula (P_H) defined in example 2 in $P_\nu E\, P_H E$ mode. Use the Schur or Routh–Hurwitz criterion to show that the interval of absolute stability is $(-\frac{1}{2}, 0)$.

5.5 Generalizations

There is no reason why hybrid formulae should make use of only one off-step point. Formulae involving two off-step points have been derived and studied by Butcher[21] and Brush, Kohfeld, and Thompson.[10] Multistep methods with an arbitrary number of off-step points have been considered by Lyche.[125] In another development, Kohfeld and Thompson[88] consider the implementation of hybrid methods in a Nordsieck formulation (see section 3.14). Space does not permit us to present these generalizations here.

In this chapter, we have not proved any theorems on the convergence (in the sense of section 2.5) of hybrid methods. Nor have we done so, in chapter 3, for predictor–corrector methods in modes other than that of correcting to convergence. Such theorems may be found in the work of Butcher[20] and Watt,[180] who consider general classes of methods which include as special cases all the methods considered so far. Gear[51] also presents a general matrix formulation which facilitates examination of the equivalence of methods of different classes.

5.6 Comparison with linear multistep and Runge–Kutta methods

If a comparison on the basis of attainable order for a given number of function evaluations were to be made, in the spirit of section 4.9, hybrid methods would not fare particularly well, since most of the hybrid modes we have discussed call for three or four evaluations per step. However, such an approach ignores a major advantage of hybrid methods—one which we have emphasized in this chapter—namely that they can possess remarkably small error constants. Meaningful comparisons must therefore rest on numerical experiments. To date there have been no really comprehensive numerical tests comparing hybrid with other methods. Hybrid methods present considerable difficulties when the steplength has to be changed—the Nordsieck approach considered by Kohfeld and Thompson[88] is relevant here—and their implementation has not yet been developed to a stage comparable with that of conventional predictor–corrector methods.

6

Extrapolation methods

6.1 Polynomial extrapolation

In many situations in numerical analysis† we wish to evaluate a number A_0, but are able to compute only an approximation $A(h)$, where h is a positive discretization parameter (typically steplength) and where $A(h) \to A_0$ as $h \to 0$. Let us suppose that, for every fixed N, $A(h)$ possesses an asymptotic expansion of the form

$$A(h) = A_0 + A_1 h + A_2 h^2 + \dots + A_N h^N + R_N(h),$$
$$R_N(h) = O(h^{N+1}) \quad \text{as } h \to 0,$$

where the coefficients $A_0, A_1, \dots, A_N$ are independent of h. In future we shall summarize this statement by writing

$$A(h) \sim A_0 + A_1 h + A_2 h^2 + \dots. \tag{1}$$

Suppose that we have calculated $A(h_0)$ and $A(\frac{1}{2}h_0)$. Clearly, $A(h_0) = A_0 + O(h_0)$ and $A(\frac{1}{2}h_0) = A_0 + O(h_0)$ as $h \to 0$, whereas there exists a linear combination of $A(h_0)$ and $A(\frac{1}{2}h_0)$ which differs from A_0 by an $O(h_0^2)$ term. Specifically,

$$2A(\tfrac{1}{2}h_0) - A(h_0) = A_0 - \tfrac{1}{2}A_2 h_0^2 + \dots = A_0 + O(h_0^2). \tag{2}$$

The left-hand side of (2) is thus a "better" approximation than either $A(h_0)$ or $A(\frac{1}{2}h_0)$. This is the basic idea of *Richardson*[151] *extrapolation* (which we have already employed in section 4.6 to estimate the principal local truncation error of Runge–Kutta methods). It can be extended in several ways. Thus, if, in addition to $A(h_0)$ and $A(\frac{1}{2}h_0)$, we compute $A(\frac{1}{4}h_0)$, then we can find a linear combination of these three values which differs from A_0

† A recent survey of the application of extrapolation processes in numerical analysis has been compiled by Joyce.[83]

by an $O(h_0^3)$ term. Moreover, we need not consider only the sequence $h_0, \frac{1}{2}h_0, \frac{1}{4}h_0, \ldots$, but a general sequence $h_0, h_1, h_2, \ldots$ of values of h, where

$$h_0 > h_1 > h_2 > \ldots > h_S > 0. \tag{3}$$

In general, we can then find a linear combination with the property

$$\sum_{s=0}^{S} c_{s,S} A(h_s) = A_0 + O(h_0^{S+1}), \qquad h \to 0. \tag{4}$$

The forming of such *linear* combinations is essentially equivalent to *polynomial* interpolation at $h = 0$ of the data $(h_s, A(h_s))$, $s = 0, 1, \ldots, S$. (Since $h_s > 0$, $s = 0, 1, \ldots, S$, the process is strictly one of *extrapolation*.) If the extrapolation is performed in an iterative manner due originally to Aitken[2] and Neville,[136] it is possible to avoid computing the coefficients $c_{s,S}$ in (4). For each value of h_s (subject to (3)) compute $A(h_s)$ and denote the result by $a_s^{(0)}$. Let $I_{01}(h)$ be the unique polynomial of degree 1 in h which interpolates the points $(h_0, a_0^{(0)})$ and $(h_1, a_1^{(0)})$ in the h–$A(h)$ plane. This polynomial may be conveniently represented in terms of a 2×2 determinant as follows:

$$I_{01}(h) = \frac{1}{h_1 - h_0} \begin{vmatrix} a_0^{(0)} & h_0 - h \\ a_1^{(0)} & h_1 - h \end{vmatrix}. \tag{5}$$

(Clearly $I_{01}(h)$ so defined is indeed a polynomial of degree 1 in h and, moreover, $I_{01}(h_0) = a_0^{(0)}$, $I_{01}(h_1) = a_1^{(0)}$.) Let us denote by $a_0^{(1)}$ the result of extrapolating to $h = 0$ using this polynomial; that is, set $I_{01}(0) = a_0^{(1)}$. It follows easily from (1) that $a_0^{(1)} = A_0 + O(h_0^2)$ (and that, in the case when $h_1 = \frac{1}{2}h_0$, $a_0^{(1)}$ coincides with the left-hand side of (2)). Similarly we obtain a value $a_1^{(1)}$ $(= A_0 + O(h_1^2))$ by extrapolating to zero using the linear interpolant of the data $(h_1, a_1^{(0)})$ and $(h_2, a_2^{(0)})$, where

$$a_1^{(1)} = I_{12}(0), \qquad I_{12}(h) = \frac{1}{h_2 - h_1} \begin{vmatrix} a_1^{(0)} & h_1 - h \\ a_2^{(0)} & h_2 - h \end{vmatrix}.$$

Suppose we now denote by $I_{012}(h)$ the unique polynomial of degree 2 which interpolates the points $(h_0, a_0^{(0)}), (h_1, a_1^{(0)}), (h_2, a_2^{(0)})$ in the h–$A(h)$ plane. Then we may write

$$I_{012}(h) = \frac{1}{h_2 - h_0} \begin{vmatrix} I_{01}(h) & h_0 - h \\ I_{12}(h) & h_2 - h \end{vmatrix}$$

since (i) $I_{012}(h)$ is clearly a polynomial of degree 2 in h, (ii) $I_{012}(h_0) = I_{01}(h_0) = a_0^{(0)}, I_{012}(h_2) = I_{12}(h_2) = a_2^{(0)}$, and (iii) $I_{012}(h_1) = [(h_2 - h_1)a_1^{(0)} - (h_0 - h_1)a_1^{(0)}]/(h_2 - h_0) = a_1^{(0)}$. Extrapolating to zero by this polynomial, we define $a_0^{(2)} = I_{012}(0)$, and find that $a_0^{(2)} = A_0 + O(h_0^3)$. Note that $a_0^{(2)}$ is a *linear* combination of $a_0^{(0)}$, $a_1^{(0)}$, and $a_2^{(0)}$, so that we have found the required linear combination (4) in the case $S = 2$. The process can be continued to give higher-order approximations to A_0, and may be summarized by the following tableau.

$$\begin{array}{lllllll} h_0 & a_0^{(0)} & & & & & \\ h_1 & a_1^{(0)} & a_0^{(1)} & & & & \\ h_2 & a_2^{(0)} & a_1^{(1)} & a_0^{(2)} & & & \\ h_3 & a_3^{(0)} & a_2^{(1)} & a_1^{(2)} & a_0^{(3)} & & \\ \vdots & \vdots & \vdots & \vdots & \vdots & \vdots & \vdots \end{array}, \tag{6}$$

where

$$\left.\begin{aligned} a_s^{(0)} &= A(h_s), \\ a_s^{(m)} &= \frac{1}{h_{m+s} - h_s}\begin{vmatrix} a_s^{(m-1)} & h_s \\ a_{s+1}^{(m-1)} & h_{m+s} \end{vmatrix}, m = 1, 2, \ldots \end{aligned}\right\} s = 0, 1, 2, \ldots. \tag{7}$$

An equivalent form of (7), more suitable for computation, is

$$a_s^{(0)} = A(h_s), \qquad a_s^{(m)} = a_{s+1}^{(m-1)} + \frac{a_{s+1}^{(m-1)} - a_s^{(m-1)}}{h_s/h_{m+s} - 1}, \qquad m = 1, 2, \ldots, \quad s = 0, 1, 2, \ldots. \tag{8}$$

It is straightforward to show that $a_s^{(m)} = A_0 + O(h_s^{m+1})$.

The benefits of repeated extrapolation are clearly greatly enhanced if it happens that the asymptotic expansion for $A(h)$ contains only even powers of h (and this will be the case for some important applications to ordinary differential equations). For example, corresponding to (2), we will then have

$$\tfrac{4}{3}A(\tfrac{1}{2}h_0) - \tfrac{1}{3}A(h_0) = A_0 + O(h_0^4).$$

If the asymptotic expansion for $A(h)$ has the form

$$A(h) \sim A_0 + A_2h^2 + A_4h^4 + \ldots, \tag{9}$$

then the process of repeated polynomial extrapolation is described by (6), where now, in place of (8),

$$a_s^{(0)} = A(h_s), \qquad a_s^{(m)} = a_{s+1}^{(m-1)} + \frac{a_{s+1}^{(m-1)} - a_s^{(m-1)}}{(h_s/h_{m+s})^2 - 1}, \qquad m = 1, 2, \ldots, \quad s = 0, 1, \ldots. \tag{10}$$

We then have $a_s^{(m)} = A_0 + O(h_s^{2m+2})$. The algorithm defined by (6) and (10) (where (9) is assumed) is analysed by Gragg[59,60] who shows that if $A(h)$ is continuous from the right at $h = 0$, then a necessary and sufficient condition for the convergence of $\{a_0^{(n)}\}$ to A_0 as $n \to \infty$ is that $\sup_{n \geqslant 0} (h_{n+1}/h_n) < 1$. Further results by Gragg show that each column of (6) then converges to A_0 faster than the one to its left, and that if, in addition, $\inf_{n \geqslant 0} (h_{n+1}/h_n) > 0$, the principal diagonal $a_0^{(0)}, a_0^{(1)}, a_0^{(2)}, \ldots$ converges to A_0 faster than any column. Indeed, under mild conditions on $A(h)$, $\{a_0^{(n)}\}$ converges to A_0 *superlinearly*, in the sense that $|a_0^{(n)} - A_0| \leqslant K_n$ and $\lim_{n \to \infty} (K_{n+1}/K_n) = 0$.

Exercise

1. (Archimedes.) Let $A(h)$ be half the perimeter of a regular n-gon inscribed in a circle of unit radius, and let $nh = 1$. Show that $A(h)$ may be written in the form

$$A(h) = \pi + A_2 h^2 + A_4 h^4 + \ldots.$$

Hence obtain approximations to π by computing $A(h)$ for $h = \frac{1}{4}, \frac{1}{6}, \frac{1}{8}$.

Apply repeated polynomial extrapolation using (6) and (10) to obtain a better approximation to π.

6.2 Application to initial value problems in ordinary differential equations

The use of polynomial extrapolation in the numerical solution of both boundary and initial value problems in ordinary differential equations has been studied by several authors, notably Fox,[44] Fox and Goodwin,[45] Gragg,[59,60] and Pereyra.[143,144,145] For initial value problems, we may proceed as follows. For a given (fixed) discrete numerical method (linear multistep, Runge–Kutta, etc.), let $y(x; h)$ denote the approximation at x, given by the numerical method with steplength h, to the theoretical solution $y(x)$ of the initial value problem $y' = f(x, y)$, $y(x_0) = y_0$. We intend to use polynomial extrapolation to furnish approximations to $y(x)$ at the *basic points* $x_0 + jH, j = 0, 1, \ldots,$ where H is the *basic steplength.* (For a given problem and required accuracy, H will typically be large compared with appropriate steplengths for previously discussed methods.) We first choose a steplength $h_0 = H/N_0$, where N_0 is a positive integer (possibly 1), and apply the numerical method N_0 times starting from $x = x_0$ to obtain an approximation $y(x_0 + H; h_0)$ to the theoretical solution $y(x_0 + H)$. A second steplength $h_1 = H/N_1, N_1$ a positive integer greater than N_0, is chosen, and the method applied N_1 times, again starting from x_0, to yield the approximation $y(x_0 + H; h_1)$. Proceeding in this fashion for the sequence of steplengths $\{h_s\}$, where $h_s = H/N_s, \{N_s | s = 0, 1, \ldots, S\}$

being an increasing sequence of positive integers, we obtain the sequence of approximations $\{y(x_0 + H; h_s)|s = 0, 1, \ldots, S\}$ to $y(x_0 + H)$. (In practice, S is typically in the range 4 to 7.) *Provided that there exists, for the given numerical method, an asymptotic expansion of the form*†

$$y(x; h) \sim y(x) + A_1 h + A_2 h^2 + A_3 h^3 + \ldots, \tag{11}$$

then we can set $a_s^{(0)} = y(x_0 + H; h_s)$ in (6) and apply the process of repeated polynomial extrapolation using (8). Equation (10) of course replaces (8) in the case when the numerical method possesses an asymptotic expansion of the form

$$y(x; h) \sim y(x) + A_2 h^2 + A_4 h^4 + A_6 h^6 + \ldots. \tag{12}$$

We then take the last entry in the main diagonal of the tableau (6) as our final approximation to $y(x_0 + H)$, and denote it by $y^*(x_0 + H; H)$. To obtain a numerical solution at the next basic point $x_0 + 2H$, we apply the whole of the above procedure to the new initial value problem $y' = f(x, y)$, $y(x_0 + H) = y^*(x_0 + H; H)$. Note that (11) and (12) are tantamount to asymptotic expansions for the *global* truncation error (cf. our application of Richardson extrapolation in section 4.6, where only *local* truncation error was involved).

6.3 Existence of asymptotic expansions: Gragg's method

The success of extrapolation methods thus hinges on the existence of numerical methods which yield asymptotic expansions of the form (11) or, preferably, (12). The existence of such expansions had frequently, in the past, been tacitly assumed, but was first rigorously investigated by Gragg.[59,60] Among Gragg's findings is the following. Every explicit one-step method and every linear multistep method whose first characteristic polynomial $\rho(\zeta)$ has only one root (the principal root) on the unit circle yields an asymptotic expansion of the form (11). (Difficulties can arise when $\rho(\zeta)$ has more than one root on the unit circle.)

In the interests of computational efficiency, we are, however, much more interested in finding methods which yield expansions of the form (12). Gragg shows that linear k-step methods whose characteristic polynomials $\rho(\zeta)$ and $\sigma(\zeta)$ satisfy the symmetry requirement

$$\rho(\zeta) + \zeta^k \rho(\zeta^{-1}) \equiv 0, \qquad \sigma(\zeta) - \zeta^k \sigma(\zeta^{-1}) \equiv 0 \tag{13}$$

† Note that as $h \to 0$ (11) implies the convergence of the numerical method. We shall assume throughout the remainder of this chapter that all numerical methods considered are consistent, and, in the case of linear multistep methods, zero-stable.

can yield such expansions, provided that the necessary additional starting values y_μ, $\mu = 1, 2, \ldots, k - 1$, can be supplied in a way which preserves the symmetry. This proviso does not, of course, apply in the case of a linear one-step method; we easily find that the only such (consistent) method satisfying (13) is the Trapezoidal rule, which does indeed possess an asymptotic expansion of the form (12). Unfortunately, this method is implicit, and it is necessary to solve it *exactly* at each step if the expansion (12) is to be valid. An explicit two-step method (of Nyström type) which satisfies (13) is the mid-point rule, $y_{n+2} - y_n = 2hf_{n+1}$, which has order 2. An important result of Gragg's shows that if, for this method, the additional starting value y_1 is chosen to be $y_0 + hf_0$, then the resulting method has an asymptotic expansion of the form (12), provided that the number of times the method is applied is always even or always odd; that is, in the terminology of section 6.2, provided that the increasing sequence $\{N_s | s = 0, 1, 2, \ldots, S\}$ contains only even or only odd positive integers. There is some advantage (see Gragg[60]) in choosing the N_s to be even, and this we shall do. (Note that the mid-point rule, being of order 2, has *local* truncation error of the form $C_3 h^3 y^{(3)}(x_n) + O(h^4)$. That the *global* truncation error is, by (12), of the form $-A_2 h^2 + O(h^4)$ is consistent with our discussion in section 3.5.)

It is easily checked (see exercise 2 below) that the mid-point rule has no interval of absolute stability. As a result, the coefficients A_{2m}, $m = 1, 2, \ldots$ in (12) (which are functions of x) contain components of the truncation error which can increase exponentially, even when the solution $y(x)$ decays exponentially. This potential instability can be controlled (to the extent that the leading coefficient A_2 becomes independent of such unstable components) by applying, at the end of the basic step, a smoothing procedure similar to that used by Milne and Reynolds[132,133] in conjunction with Milne's method (see section 3.13). Moreover, this can be done without destroying the form (12) of the asymptotic expansion. *Gragg's method* (or, as it is sometimes called, the *modified mid-point method*) is thus defined as follows.

$$
\begin{aligned}
h_s &= H/N_s, \qquad N_s \text{ even}, \\
y_0 &= y(x_0), \\
y_1 &= y_0 + h_s f(x_0, y_0), \qquad\qquad (14) \\
y_{m+2} - y_m &= 2h_s f(x_{m+1}, y_{m+1}), \qquad m = 0, 1, 2, \ldots, N_s - 1, \\
y(x_0 + H; h_s) &= \tfrac{1}{4} y_{N_s+1} + \tfrac{1}{2} y_{N_s} + \tfrac{1}{4} y_{N_s-1}.
\end{aligned}
$$

If (14) is repeated for an increasing sequence N_s, $s = 0, 1, \ldots, S$, of even integers, polynomial extrapolation, using (6) and (10), can be applied as described in section 6.2. Two popular choices for the sequence $\{N_s\}$ are $\{2, 4, 6, 8, 12, 16, 24, \ldots\}$ and $\{2, 4, 8, 16, 32, 64, \ldots\}$.

Recently, Stetter[174] has investigated the existence of asymptotic expansions of the form (12) for general implicit one-step methods of the form

$$y_{n+1} - y_n = h\phi(x_n, x_{n+1}, y_n, y_{n+1}, h), \tag{15}$$

and shows that if the function ϕ satisfies the symmetry requirement

$$\phi(s, t, \eta, \zeta, h) = \phi(t, s, \zeta, \eta, -h) \tag{16}$$

then (15) possesses an asymptotic expansion of the form (12). Clearly, the Trapezoidal rule satisfies this requirement, as does the *implicit mid-point method*,

$$y_{n+1} - y_n = hf\left(\frac{x_n + x_{n+1}}{2}, \frac{y_n + y_{n+1}}{2}\right).$$

This method, like the Trapezoidal rule, suffers the disadvantage that it must be solved exactly for y_{n+1} at each step, if the asymptotic expansion is to remain valid. Stetter also shows that it is possible to interpret the two-step Gragg method for a scalar y as a one-step method for a two-dimensional vector variable $\mathbf{w} = [u, v]^T$ which satisfies the differential system $u' = f(x, v)$, $v' = f(x, u)$, $u(x_0) = v(x_0) = y_0$, which clearly has the solution $u(x) = v(x) = y(x)$. This one-step method is of the form (15) and satisfies (16), whence the existence of an asymptotic expansion of the form (12). This approach can be generalized to produce a two-step method analogous to (14) but which has order 4. The new method is, unfortunately, implicit, and the result is of more theoretical than practical interest. For practical purposes, (14) still remains easily the most appropriate numerical method on which to base an extrapolation algorithm.

Example 1 *Apply Gragg's method (14), with polynomial extrapolation for the sequence* $N_s = 2, 4, 6, 8, 12$, *to solve the initial value problem* $y' = -y$, $y(0) = 1$, *for one basic step of length* $H = 1{\cdot}0$. *Compare with the fourth-order Runge–Kutta method (equation (20) of chapter 4) on the basis of accuracy for a given number of function evaluations.*

Applying (14) with steplengths $h_s = H/N_s$, we obtain the following values:

$y(1; \frac{1}{2}) = 0{\cdot}375{,}000{,}00$, $y(1; \frac{1}{4}) = 0{\cdot}371{,}093{,}75$, $y(1; \frac{1}{6}) = 0{\cdot}369{,}455{,}88$,

$y(1; \frac{1}{8}) = 0{\cdot}368{,}796{,}83$, $y(1; \frac{1}{12}) = 0{\cdot}368{,}297{,}12$.

The tableau (6) can now be constructed using (10); it is displayed in table 22. Each of the entries $a_s^{(m)}$ is an approximation to the theoretical solution $y(1) = \exp(-1)$; the entries in brackets denote the global error, $y(1) - a_s^{(m)}$, multiplied by 10^8.

Table 22

$h_0 = \frac{1}{2}$	0·375,000,00 (−712,056)				
$h_1 = \frac{1}{4}$	0·371,093,75 (−321,431)	0·369,791,67 (−191,223)			
$h_2 = \frac{1}{6}$	0·369,455,88 (−157,644)	0·368,145,58 (−26,614)	0·367,939,82 (−6038)		
$h_3 = \frac{1}{8}$	0·368,796,83 (−91,739)	0·367,949,48 (−7004)	0·367,884,11 (−467)	0·367,880,40 (−96)	
$h_4 = \frac{1}{12}$	0·368,297,12 (−41,768)	0·367,897,35 (−1791)	0·367,879,98 (−54)	0·367,879,46 (−2)	0·367,879,43 (1)

The results are clearly consistent with the assertions of section 6.1 on the rates of convergence of the columns and the principal diagonal. Note that the error in the final extrapolated value is roughly 4×10^4 times smaller than the error in the most accurate of the original computed solutions $y(1; h_s)$.

The first application of (14) clearly costs $N_0 + 1$ evaluations of the function $f(x, y)$. Subsequently, each application of (14) for given N_s will cost N_s evaluations (since the evaluation of $f(x_0, y_0)$ need not be repeated). Thus, to compute $y(1; h_s)$ for the first 3, 4, and 5 members of the sequence $\{\frac{1}{2}, \frac{1}{4}, \frac{1}{6}, \frac{1}{8}, \frac{1}{12}\}$ of steplengths, costs respectively 13, 21, and 33 evaluations of f. Since the fourth-order Runge–Kutta method costs 4 evaluations per step, we shall compute solutions by it, using steplengths $\frac{1}{3}, \frac{1}{5}$, and $\frac{1}{8}$, which will cost 12, 20, and 32 evaluations respectively. The errors in the two processes are compared below.

Polynomial extrapolation		*Runge–Kutta*	
Evaluations	*Error* $\times 10^8$	*Evaluations*	*Error* $\times 10^8$
13	−6038	12	−5002
21	−96	20	−580
33	1	32	−83

The superiority of the extrapolation method asserts itself only when $S \geqslant 2$. (Recall our earlier remark that, in practice, S is typically in the range 4 to 7.)

Exercises

2. Show that the mid-point rule has no interval of absolute stability. Show also that all linear multistep methods which satisfy (13) are absolutely stable either for no $\bar{h}$ or for all $\bar{h} < 0$ (see exercise 11 of chapter 3, page 83).

3. For $N_s = 2$, write (14) in the form of an explicit Runge–Kutta method with steplength H. Use equations (17) of chapter 4 to corroborate that the method has order 2.

4. Use the results of section 4.7 applied to the Runge–Kutta method derived in exercise 3 to establish that the interval of absolute stability of (14) with $N_s = 2$ is approximately $(-3{\cdot}1, 0)$. (Comparing this result with that of exercise 2 demonstrates the stabilizing effect of the smoothing procedure incorporated in the last line of (14).)

5. Demonstrate the necessity of the special starting procedure of Gragg's method by repeating the calculations of example 1, but replacing the value for y_1 given in (14) by a more accurate starting value given by a Runge–Kutta or a Taylor series method.

6. Euler's rule possesses an asymptotic expansion of the form (11), not (12). Use Euler's rule followed by polynomial extrapolation to solve the problem of example 1, constructing the tableau (6) according to (i) the appropriate equations (8), and (ii) the inappropriate equations (10). Compare the results.

6.4 Weak stability

As we have seen, the mid-point method has no interval of absolute stability; but, due to the smoothing procedure incorporated in the last line, the algorithm (14) (regarded as a one-step method with steplength H and parameter N_s) has a non-vanishing interval of absolute stability (see exercise 4 above). However, we cannot deduce from these intervals any useful information on the weak stability properties of the overall method consisting of (14), followed by polynomial extrapolation, since the latter forms linear combinations of the $y(x_0 + H; h_s)$, $s = 0, 1, \ldots, S$, the coefficients in the combinations sometimes being negative.

Stetter[173] adopts a new approach by computing the perturbation at $x_0 + H$ which results on introducing unit perturbations in each step of (14) applied to the linear differential equation $y' = \lambda y$, and then forming the linear combinations (but using absolute values of the coefficients) which correspond to the extrapolation process. This is done for a 'worst possible' and a 'natural' propagation of error. (The latter corresponds to the 1 in the requirement $|r_s| < 1$ for absolute stability as defined in section 3.6. Note that the use of the linear equation $y' = \lambda y$ is consistent with our treatment of weak stability in section 3.6.) For a given sequence $\{N_s\}$, the overall perturbations at $x_0 + H$ are functions of $S + 1$, the number of elements of the sequence $\{N_s\}$ employed, and $\bar{H} = \lambda H$. If these perturbations are denoted by $M(\bar{H}, S)$ and $M_0(\bar{H}, S)$ in the 'worst possible' and the 'natural' case respectively, then an interval of absolute stability may be defined to be an interval of $\bar{H}$ for which $M(\bar{H}, S) \leqslant M_0(\bar{H}, S)$. Stetter has computed such intervals for Gragg's method with polynomial extrapolation for the two popular choices $F_1 = \{2, 4, 6, 8, 12, \ldots\}$ and $F_2 = \{2, 4, 8, 16, \ldots\}$ for the sequence $\{N_s\}$. For these two sequences, we quote the approximate value of the left-hand end point, α, of the interval (which in all cases extends to $\bar{H} = 0$).

F_1:	N_s:	2	4	6	8	12	16	24	32	48	64
	$-\alpha$:	3·3	3·9	4·3	4·7	5·1	5·6	6·1	6·7	7·1	7·8

F_2:	N_s:	2	4	8	16	32	64	128
	$-\alpha$;	3·3	3·9	4·7	5·6	6·7	7·8	9·1

At first sight, these results might suggest that the weak stability properties of extrapolation methods compare very favourably with those of other methods we have studied; but we must remember that the intervals quoted are for $\bar{H} = \lambda H$, H being the *basic* steplength, not the individual steplength h_s. The motivation for extrapolation methods, after all, depends heavily on the possibility of choosing H to be large. Nevertheless, these results can be fairly described as generally encouraging, particularly since the numerical method behind the algorithm, the mid-point rule, has no interval of absolute stability. Thus, in example 1, our choice of $H = 1$ ($\bar{H} = -1$) is clearly acceptable; but $H = 10$ would have been dangerous, unless we were prepared to choose N_s very large (and consequently the individual steplength h_s very small).

Exercise

7. In some ways it would be fairer to consider intervals of absolute stability for $\bar{h}_S = \lambda h_S$, where h_S is the smallest individual steplength which will be called for in a given sequence $\{N_s | s = 0, 1, \ldots, S\}$. On this basis, compare the weak stability properties of Gragg's method, followed by polynomial extrapolation, with those of previously considered methods.

6.5 Rational extrapolation; the GBS method

When $A(h)$ possesses an asymptotic expansion of the form (1), the tableau (6) with elements defined by (8) corresponds to extrapolation to zero in the h–$A(h)$ plane by polynomials, the first extrapolated column of the tableau corresponding to polynomials of degree one, the second to polynomials of degree two, and so on. Bulirsch and Stoer[11] (see also Stoer[176]) have devised a similar tableau which corresponds to extrapolation to zero by rational functions of the form $P(h)/Q(h)$, $P(h)$ and $Q(h)$ being polynomials. In the new tableau, the first extrapolated column corresponds to P having degree zero and Q degree one, and subsequent columns correspond to the degree of $P(h)$ and of $Q(h)$ being alternately increased by one. The algorithm is adapted to the case when $A(h)$ has an asymptotic expansion of the form (9) in the same way as the polynomial extrapolation algorithm was adapted by replacing (8) by (10). In the case when the expansion (9) is assumed—the relevant case for our purposes—the Bulirsch–Stoer rational extrapolation algorithm is defined by the following tableau.

$$\begin{array}{ccccccc} h_0 & b_0^{(0)} & & & & & \\ h_1 & b_1^{(0)} & b_0^{(1)} & & & & \\ h_2 & b_2^{(0)} & b_1^{(1)} & b_0^{(2)} & & & \\ h_3 & b_3^{(0)} & b_2^{(1)} & b_1^{(2)} & b_0^{(3)} & & \\ \vdots & \vdots & \vdots & \vdots & \vdots & \vdots & \vdots, \end{array} \tag{17}$$

where

$$b_s^{(0)} = A(h_s), \; b_s^{(-1)} = 0,$$

$$b_s^{(m)} = b_{s+1}^{(m-1)} + \frac{b_{s+1}^{(m-1)} - b_s^{(m-1)}}{(h_s/h_{m+s})^2[1 - (b_{s+1}^{(m-1)} - b_s^{(m-1)})/(b_{s+1}^{(m-1)} - b_{s+1}^{(m-2)})] - 1}, \quad m = 1, 2, \ldots \quad s = 0, 1, 2, \ldots . \tag{18}$$

(The appropriate tableau for the case when $A(h)$ has an expansion of the form (1) is obtained by replacing $(h_s/h_{m+s})^2$ in (18) by h_s/h_{m+s}.)

This algorithm is supported by fewer theoretical results than is the polynomial extrapolation algorithm. Gragg[60] shows that for the Bulirsch–Stoer tableau to exist and for each column to converge faster than the one to its left, it is necessary to assume, in addition to the assumptions made in the polynomial extrapolation case, that h_0 is sufficiently small and that a certain determinant is non-vanishing. If this last condition is not satisfied, it is possible that some columns will not be accelerated. There exist no known general results on the rate of convergence of the principal diagonal of the tableau. Nevertheless, in most practical cases the Bulirsch–Stoer tableau does converge and usually (but not always) gives better approximations than does the polynomial extrapolation algorithm.

A particularly successful algorithm consists of applying the rational extrapolation scheme, (17), (18) of Bulirsch and Stoer to the method (14) of Gragg. The resulting method—dubbed by Stetter[173] the GBS method—is one of the most efficient general purpose methods for the numerical solution of initial value problems. A full description of the method, including a step-control procedure, an ALGOL programme, and numerical tests, is given by Bulirsch and Stoer.[12] In the recent survey by Hull *et al.*[73] referred to in section 3.14, the GBS method emerges as the best of those tested when function evaluations are relatively inexpensive (quantitatively, when each function evaluation costs less than roughly twenty-five arithmetic operations per component).

Example 2 Use the GBS method to solve the problem of example 1 (page 192).

In (18), we set $b_s^{(0)} = y(1; h_s)$, $s = 0, 1, 2, 3, 4$, these values having been computed in example 1. The resulting tableau (17) is displayed in table 23, where the bracketed entries are the global errors multiplied by 10^8.

Table 23

$h_0 = \frac{1}{2}$	0·375,000,00 (−712,056)				
$h_1 = \frac{1}{4}$	0·371,093,75 (−321,431)	0·369,809,69 (−193,025)			
$h_2 = \frac{1}{6}$	0·369,455,88 (−157,644))	0·368,155,96 (−27,652)	0·367,592,27 (28,717)		
$h_3 = \frac{1}{8}$	0·368,796,83 (−91,739)	0·367,952,93 (−7349)	0·367,870,39 (905)	0·367,879,48 (−4)	
$h_4 = \frac{1}{12}$	0·368,297,12 (−41,768)	0·367,898,33 (−1889)	0·367,878,53 (91)	0·367,879,44 (0)	0·367,879,44 (0)

The superiority of rational over polynomial extrapolation when $S \geqslant 3$ is clear.

Exercise

8. (Gragg.) A variant of the rational extrapolation algorithm consists of setting $b_s^{(-1)} = \infty$ in (18). Show that the first extrapolated column $b_0^{(1)}, b_1^{(1)}, b_2^{(1)}, \ldots$ in (17) then coincides with the first extrapolated column of the polynomial extrapolation tableau defined by (6) and (10). Investigate the effect of incorporating this variant in the GBS method in example 2.

7

Methods for special problems

7.1 Introduction

In preceding chapters we have discussed various families of methods, all of which are intended for application to the general initial value problem $y' = f(x, y), y(x_0) = y_0$. When the initial value problem, or its solution, is known in advance to have some special property, then it is natural to ask whether we cannot devise some special numerical method which makes use of this property. The special property may be some analytical property of the function $f(x, y)$, or it may be that the solution of the real life problem, which the initial value problem models, is known to have a solution with special characteristics—is known to be periodic, for example. In this chapter we shall briefly describe some such methods for special problems. These will be presented with less detail and analytical examination than hitherto. (In some cases, analytical treatment analogous to that employed in earlier chapters is not even possible.) One of the most important classes of initial value problems with special properties is the class which exhibits 'stiffness'. However, since this phenomenon is essentially concerned with systems of differential equations, we shall not treat it in this chapter; we shall devote much of chapter 8 to this topic.

At the present time, general purpose methods have been developed to a high degree of efficiency, and it may well be that in the future the significant new results will arise in the area of special methods for special problems. (Indeed, the most significant new results of the last few years have been concerned with the problem of stiffness.) It is interesting to compare the development of numerical methods for ordinary differential equations with that of those for partial differential equations. (See, for example, Mitchell.[134]) In the latter, the development has been from the particular to the general, most of the early work being concerned with specific problems, frequently linear, with constant coefficients. Successful numerical methods for such problems frequently depend heavily on analytical properties of the partial differential equation. (One reason for this emphasis is, of course, that a linear, constant coefficient, partial differential equation satisfying given conditions on a boundary other than a geometrically simple one cannot, in general, be solved analytically, whereas a linear, constant coefficient, ordinary differential equation can.) As numerical analysts working in partial differential equations have broadened their interests to include classes of linear variable coefficient and non-linear problems, those working in ordinary differential equations, having found reasonably efficient methods for general non-linear problems, have been trying to identify, and treat more efficiently, specific classes of problems.

7.2 Obrechkoff methods

All of the methods we have discussed so far have called for us to do nothing more elaborate with the function $f(x, y)$ than evaluate it. The most obvious and direct way of incorporating more analytical information into a numerical method is to let it depend on the total derivatives, with respect to x, of $f(x, y)$, that is, on the higher derivatives of y, which can be obtained explicitly by formal differentiation. Thus, in the notation of section 4.3,

$$\begin{aligned} y^{(1)} &= f(x, y) = f, \\ y^{(2)} &= f_x + ff_y, \\ y^{(3)} &= f_{xx} + 2ff_{xy} + f^2 f_{yy} + f_y(f_x + ff_y), \\ &\vdots \qquad\qquad \vdots \end{aligned} \tag{1}$$

For many problems, such explicit differentiation is intolerably complicated (see exercise 1 of chapter 3); but when it is feasible to evaluate the first few total derivatives of y, then generalizations of linear multistep methods which employ such derivatives can be very efficient. Such methods are called *Obrechkoff methods*, although the original work of Obrechkoff[139] was concerned only with numerical quadrature, and it would appear that Milne[130] was the first to advocate the use of Obrechkoff formulae for the numerical solution of differential equations. (In fact, quadrature formulae involving higher derivatives go back to Hermite[69] (pages 438–439).)

The k-step Obrechkoff method using the first l derivatives of y may be written

$$\sum_{j=0}^{k} \alpha_j y_{n+j} = \sum_{i=1}^{l} h^i \sum_{j=0}^{k} \beta_{ij} y_{n+j}^{(i)}; \qquad \alpha_k = +1. \tag{2}$$

Order, error constant, and local truncation error are defined exactly as for linear multistep methods, through the operator $\mathscr{L}$, where

$$\mathscr{L}[y(x); h] = \sum_{j=0}^{k} \left[\alpha_j y(x + jh) - \sum_{i=1}^{l} h^i \beta_{ij} y^{(i)}(x + jh) \right].$$

A count of the available coefficients shows that order $kl + k + l - 1$ is attainable if the method is implicit, and order $kl + k - 1$ if it is explicit (that is, if $\beta_{ik} = 0, i = 1, 2, \ldots, l$). However, the coefficients $\alpha_j, j = 0, 1, \ldots, k$, must satisfy the usual zero-stability condition, and this may prevent the above orders being attained for some k. The coefficients α_j, β_{ij}, $i = 1, \ldots, l$, $j = 0, 1, \ldots, k$, and error constants are listed by Lambert and Mitchell[102] for explicit and implicit methods with $l = 1, 2, \ldots, 5 - k, k = 1, 2, 3, 4$. These are quoted in terms of parameters $a, b, c, \ldots$, after the style of the linear multistep methods quoted in section 2.10; by choosing the parameters appropriately, zero-stability can be satisfied. Obrechkoff methods are also quoted by Jankovič[81] and by Krückeberg.[96] The expression for maximal order, $kl + k + l - 1$, is clearly symmetric in k and l, and it might appear that a required order is as readily obtained by increasing k as by increasing l. However, apart from the difficulty of satisfying the zero-stability condition when k is large, there is a definite advantage in increasing l rather than k, namely that the error constants decrease more rapidly with increasing l than with increasing k. Thus, eighth order can be attained by an implicit method with $k = 4$, $l = 1$, in which case the error constant is $-1/2625$, or with $k = 1$, $l = 4$, when the error constant is $+1/25{,}401{,}600$.

Explicit and implicit Obrechkoff methods can be combined in predictor–corrector modes in a manner analogous to that discussed in section 3.9. Thus, for example, if we indicate by $y^{(i)}(x, y)$ the ith total derivative given by (1) expressed as a function of x and y only, then, with the usual convention that k is the steplength of the overall method and that coefficients of the predictor are marked *, we may define a general Obrechkoff *PECE* method by

$$\left.\begin{aligned} y_{n+k}^{[0]} + \sum_{j=0}^{k-1} \alpha_j^* y_{n+j}^{[1]} &= \sum_{i=1}^{l^*} h^i \sum_{j=0}^{k-1} \beta_{ij}^* [y_{n+j}^{(i)}]^{[1]}, \\ [y_{n+k}^{(i)}]^{[0]} &= y^{(i)}(x_{n+k}, y_{n+k}^{[0]}), \qquad i = 1, 2, \ldots, l, \\ \sum_{j=0}^{k} \alpha_j y_{n+j}^{[1]} &= \sum_{i=1}^{l} h^i \left\{ \beta_{ik} [y_{n+k}^{(i)}]^{[0]} + \sum_{j=0}^{k-1} \beta_{ij} [y_{n+j}^{(i)}]^{[1]} \right\}, \\ [y_{n+k}^{(i)}]^{[1]} &= y^{(i)}(x_{n+k}, y_{n+k}^{[1]}), \qquad i = 1, 2, \ldots, \max(l, l^*). \end{aligned}\right\} \quad (3)$$

Note that, as indicated in (3), it is not necessary for the number l^* of higher derivatives employed in the predictor to be the same as the number l employed in the corrector. Note also that the number of evaluations per step of the above *PECE* method is $l + \max(l, l^*)$. However, for the sort of problem for which Obrechkoff methods are appropriate, a count of the number of function evaluations per step is not a good estimate of the computing effort involved. Such problems will have fairly simple functions $f(x, y)$, and in some cases the higher derivatives will be easily evaluated in terms of lower derivatives. In such circumstances the major computing effort may well be in the overheads rather than in the function evaluations.

The results of section 3.10 on the local truncation error of conventional predictor–corrector methods apply without modification to Obrechkoff predictor–corrector methods. In particular, if the order of the predictor is at least that of the corrector, then the principal local truncation error of the overall method is that of the corrector alone. Moreover, if the predictor and the corrector have the same order p, then Milne's device is applicable, and the principal local truncation error may be estimated by equation (63) of chapter 3.

We now quote some specific Obrechkoff methods, from which fourth- and sixth-order predictor–corrector methods may be constructed. (Note that Taylor algorithms—that is, explicit Obrechkoff methods with $k = 1$—may also be used as predictors.)

Implicit, $k = 1, l = 2$

$$y_{n+1} - y_n = \frac{h}{2}(y^{(1)}_{n+1} + y^{(1)}_n) - \frac{h^2}{12}(y^{(2)}_{n+1} - y^{(2)}_n), \tag{4}$$

$$p = 4, \qquad C_5 = +\tfrac{1}{720}.$$

Implicit, $k = 1, l = 3$

$$y_{n+1} - y_n = \frac{h}{2}(y^{(1)}_{n+1} + y^{(1)}_n) - \frac{h^2}{10}(y^{(2)}_{n+1} - y^{(2)}_n) + \frac{h^3}{120}(y^{(3)}_{n+1} + y^{(3)}_n), \tag{5}$$

$$p = 6, \qquad C_7 = -\tfrac{1}{100800}.$$

Implicit, $k = 2, l = 2$ (*Adams type*)

$$\begin{aligned} y_{n+2} - y_{n+1} &= \frac{h}{240}(101y^{(1)}_{n+2} + 128y^{(1)}_{n+1} + 11y^{(1)}_n) \\ &\quad + \frac{h^2}{240}(-13y^{(2)}_{n+2} + 40y^{(2)}_{n+1} + 3y^{(2)}_n), \end{aligned} \tag{6}$$

$$p = 6, \qquad C_7 = +\tfrac{1}{9450}.$$

Explicit, $k = 2, l = 2$ (*Adams type*)

$$y_{n+2} - y_{n+1} = \frac{h}{2}(-y_{n+1}^{(1)} + 3y_n^{(1)}) + \frac{h^2}{12}(17y_{n+1}^{(2)} + 7y_n^{(2)}), \tag{7}$$

$$p = 4, \qquad C_5 = +\tfrac{31}{720}.$$

Explicit, $k = 2, l = 3$ (*Adams type*)

$$y_{n+2} - y_{n+1} = \frac{h}{2}(15y_{n+1}^{(1)} - 13y_n^{(1)}) - \frac{h^2}{10}(31y_{n+1}^{(2)} + 29y_n^{(2)})$$

$$+ \frac{h^3}{120}(111y_{n+1}^{(3)} - 49y_n^{(3)}), \tag{8}$$

$$p = 6, \qquad C_7 = +\tfrac{209}{100800}.$$

The weak stability of Obrechkoff methods may be investigated through an obvious extension of section 3.11. Thus, if we define the characteristic polynomials

$$\rho(r) = \sum_{j=0}^{k} \alpha_j r^j; \qquad \sigma_i(r) = \sum_{j=0}^{k} \beta_{ij} r^j, \qquad i = 1, 2, \ldots, l;$$

$$\rho^*(r) = \sum_{j=0}^{k} \alpha_j^* r^j; \qquad \sigma_i^*(r) = \sum_{j=0}^{k-1} \beta_{ij}^* r^j, \qquad i = 1, 2, \ldots, l^*,$$

then the stability polynomial for the *PECE* mode defined by (3) turns out to be

$$\pi(r, \bar{h}) = \rho(r) - \sum_{i=1}^{l} \bar{h}^i \sigma_i(r) + \left(\sum_{i=1}^{l} \bar{h}^i \beta_{ik}\right)\left[\rho^*(r) - \sum_{i=1}^{l^*} \bar{h}^i \sigma_i^*(r)\right], \tag{9}$$

an obvious extension of the result for linear multistep methods given in section 3.11. The literature contains very little information on the size of stability intervals for Obrechkoff methods, but they would appear to be rather small.

Obrechkoff methods, with steplengths comparable with those used with linear multistep methods of roughly the same stepnumber, will yield results of much higher accuracy, frequently necessitating the use of double-length arithmetic; they are thus appropriate to problems which require such accuracy, and which allow the differentiation (1) to be performed. Attempts to utilize the small error constants of Obrechkoff methods by applying them with large steplengths may be frustrated by the restrictions of weak stability; moreover, when h is large, Milne's device may seriously underestimate the local error (see Example 2 below).

Obrechkoff methods using predictor–corrector modes not directly analogous to those discussed in section 3.9 are proposed by Ceschino[24] (see exercise 5 below). Methods of Runge–Kutta type which involve higher derivatives are derived by Fehlberg.[42] Space does not permit us to quote these methods here.

Example 1 Construct a sixth-order PECE method based on (5) and (8). Use it to solve the problem $y' = -y$, $y(0) = 1$ from $x = 0$ to $x = 3$, using a steplength $h = 0{\cdot}375$, first checking that this choice corresponds to $\bar{h}$ lying within an interval of absolute stability. Compare the results with those obtained by the Runge–Kutta method (equation (20) of chapter 4) with steplength $h = 0{\cdot}25$. (Note that the Obrechkoff method costs six evaluations per step, and the Runge–Kutta four: thus the above choice of steplengths equalizes the number of evaluations over a given interval, in the spirit of example 7 of chapter 4, page 146).

The algorithm is readily constructed from (3), (5), and (8). When $h = 0{\cdot}375$, $\bar{h} = -0{\cdot}375$, and, using (9), we find that

$$\pi(r, -0{\cdot}375) = r^2 - 1{\cdot}2902r + 0{\cdot}3524.$$

whose roots are 0·898 and 0·393. Thus the method is absolutely stable when $h = 0{\cdot}375$.

Table 24

x	Errors $\times 10^8$	
	Obrechkoff: $h = 0{\cdot}375$	Runge–Kutta: $h = 0{\cdot}25$
0·75	34	−1421
1·50	90	−1343
2·25	98	−951
3·00	80	−599

Table 24 displays global errors in the solutions by the Obrechkoff method and by the Runge–Kutta method; the former is more accurate than the latter, even when steplengths have been adjusted to equalize the number of function evaluations. It is noticeable that the Runge–Kutta method damps out the global error more rapidly than does the Obrechkoff method. This is because the value $\bar{h} = -0{\cdot}25$ lies well within the interval, $(-2.78, 0)$, of absolute stability of the Runge–Kutta method, whereas the value $\bar{h} = -0{\cdot}375$ is close to the end-point of the interval of absolute stability of the Obrechkoff method (see exercise 2 below).

Example 2 Use Milne's device to estimate the principal local truncation error at the first step of the numerical solution, in example 1, by the Obrechkoff PECE method, and compare it with the actual error. Why is the estimate so bad?

Assuming that the necessary additional starting value y_1 coincides with the theoretical solution $y(h) = \exp(-0{\cdot}375)$, we find that $y_2^{[0]} = 0{\cdot}472{,}368{,}227$,

$y_2^{[1]} = 0{\cdot}472{,}366{,}209$, and $y(2h) = 0{\cdot}472{,}366{,}553$. Thus the actual local error (which coincides with the global error, since the localizing assumptions are satisfied) is $+344 \times 10^{-9}$. Using equation (63) of chapter 3, the estimate by Milne's device is

$$\frac{C_7}{C_7^* - C_7}(y_2^{[1]} - y_2^{[0]}) = \frac{-1}{210}(-0{\cdot}000{,}002{,}018) \approx 1 \times 10^{-8}.$$

On adapting the analysis of section 3.10 to the case of an Obrechkoff method in *PECE* mode, it emerges that

$$y(x_{n+k}) - y_{n+k}^{[1]} = \left[\sum_{i=1}^{l} h^i \beta_{ik} \frac{\partial y^{(i)}}{\partial y}\right][C_{p^*+1}^* h^{p^*+1} y^{(p^*+1)}(x_n) + O(h^{p^*+2})]$$
$$+ C_{p+1} h^{p+1} y^{(p+1)}(x_n) + O(h^{p+2}).$$

For the case in hand, $p = p^* = 6$, so that $y(x_{n+k}) - y_{n+k}^{[1]} = C_7 h^7 y^{(7)}(x_n) + O(h^8)$. However, the $O(h^8)$ term contains a term $h\beta_{1k}(\partial f/\partial y)C_7^* h^7 y^{(7)}(x_n)$, and since $h = 0{\cdot}375$, $C_7^*/C_7 = -209$, $\partial f/\partial y = -1$, $y^{(7)} = -y$, this term is not at all negligible compared with $C_7 h^7 y^{(7)}(x_n)$. Note that this phenomenon is not specifically concerned with Obrechkoff methods; it could arise in any application of Milne's device in which hC_{p+1}^* and C_{p+1} are of comparable magnitude. Thus Milne's device is always suspect when h is large. This example serves as a reminder that the principal local truncation error approximates the local truncation error only asymptotically as $h \to 0$.

Exercises

1. Repeat example 1, using a fourth-order Obrechkoff method based on (4) and (7). Choose steplengths for this and for the Runge–Kutta method so as to equalize the number of function evaluations per step.

2. Show that the Obrechkoff *PECE* method of example 1 is absolutely unstable when $\bar{h} = -\frac{1}{2}$.

3. Repeat example 2, but with (6) replacing (5) in the Obrechkoff method.

4. Let the solution of the initial value problem be locally represented by the polynomial $I(x) = ax^4 + bx^3 + cx^2 + dx + e$. It was shown in section 2.4 that Simpson's rule could be derived by eliminating a, b, c, d, and e between the six equations

$$I(x_{n+j}) = y_{n+j}, \qquad I'(x) = f_{n+j}, \qquad j = 0, 1, 2.$$

Show that similar elimination between the six equations

$$I(x_{n+j}) = y_{n+j}, \qquad I'(x_{n+j}) = y_{n+j}^{(1)}, \qquad I''(x_{n+j}) = y_{n+j}^{(2)}, \qquad j = 0, 1,$$

yields (4).

5. (Ceschino.[24]) Show that the following method has order six:

$$\bar{\bar{\bar{y}}}_{n+1} = y_n + h\bar{y}^{(1)}_n + \frac{h^2}{2}\bar{\bar{y}}^{(2)}_n + \frac{h^3}{6}\bar{\bar{\bar{y}}}^{(3)}_n,$$

$$\bar{\bar{y}}_{n+1} = y_n + h\bar{y}^{(1)}_n + \frac{h^2}{2}\bar{\bar{y}}^{(2)}_n + \frac{h^3}{24}(\bar{\bar{\bar{y}}}^{(3)}_{n+1} + 3\bar{\bar{\bar{y}}}^{(3)}_n),$$

$$\bar{y}_{n+1} = y_n + h\bar{y}^{(1)}_n + \frac{h^2}{20}(3\bar{\bar{y}}^{(2)}_{n+1} + 7\bar{\bar{y}}^{(2)}_n) + \frac{h^3}{60}(-2\bar{\bar{\bar{y}}}^{(3)}_{n+1} + 3\bar{\bar{\bar{y}}}^{(3)}_n),$$

$$y_{n+1} = y_n + \frac{h}{2}(\bar{y}^{(1)}_{n+1} + \bar{y}^{(1)}_n) + \frac{h^2}{10}(-\bar{\bar{y}}^{(2)}_{n+1} + \bar{\bar{y}}^{(2)}_n) + \frac{h^3}{120}(\bar{\bar{\bar{y}}}^{(3)}_{n+1} + \bar{\bar{\bar{y}}}^{(3)}_n),$$

where $\bar{y}^{(1)}_n = y^{(1)}(x_n, \bar{y}_n)$, etc. How many function evaluations per step are required? What is the interval of absolute stability?

7.3 Problems with oscillatory solutions

An interesting and important class of initial value problems which can arise in practice consists of problems whose solutions are known to be periodic, or to oscillate with a known frequency. If this frequency, or a reasonable estimate for it, is known *in advance*, then a class of methods based on trigonometrical polynomials, developed by Gautshi,[49] is particularly appropriate.

We recall (see exercise 6 of chapter 2, page 27) that the order p of a linear multistep method may be defined by the requirement

$$\mathscr{L}[x^r; h] \equiv 0, \qquad r = 0, 1, \ldots, p; \qquad \mathscr{L}[x^{p+1}; h] \not\equiv 0, \tag{10}$$

where $\mathscr{L}$, the associated linear difference operator, is defined by equation (17) of chapter 2. In other words, the linear difference operator $\mathscr{L}$, of order p, annihilates all algebraic polynomials of order $\leqslant p$. The methods developed by Gautschi similarly annihilate trigonometric polynomials up to a certain degree. Let the known or estimated period of the solution be T, and define the frequency $\omega = 2\pi/T$. Then the method

$$\sum_{j=0}^{k} \alpha_j y_{n+j} = h \sum_{j=0}^{k} \beta_j(v) f_{n+j}, \qquad v = \omega h, \qquad \alpha_k = +1 \tag{11}$$

is said to be of *trigonometric order q relative to the frequency ω* if the associated linear difference operator

$$\mathscr{L}_\omega[y(x); h] = \sum_{j=0}^{k} [\alpha_j y(x + jh) - h\beta_j(v) y'(x + jh)]$$

satisfies

$$\left.\begin{aligned}&\mathscr{L}_\omega[1;h] \equiv 0, \qquad \mathscr{L}_\omega[\cos(r\omega x);h] \equiv \mathscr{L}_\omega[\sin(r\omega x);h] \equiv 0,\\ &\qquad\qquad\qquad\qquad\qquad\qquad\qquad\qquad r = 1, 2, \ldots, q,\\ &\mathscr{L}_\omega[\cos((q+1)\omega x);h] \quad \text{and} \quad \mathscr{L}_\omega[\sin((q+1)\omega x);h]\\ &\qquad\qquad\qquad\qquad\qquad\qquad \text{not both identically zero.}\end{aligned}\right\} \tag{12}$$

Note that (11) is not a linear multistep method as defined in chapter 2, since the coefficients $\beta_j(v)$, $j = 0, 1, \ldots, k$ are functions of h.

Let us consider an implicit two-step method of the class (11). If we require it to be of trigonometric order 1, then the three unspecified coefficients α_0, $\beta_1(v)$, and $\beta_0(v)$ must satisfy the conditions

$$\begin{aligned}\mathscr{L}_\omega[1;h] &= 1 + \alpha_0 = 0,\\ \mathscr{L}_\omega[\cos(\omega x);h] &= \cos[\omega(x+h)] + \alpha_0 \cos(\omega x)\\ &\quad + h\omega\{\beta_1 \sin[\omega(x+h)] + \beta_0 \sin(\omega x)\} \equiv 0,\\ \mathscr{L}_\omega[\sin(\omega x);h] &= \sin[\omega(x+h)] + \alpha_0 \sin(\omega x)\\ &\quad - h\omega\{\beta_1 \cos[\omega(x+h)] + \beta_0 \cos(\omega x)\} \equiv 0.\end{aligned}$$

After some elementary manipulation, the solution of this set of equations is found to be

$$\alpha_0 = -1, \qquad \beta_0 = \beta_1 = [\tan(v/2)]/v, \quad \text{where} \quad v = \omega h,$$

yielding the following method, which can be regarded as the trigonometric equivalent of the Trapezoidal rule:

$$y_{n+1} - y_n = \frac{h\tan(v/2)}{v}(f_{n+1} + f_n). \tag{13}$$

On expanding $\tan(v/2)$ as a power series, we obtain the following alternative form:

$$y_{n+1} - y_n = h(\tfrac{1}{2} + \tfrac{1}{24}v^2 + \tfrac{1}{240}v^4 + \ldots)(f_{n+1} + f_n); \qquad v = \omega h. \tag{14}$$

We now establish the principal local truncation error of (14). Although the method is not a linear multistep method, we can proceed along lines similar to those of section 2.7 and define the local truncation error at x_n to be $\mathscr{L}_\omega[y(x_n);h]$, when $y(x)$ is the theoretical solution of the problem.

Then,

$$\mathscr{L}_\omega[y(x_n);h] = y(x_n+h) - y(x_n) - h(\tfrac{1}{2} + \tfrac{1}{24}v^2 + \tfrac{1}{240}v^4 + \ldots) \times [y'(x_n+h) + y'(x_n)]$$

$$= hy^{(1)}(x_n) + \frac{h^2}{2!}y^{(2)}(x_n) + \frac{h^3}{3!}y^{(3)}(x_n) + \ldots$$

$$-h[\tfrac{1}{2} + \tfrac{1}{24}\omega^2h^2 + \tfrac{1}{240}\omega^4h^4 + \ldots]$$

$$\times [2y^{(1)}(x_n) + hy^{(2)}(x_n) + \frac{h^2}{2!}y^{(3)}(x_n) + \ldots]$$

$$= -\tfrac{1}{12}h^3[y^{(3)}(x_n) + \omega^2 y^{(1)}(x)] + O(h^4).$$

Thus the 'order' of (14), in a sense analogous to that of a linear multistep method, is two. Following Gautschi, we shall refer to order in this sense as *algebraic order*. (That the trigonometric order is lower than the algebraic order is a consequence of the fact that each value of r ($\geqslant 1$) corresponds to two conditions in (12), but to only one in (10).)

Gautschi[49] proves the following result on the trigonometric order attainable by a method of class (11) of given stepnumber. Let $v = \omega h$ be given, and let k be a given even/odd integer: then, for any given set of coefficients $\alpha_j, j = 0, 1, \ldots, k$, subject to $\sum_{j=0}^k \alpha_j = 0$, there exists a unique explicit/implicit method of the class (11) whose trigonometric order is $q = \frac{1}{2}k/q = \frac{1}{2}(k+1)$. In both cases, when $\omega = 0$ the method will reduce to a linear multistep method of (algebraic) order $2q$. Coefficients of such methods, when the left-hand side is specified to be of Adams type, are derived by Gautschi for $k = 1, 2, 3, \ldots, 6$. The Adams type method in the case $k = 1$ has already been derived in (14); similar methods for the cases $k = 2, 3, 4$ are quoted below.

$k = 2$ (explicit):

$$\begin{aligned} y_{n+2} - y_{n+1} &= h(\beta_1 f_{n+1} + \beta_0 f_n), \\ \beta_1 &= \tfrac{3}{2}(1 - \tfrac{1}{4}v^2 + \tfrac{1}{120}v^4 + \ldots), \\ \beta_0 &= -\tfrac{1}{2}(1 + \tfrac{1}{12}v^2 + \tfrac{1}{120}v^4 + \ldots). \end{aligned} \tag{15}$$

Trigonometric order $q = 1$, algebraic order $p = 2$.
Local truncation error $= \frac{5}{12}h^3[y^{(3)} + \omega^2 y^{(1)}] + O(h^4)$.

$k = 3$ (implicit):

$$
\begin{aligned}
y_{n+3} - y_{n+2} &= h(\beta_3 f_{n+3} + \beta_2 f_{n+2} + \beta_1 f_{n+1} + \beta_0 f_n),\\
\beta_3 &= \tfrac{9}{24}(1 + \tfrac{1}{4}v^2 + \tfrac{11}{120}v^4 + \ldots),\\
\beta_2 &= \tfrac{19}{24}(1 - \tfrac{43}{228}v^2 + \tfrac{13}{360}v^4 + \ldots),\\
\beta_1 &= -\tfrac{5}{24}(1 - \tfrac{1}{12}v^2 - \tfrac{7}{72}v^4 + \ldots),\\
\beta_0 &= \tfrac{1}{24}(1 + \tfrac{11}{12}v^2 + \tfrac{193}{360}v^4 + \ldots).
\end{aligned}
\tag{16}
$$

Trigonometric order $q = 2$, algebraic order $p = 4$.
Local truncation error $= -\frac{19}{720}h^5[y^{(5)} + 5\omega^2 y^{(3)} + 4\omega^4 y^{(1)}] + O(h^6)$.

$k = 4$ (explicit):

$$
\begin{aligned}
y_{n+4} - y_{n+3} &= h(\beta_3 f_{n+3} + \beta_2 f_{n+2} + \beta_1 f_{n+1} + \beta_0 f_n),\\
\beta_3 &= \tfrac{55}{24}(1 - \tfrac{95}{132}v^2 + \tfrac{79}{792}v^4 + \ldots),\\
\beta_2 &= -\tfrac{59}{24}(1 - \tfrac{923}{708}v^2 + \tfrac{15647}{21240}v^4 + \ldots),\\
\beta_1 &= \tfrac{37}{24}(1 - \tfrac{421}{444}v^2 + \tfrac{1921}{13320}v^4 + \ldots),\\
\beta_0 &= -\tfrac{9}{24}(1 + \tfrac{1}{4}v^2 + \tfrac{11}{120}v^4 + \ldots).
\end{aligned}
\tag{17}
$$

Trigonometric order $q = 2$, algebraic order $p = 4$.
Local truncation error $= \frac{251}{720}h^5[y^{(5)} + 5\omega^2 y^{(3)} + 4\omega^4 y^{(1)}] + O(h^6)$.

If one compares the above methods with conventional linear multistep methods of Adams type (see section 2.10 for lists of coefficients), it is apparent that the coefficients of the trigonometric methods are $O(v^2)$ perturbations of those of the corresponding linear multistep methods, where $v = \omega h$. If we do not know ω precisely, it would thus appear to be safer to underestimate rather than overestimate it. A discussion of the effects of uncertainty in estimating ω can be found in Gautschi.[49]

Exercises

6. In the manner of section 2.4, the Trapezoidal rule may be derived by eliminating a, b, and c between the four equations

$$I(x_{n+j}) = y_{n+j}, \qquad I'(x_{n+j}) = f_{n+j}, \qquad j = 0, 1,$$

where $I(x) = ax^2 + bx + c$ is a local polynomial interpolant. Show that a similar approach applied to the local trigonometric interpolant $I(x) = a + b\cos\omega x + c\sin\omega x$ yields the method (13).

7. (Gautschi.[49]) Derive the two-step explicit Adams type trigonometric method in a form analogous to (13). Using expansions in powers of v for the appropriate trigonometric functions, show that the method reduces to the form given in (15).

8. Corroborate that the local truncation error of (17) is as stated, and show that the coefficient of h^5 in this error is independent of the origin of the Taylor expansions.

9. Let y^*_{n+3} be the estimate for $y(x_{n+3})$ given by the implicit three-step Adams–Moulton method. Write (16) in the form

$$y_{n+3} = y^*_{n+3} + v^2A + v^4B,$$

and, by using Taylor expansions, express A and B in the form $ah^r + O(h^{r+1})$. Hence, corroborate the local truncation error quoted in (16).

7.4 Problems whose solutions possess singularities

We saw in section 2.4 that linear multistep methods are essentially linked with local representation, by a *polynomial*, of the theoretical solution $y(x)$ of the initial value problem. The Obrechkoff methods discussed in section 7.2 are similarly linked with polynomial representation (see exercise 4, page 204). Specifically, if we seek an explicit k-step Obrechkoff method involving the first l total derivatives of y, we demand that $y(x)$ be locally represented by a polynomial $I(x)$ satisfying

$$\begin{aligned} I(x_{n+j}) &= y_{n+j}, \qquad j = 0, 1, \ldots, k, \\ I^{(i)}(x_{n+j}) &= y^{(i)}_{n+j}, \qquad j = 0, 1, \ldots, k-1, \quad i = 1, 2, \ldots, l. \end{aligned} \tag{18}$$

That is, we impose $k(l + 1) + 1$ conditions on $I(x)$. If $I(x)$ has degree r, it has $r + 1$ undetermined coefficients, and if we choose r such that

$$k(l + 1) + 1 = r + 2, \tag{19}$$

then we can eliminate the $k(l + 1)$ undetermined coefficients between the $k(l + 1) + 1$ conditions (18), yielding the appropriate Obrechkoff method. If a similar implicit Obrechkoff method is sought, then the l additional requirements $I^{(i)}(x_{n+k}) = y^{(i)}_{n+k}$, $i = 1, 2, \ldots, l$ are added to (18), giving $(k + 1)(l + 1)$ conditions in all; the degree r of $I(x)$ must then be chosen such that

$$(k + 1)(l + 1) = r + 2. \tag{20}$$

Let us now suppose that the theoretical solution $y(x)$ possesses a singularity. It is then particularly inappropriate to attempt to represent $y(x)$, in the neighbourhood of the singularity, by a polynomial. As a result, linear multistep methods and Obrechkoff methods give very poor results if we attempt to use them to pursue the numerical solution close to a singularity. (Runge–Kutta methods also perform poorly in such circumstances.) A natural step would appear to be the replacement of the *polynomial* $I(x)$ by a *rational* function $R(x)$, a rational function being far more appropriate for the representation of a function close to a singularity. Let us consider the rational function

$$R(x) = S(x)/T(x),$$

where $S(x)$ and $T(x)$ are polynomials of degree s and t respectively, and the coefficient of the highest power of x in $T(x)$ is taken to be unity. Clearly $R(x)$ contains $s + t + 1$ undetermined coefficients, and if we choose s and t such that

$$k(l + 1) + 1 = s + t + 2, \tag{21}$$

then the elimination process can be repeated, with $I(x)$ replaced by $R(x)$ in (18), to yield an explicit k-step *rational method* involving the first l total derivatives of y. Implicit rational methods can be derived in a similar fashion, but (21) will be replaced by

$$(k + 1)(l + 1) = s + t + 2 \tag{22}$$

for the same reason that (20) replaces (19) in the case of Obrechkoff methods.

The class of one-step explicit rational methods for which $t = 1$ will, by (21), involve the first $s + 1$ derivatives of y, and is given by

$$y_{n+1} - y_n = \sum_{i=1}^{s} \frac{h^i}{i!} y_n^{(i)} + \frac{h^{s+1}}{s!} \frac{y_n^{(s)} y_n^{(s+1)}}{(s + 1)y_n^{(s)} - h y_n^{(s+1)}}, \quad s = 0, 1, 2, \ldots, \tag{23}$$

where, in the case $s = 0$, the term $\sum_{i=1}^{s} (h^i/i!) y_n^{(i)}$ is taken to be zero. The local truncation error of (23), defined in an obvious manner, is

$$\frac{h^{s+2}}{(s + 2)!}\left[y^{(s+2)} - \frac{s + 2}{s + 1} \frac{(y^{(s+1)})^2}{y^{(s)}}\right] + O(h^{s+3}), \tag{24}$$

and we can thus say that (23) has order $s + 1$. Each method of the class (23) is seen to be a truncated Taylor series with a rational correcting term. Two-step explicit rational methods with $t = 1$ will, by (21) demand that s be even. The general method of this class is unwieldy to state, and we quote only the third-order method for which $s = 2$. It is

$$3y_{n+2} - 4y_{n+1} + y_n = \frac{2h}{3}(y_{n+1}^{(1)} + y_n^{(1)}) + \frac{4h^2}{3} \frac{(y_{n+1}^{(1)} - y_n^{(1)})^2}{3(y_{n+1} - y_n) + h(y_{n+1}^{(1)} + 2y_n^{(1)})},$$

and its local truncation error is

$$h^4[\tfrac{1}{2}y^{(4)} - \tfrac{2}{3}(y^{(3)})^2/y^{(2)}] + O(h^5).$$

We quote only one implicit rational method, namely the second-order method given by $k = l = s = t = 1$ (which values satisfy (22)). It is

$$y_{n+1} - y_n = h^2 \frac{y_{n+1}^{(1)} y_n^{(1)}}{y_{n+1} - y_n}, \tag{25}$$

and its local truncation error is

$$h^3\left[-\frac{y^{(3)}}{6} + \frac{(y^{(2)})^2}{4y^{(1)}}\right].$$

Rational methods of higher accuracy than those quoted above can be very unwieldy to write down; a selection may be found in Lambert and Shaw.[103]

It is important to appreciate that the methods discussed in this section are non-linear and, as such, are unsupported by any body of theory such as exists for linear multistep methods. It would appear that they can nonetheless give good results in practice, provided that they are applied with care. In particular, bizarre results may emerge if the steplength is such that a pole of the local interpolant $R(x)$ falls within the local range of application of the method. Fortunately, such an occurrence is easily detected, since it is accompanied by a change of sign of the denominator of the rational term on the right-hand side of the method.

Example 3 (Shaw.[163]) *The initial value problem $y' = 1 + y^2$, $y(0) = 1$, has the theoretical solution $y(x) = \tan(x + \pi/4)$; it thus has a singularity (a simple pole) at $x = \pi/4 \approx 0{\cdot}785$. Use the explicit rational method obtained by setting $s = 3$ in (23) to obtain a numerical solution for this problem in the range $0 \leqslant x \leqslant 0{\cdot}75$, and compare it with that obtained by using a comparable explicit one-step method based on polynomial representation. Repeat the comparison, using now the implicit rational method (25) and a comparable implicit linear multistep method. (Use the steplength $h = 0{\cdot}05$ in all cases.)*

Setting $s = 3$ in (23) yields the fourth-order method

$$y_{n+1} - y_n = hy_n^{(1)} + \frac{h^2}{2}y_n^{(2)} + \frac{h^3}{6}y_n^{(3)} + \frac{h^4}{6}\frac{y_n^{(3)}y_n^{(4)}}{4y_n^{(3)} - hy_n^{(4)}}. \tag{26}$$

The comparable method based on polynomial representation and using the same total derivatives of y is the fourth-order explicit Obrechkoff method

$$y_{n+1} - y_n = hy_n^{(1)} + \frac{h^2}{2}y_n^{(2)} + \frac{h^3}{6}y_n^{(3)} + \frac{h^4}{24}y_n^{(4)}, \tag{27}$$

For the given problem, there is no difficulty in computing the higher derivatives of y.

The second-order implicit rational method (25) involves no derivatives of y higher than the first, and may be written

$$y_{n+1} - y_n = h^2 \frac{f_{n+1} f_n}{y_{n+1} - y_n}. \tag{28}$$

The comparable second-order linear multistep method is, of course, the Trapezoidal rule

$$y_{n+1} - y_n = \frac{h}{2}(f_{n+1} + f_n). \tag{29}$$

Despite the fact that both (28) and (29) are implicit, it is possible, for this problem, to solve for y_{n+1} exactly at each step, since $f(x, y)$ is quadratic in y, and the usual formula for the roots of a quadratic may be used.

Table 25

x	Theoretical solution	Errors in numerical solutions			
		(26) (rational)	(27) (polynomial)	(28) (rational)	(29) (polynomial)
0	1·000,000,000	—	—	—	—
0·1	1·223,048,880	8×10^{-8}	3×10^{-6}	1×10^{-4}	5×10^{-4}
0·2	1·508,497,647	2×10^{-7}	1×10^{-5}	3×10^{-4}	2×10^{-3}
0·3	1·895,765,123	4×10^{-7}	5×10^{-5}	6×10^{-4}	4×10^{-3}
0·4	2·464,962,757	7×10^{-7}	2×10^{-4}	1×10^{-3}	1×10^{-2}
0·5	3·408,223,442	1×10^{-6}	7×10^{-4}	3×10^{-3}	3×10^{-2}
0·6	5·331,855,223	4×10^{-6}	5×10^{-3}	7×10^{-3}	2×10^{-1}
0·65	7·340,436,575	8×10^{-6}	2×10^{-2}	1×10^{-2}	5×10^{-1}
0·70	11·681,373,800	2×10^{-5}	1×10^{-1}	4×10^{-2}	3·4
0·75	28·238,252,850	1×10^{-4}	3·0	3×10^{-1}	—

Table 25 displays the theoretical solution and errors in the numerical solutions by each of (26), (27), (28), and (29). No entry is quoted for $x = 0{\cdot}75$ when method (29) is used; the solution here is so inaccurate that the square root involved in solving the quadratic becomes imaginary. The overall superiority of the rational methods over the corresponding polynomial methods is apparent. (Note that this superiority is not limited to some small neighbourhood of the singularity.)

By basing methods on more general non-polynomial interpolants, it is possible further to extend the idea of tailoring methods to fit particular problems. Lambert and Shaw[104] propose the local interpolant

$$Q(x) = \begin{cases} \sum_{i=0}^{L} a_i x^i + b|A + x|^N, & N \notin \{0, 1, \dots, L\}, \quad (30i) \\ \sum_{i=0}^{L} a_i x^i + b|A + x|^N \log|A + x|, & N \in \{0, 1, \dots, L\}, \quad (30ii) \end{cases}$$

where L is a positive integer, a_i, $i = 0, 1, \ldots, L$, and b are regarded as undetermined coefficients, and A and N are regarded as parameters which will be adapted to suit the particular problem in hand. Methods based on (30) will thus be called *adaptive methods*. If N is negative, $Q(x)$ possesses a singularity at $x = -A$, but since N is not necessarily an integer the class of possible singularities is larger than that produced by rational interpolants. If N is a positive integer $> L$, $Q(x)$ is a polynomial, while positive non-integer values for N give a new type of non-polynomial interpolant. Finally, if $N \in \{0, 1, \ldots, L\}$, the interpolant $Q(x)$ given by (30*i*) is degenerate in the sense that it can be written, in either of the intervals $x < -A$, $x > -A$ as a polynomial with less than $L + 2$ undetermined coefficients—whence the need for the alternative form (30*ii*).

Elimination of the $L + 2$ undetermined coefficients in (30) between the $L + 3$ conditions $y_{n+1} = Q(x_{n+1})$, $y_n^{(i)} = Q^{(i)}(x_n)$, $i = 0, 1, \ldots, L + 1$, yields the following class of one-step explicit adaptive methods, which it is assumed will be applied entirely in one of the intervals $x < -A, x > -A$.

$$y_{n+1} - y_n = \sum_{i=1}^{L} \frac{h^i}{i!} y_n^{(i)} + \frac{(A + x_n)^{L+1}}{\alpha_L^N} y_n^{(L+1)} \times \left[\left(1 + \frac{h}{A + x_n}\right)^N - 1 - \sum_{i=1}^{L} \frac{\alpha_{i-1}^N}{i!}\left(\frac{h}{A + x_n}\right)^i\right], \quad N \notin \{0, 1, \ldots, L\}, \tag{31i}$$

$$y_{n+1} - y_n = \sum_{i=1}^{L} \frac{h^i}{i!} y_n^{(i)} + \frac{(-1)^{L-N}(A + x_n)^{L+1} y_n^{(L+1)}}{N!(L - N)!} \times \left[\left(1 + \frac{h}{A + x_n}\right)^N \log\left(1 + \frac{h}{A + x_n}\right) - \sum_{i=1}^{L} \left\{\frac{h^i \alpha_{i-1}^N}{i!(A + x_n)^i} \sum_{j=0}^{i-1} \frac{1}{N - j}\right\}\right], \quad N \in \{0, 1, \ldots, L\}, \tag{31ii}$$

where $\alpha_r^m = m(m-1)\ldots(m-r)$, r a non-negative integer. (It can be shown that if $\tilde{N}$ is one of the integers $0, 1, \ldots, L$, then as $N \to \tilde{N}$, the right-hand side of (31*i*) tends to that of (31*ii*); this justifies the form chosen for (30*ii*).) Both (31*i*) and (31*ii*) have order $L + 1$ and local truncation error

$$\sum_{q=L+2}^{\infty} T_q \frac{h^q}{q!},$$

where

$$T_q = y_n^{(q)} - \frac{\alpha_{q-L-2}^{N-L-1}}{(A + x_n)^{q-L-1}} y_n^{(L+1)}. \tag{32}$$

If neither A nor N is known in advance, then we choose for them values for which $T_{L+2} = T_{L+3} = 0$. These values are, from (32),

$$\left.\begin{aligned} -A_{(n)} &= x_n - \frac{y_n^{(L+2)} y_n^{(L+1)}}{(y_n^{(L+2)})^2 - y_n^{(L+1)} y_n^{(L+3)}}, \\ N_{(n)} &= L + 1 + \frac{(y_n^{(L+2)})^2}{(y_n^{(L+2)})^2 - y_n^{(L+1)} y_n^{(L+3)}}. \end{aligned}\right\} \tag{33}$$

The subscripts indicate that $-A_{(n)}$ and $N_{(n)}$ are estimates for $-A$ and N respectively, based on derivatives of y evaluated at $x = x_n$, $y = y_n$.

The method can be made fully adaptive by proceeding as follows. The given initial condition $y(x_0) = y_0$ is used to evaluate $-A_{(0)}$ and $N_{(0)}$, using (33). These values are substituted for $-A$ and N respectively in (31) which is now solved to give y_1. With this value available, (33) affords newer estimates $-A_{(1)}$ and $N_{(1)}$, and these are now substituted for $-A$ and N in (31), which now yields y_2. Proceeding in this way, applying (33) and (31) alternately, a numerical solution—which we shall call the *initial solution*—is obtained together with sequences of estimates $\{-A_{(n)}\}$ and $\{N_{(n)}\}$ for $-A$ and N respectively. At each stage of the process, a test is made to see whether $N_{(n)} = \tilde{N}$, $\tilde{N} \in \{0, 1, \ldots, L\}$, in which case (31*ii*) rather than (31*i*) is used for the integration step. (In practice, this is assumed to be the case if $|N_{(n)} - \tilde{N}| < \varepsilon$, where ε is a pre-assigned tolerance.) If the sequences $\{-A_{(n)}\}$ and $\{N_{(n)}\}$ appear to converge to limits $-A^*$ and N^* respectively, then we assume that the theoretical solution $y(x)$ is locally well-represented by the interpolant $Q(x)$ defined by (30) with $-A$ and N replaced by $-A^*$ and N^* respectively. If N^* is negative, we thus assume that $y(x)$ has a singularity of the type $|A^* + x|^{N^*}$ at $x = -A^*$. (If $N_{(n)} \to -\infty$, we assume that $y(x)$ has an essential singularity at $x = -A^*$.) Thus, in addition to a numerical solution, the procedure can yield information on the nature and position of any singularities.

The integration process is terminated if at any stage $x_m + A_{(m)}$ becomes small, since the method (31) is invalid if the integration is pursued past $x = -A$. A further refinement consists of recomputing an *improved solution*, using (31) with *fixed* values for $-A$ and N, these values being taken to be the latest available estimates $-A_{(m)}$, $N_{(m)}$, obtained while finding the initial solution. Note that in computing the initial solution we have set $T_{L+2} = T_{L+3} = 0$ at each step, so that the method is effectively of order $L + 3$. In computing the improved solution, we correspondingly increase by 2 the value of L in (31), so that the same order of accuracy is maintained throughout. This will involve two more total derivatives of y, but these will already have been calculated in order to apply (33).

Example 4 Apply the procedure described above to the problem of example 3 (page 211). Use L = 1 for the initial solution, and L = 3 for the improved solution.

The steplength is chosen, as in example 3, to be $h = 0{\cdot}05$; ε is also chosen to be 0·05. For the given values of L, the overall method clearly has order four, as have the methods (26) and (27) tested for this problem in example 3. The results of the procedure outlined above are displayed in table 26. The sequences $\{-A_{(n)}\}$ and

Table 26

x	Theoretical solution y	$-A_{(n)}$	$N_{(n)}$	Initial solution (error)	Improved solution (error)
0	1·000,000,000	1·000,000,0	−2·000,000,0	—	—
0·1	1·223,048,880	0·871,052,4	−1·459,538,7	2×10^{-7}	2×10^{-8}
0·2	1·508,497,647	0·818,606,7	−1·209,581,0	5×10^{-7}	3×10^{-8}
0·3	1·895,765,123	0·797,042,8	−1·089,014,1	1×10^{-6}	6×10^{-8}
0·4	2·464,962,757	0·788,793,7	−1·032,812,0	3×10^{-6}	1×10^{-7}
0·5	3·408,223,442	0·786,114,1	−1·009,367,1	5×10^{-6}	2×10^{-7}
0·6	5·331,855,223	0·785,478,4	−1·001,612,6	1×10^{-5}	5×10^{-7}
0·65	7·340,436,575	0·785,415,0	−1·000,453,6	3×10^{-5}	1×10^{-6}
0·70	11·681,373,800	0·785,400,2	−1·000,071,2	7×10^{-5}	3×10^{-6}
0·75	28·238,252,850	0·785,398,7	−1·000,002,0	4×10^{-4}	4×10^{-5}

$\{N_{(n)}\}$ appear to be converging to 0·7854 and −1 respectively, thus indicating the presence of a simple pole at $x = 0{\cdot}7854$. The theoretical solution $y(x) = \tan(x + \pi/4)$ does indeed have a simple pole at $x = \pi/4 \approx 0{\cdot}785{,}398$. The improved solution is seen to be more accurate than that obtained in example 3, using the fourth-order rational method (26).

The main disadvantage of the adaptive methods described above is that they require total derivatives of y to be calculated. Lambert and Shaw[105] show that *multistep* methods based on (30) may also be derived, and for these it is possible to attain a suitably high order by increasing stepnumber rather than by requiring that higher derivatives be calculated. Unfortunately, high-order derivatives are still required in the equations, analogous to (33), which give estimates for $-A$ and N. However, Shaw[164] has developed a technique whereby these estimates can be made without calculating higher derivatives, and this results in the most practicable form of the adaptive procedure. Shaw's formulae are too unwieldy to quote here, and the interested reader is referred to the original paper.

Exercises

10. If either of A or N is known in advance, the other may be estimated by setting to zero T_{L+2} as defined by (32). Use this approach to adapt example 4 to the case where

(i) $y(x)$ is known in advance to have a simple pole somewhere.
(ii) $y(x)$ is known in advance to have a singularity at $x = \pi/4$.

11. If the sequence $\{N_{(n)}\}$ converges to zero, a logarithmic singularity is indicated. Illustrate this by using the procedure described above to solve the problem $xy' = y + 5x^2 \exp(y/5x)$, $y(1) = 0$, whose theoretical solution is

$$y(x) = -5x \log(2 - x).$$

12. If the sequence $\{N_{(n)}\}$ diverges to $-\infty$, an essential singularity is indicated. Illustrate this by using the procedure described above to solve the problem $(1 - x)y' = y \log y$, $y(0) = \exp(0{\cdot}2)$, whose theoretical solution is

$$y(x) = \exp[0{\cdot}2/(1 - x)].$$

13. (Lambert and Shaw.[104]) In the initial value problem $y' = f(x, y)$, $y(x_0) = y_0$, consider a change of dependent variable from y to η, where $\eta = y - b|A + x|^N$, $x \neq -A$. Consider the explicit method

$$\eta_{n+1} - \eta_n = \sum_{i=1}^{L} \frac{h^i}{i!} \eta_n^{(i)}$$

of order L for the numerical solution of the transformed problem. By direct substitution, show that this may be written in the form

$$y_{n+1} - y_n = \beta[(A + x_n + h)^N - (A + x_n)^N] + \sum_{i=1}^{L} \frac{h^i}{i!}[y_n^{(i)} - \beta\alpha_{i-1}^N(A + x_n)^{N-i}],$$

where α_r^m is defined as in the preceding section, and where $\beta = b$ if $x > -A$ and $\beta = (-1)^N b$ if $x < -A$. Evaluate β by setting the principal local truncation error to zero, and hence find an alternative derivation of (31*i*).

14. (Lambert and Shaw.[105]) Let $Q(x) = a + b|A + x|^N$, $N \neq 0$. By eliminating a, b, *and* A, derive the following method:

$$y_{n+1} - y_n = \frac{h}{N} \frac{f_{n+1}^{N/(N-1)} - f_n^{N/(N-1)}}{f_{n+1}^{1/(N-1)} - f_n^{1/(N-1)}}, \qquad N \neq 0.$$

Show that setting $N = 2, -1$, and $\frac{1}{2}$ yields a method which equates $(y_{n+1} - y_n)/h$ to the arithmetic, geometric, and harmonic means, respectively of f_{n+1} and f_n. Which *two* of these methods have we discussed previously?

8

First-order systems and the problem of stiffness

8.1 Applicability to systems

In practice, we are frequently faced with an initial value problem which involves not just a single first-order differential equation but a system of m simultaneous first-order differential equations. Such systems arise naturally in a large number of applications; as we have seen in section 1.5, they also arise in connection with the initial value problem

$$y^{(m)} = f(x, y^{(0)}, y^{(1)}, \ldots, y^{(m-1)}), \qquad y^{(t)}(a) = \eta_t, \qquad t = 0, 1, \ldots, m-1. \tag{1}$$

Indeed, with some exceptions to be discussed in chapter 9, the usual procedure for dealing with the problem (1) is to convert it, as described in section 1.5, to an equivalent initial value problem involving a system of m first-order equations.

In this chapter we consider, in place of the problem

$$y' = f(x, y), \qquad y(a) = \eta,$$

the problem

$${}^i y' = {}^i f(x, {}^1 y, {}^2 y, \ldots, {}^m y), \qquad {}^i y(a) = {}^i \eta, \qquad i = 1, 2, \ldots, m. \tag{2}$$

We shall call (2) the initial value problem for a *first-order system.* A vector notation for (2) may be introduced by defining the following vectors:

$$\mathbf{y} = [{}^1 y, {}^2 y, \ldots, {}^m y]^T, \qquad \mathbf{f} = [{}^1 f, {}^2 f, \ldots, {}^m f]^T,$$
$$\boldsymbol{\eta} = [{}^1 \eta, {}^2 \eta, \ldots, {}^m \eta]^T. \tag{3}$$

The problem (2) may now be written in the form

$$\mathbf{y}' = \mathbf{f}(x, \mathbf{y}), \qquad \mathbf{y}(a) = \boldsymbol{\eta}. \tag{4}$$

In sections 8.2 to 8.4 we shall consider the applicability to problem (4) of the methods and results discussed in previous chapters. Most of the work carries over with only simple and obvious notational changes; there are, however, a few important exceptions.

First let us consider what we mean by saying that a particular method or result is 'applicable to a (first-order) system'. Consider, for example, Euler's rule

$$y_{n+1} - y_n = hf_n \tag{5}$$

and the one-step explicit rational method

$$y_{n+1} - y_n = h\frac{y_n f_n}{y_n - hf_n} \tag{6}$$

obtained by setting $s = 0$ in equation (23) of chapter 7. Applying these to each of the scalar equations in (2) we obtain

$${}^i y_{n+1} - {}^i y_n = h\,{}^i f_n, \qquad i = 1, 2, \ldots, m, \tag{7}$$

and

$${}^i y_{n+1} - {}^i y_n = h\frac{{}^i y_n\,{}^i f_n}{{}^i y_n - h\,{}^i f_n}, \qquad i = 1, 2, \ldots, m, \tag{8}$$

where ${}^i f_n = {}^i f(x, {}^1 y_n, {}^2 y_n, \ldots, {}^m y_n)$. Augmenting (3) by the definitions $\mathbf{y}_n = [{}^1 y_n, {}^2 y_n, \ldots, {}^m y_n]^T$, $\mathbf{f}_n = [{}^1 f_n, {}^2 f_n, \ldots, {}^m f_n]^T$, we may write (7) in the form

$$\mathbf{y}_{n+1} - \mathbf{y}_n = h\mathbf{f}_n, \tag{9}$$

but we cannot write (8) in terms of vectors. (Nevertheless, we could compute a solution to (4) as easily by (8) as by (7). Indeed, when it comes to writing a programme for Euler's rule applied to (4), we essentially use (7) rather than (9).) We shall distinguish between these two cases by saying that (5) is *applicable to a system*, and that (6) is *component-applicable to a system*. A method or result which is applicable to a system may be written down for a system simply by replacing scalars by the appropriate vectors (and, occasionally, replacing moduli of scalars by norms of vectors); one which is only component-applicable cannot be so written down, but is still interpretable in terms of the components of the system. Not all of our previous results will fall into one of these two categories. An example is the equation (73*i*) of chapter 3,

$$h\frac{\partial f}{\partial y} \approx (y_{n+k}^{[m]} - y_{n+k}^{[m-1]})/\beta_k(y_{n+k}^{[m-1]} - y_{n+k}^{[m-2]}),$$

which estimates $h\,\partial f/\partial y$ in terms of iterates of a predictor–corrector method. Since, for the system (4), $\partial \mathbf{f}/\partial \mathbf{y}$ can only be interpreted as a matrix (the Jacobian), we cannot make this equation meaningful for a system; we thus say that it is *inapplicable to a system.*

Example 1 Demonstrate that (6) is indeed component-applicable to a system by setting up the difference equations which result on applying it to the first-order system arising from the problem $y'' = 2y(1 + y^2)$, $y(0) = 1$, $y'(0) = 2$.

Let $y = u$, $y' = v$; then the given initial value problem is equivalent to

$$u' = v, \qquad u(0) = 1,$$

$$v' = 2u(1 + u^2), \qquad v(0) = 2.$$

Applying (8) with ${}^1y = u$, ${}^2y = v$, ${}^1f = v$, and ${}^2f = 2u(1 + u^2)$, we obtain

$$u_{n+1} - u_n = h\frac{u_n v_n}{u_n - hv_n}, \qquad v_{n+1} - v_n = h\frac{2v_n u_n(1 + u_n^2)}{v_n - 2hu_n(1 + u_n^2)}.$$

Given $u_0 = 1$, $v_0 = 2$, we can obviously solve the above difference equations in a step-by-step manner to obtain the sequences $\{u_n\}$ and $\{v_n\}$.

8.2 Applicability of linear multistep methods

All the linear multistep and predictor–corrector methods derived in chapters 2 and 3 are applicable to systems. Our definitions of order, error constant, consistency, and zero-stability, depending as they do only on the coefficients of the method, are unaffected by the change from a single equation to a system. Note, however, that the scalar operator $\mathscr{L}$, defined in section 2.6, will be replaced by the vector-valued operator $\mathscr{L}$, where

$$\mathscr{L}[\mathbf{y}(x); h] = \sum_{j=0}^{k} [\alpha_j \mathbf{y}(x + jh) - h\beta_j \mathbf{y}'(x + jh)],$$

so that the local truncation error, defined as in section 2.7, is now a vector. The definition of convergence given in section 2.5 is adapted as follows. A linear multistep method is said to be convergent if, for all initial value problems (4) satisfying the hypotheses of the vector form of theorem 1.1, we have that

$$\lim_{\substack{h \to 0 \\ nh = x - a}} \|\mathbf{y}_n - \mathbf{y}(x_n)\| \to 0$$

holds for all $x \in [a, b]$ and for all solutions $\{\mathbf{y}_n\}$ of the (vector) difference equation resulting from applying the method to (4) with starting conditions

$\mathbf{y}_\mu = \boldsymbol{\eta}_\mu(h)$ for which

$$\lim_{h\to 0} \|\boldsymbol{\eta}_\mu(h) - \boldsymbol{\eta}\| = 0, \qquad \mu = 0, 1, \ldots, k-1.$$

The important theorem that zero-stability and consistency are together necessary and sufficient for convergence then holds for systems. A proof of this theorem for a particular vector norm can be found in Henrici.[68] (See also Butcher,[20] Spijker,[169] and Watt.[180]) A bound for the global error analogous to equation (31) of chapter 3 can also be deduced from Henrici's approach. (At several points in chapters 2 and 3 we have referred the reader to the book by Henrici;[67] when systems are involved, the appropriate reference is the later book by Henrici.[68])

An important section of chapter 3 which is, as it stands, inapplicable to systems is section 3.6, which develops a theory of weak stability. The criteria evolved there, and in section 3.11, require us to choose a steplength h such that $h\lambda$ falls within a given interval of the real line, where λ is a (fixed) estimate for the scalar $\partial f/\partial y$. Since, when a system is involved, this last quantity is replaced by the Jacobian matrix $\partial \mathbf{f}/\partial \mathbf{y}$, some modification is clearly necessary. The first few steps of section 3.6 are clearly applicable to a system, and we find that the error equation which arises when the method

$$\sum_{j=0}^{k} \alpha_j \mathbf{y}_{n+j} = h \sum_{j=0}^{k} \beta_j \mathbf{f}_{n+j}$$

is applied to the system (4) is

$$\sum_{j=0}^{k} \alpha_j \tilde{\mathbf{e}}_{n+j} = h \sum_{j=0}^{k} \beta_j [\mathbf{f}(x_{n+j}, \mathbf{y}(x_{n+j})) - \mathbf{f}(x_{n+j}, \mathbf{y}_{n+j})] + \boldsymbol{\phi}_{n+k}, \tag{10}$$

where $\tilde{\mathbf{e}}_n$ is the global error, including round-off, and $\boldsymbol{\phi}_{n+k}$ represents the total error committed at the nth step. We now apply the mean value theorem for a function of several variables (see, for example, Gillespie,[58] page 60) to each component on the right-hand side of (10) to obtain

$$\sum_{j=0}^{k} \alpha_j \tilde{\mathbf{e}}_{n+j} = h \sum_{j=0}^{k} \beta_j \frac{\partial \mathbf{f}}{\partial \mathbf{y}} \tilde{\mathbf{e}}_{n+j} + \boldsymbol{\phi}_{n+k},$$

where $\partial \mathbf{f}/\partial \mathbf{y}$ is the $m \times m$ Jacobian matrix whose i,jth element is the partial derivative of ${}^i f(x, {}^1y, {}^2y, \ldots, {}^m y)$ with respect to ${}^j y$, evaluated at an appropriate intermediate point. Corresponding to equation (37) of chapter 3, the following assumptions are made:

$$\partial \mathbf{f}/\partial \mathbf{y} = J, \quad \text{a constant } m \times m \text{ matrix,}$$

$$\boldsymbol{\phi}_n = \boldsymbol{\phi}, \quad \text{a constant } m\text{-dimensional vector.}$$

The equation for $\tilde{\mathbf{e}}_n$ now reduces to the linearized error equation

$$\sum_{j=0}^{k} (\alpha_j I - h\beta_j J)\tilde{\mathbf{e}}_{n+j} = \boldsymbol{\phi}, \tag{11}$$

where I is the $m \times m$ unit matrix. We now assume further that the eigenvalues λ_t, $t = 1, 2, \ldots, m$ of J are distinct. There then exists a non-singular matrix H such that

$$H^{-1}JH = \Lambda = \begin{bmatrix} \lambda_1 & 0 & \ldots & 0 \\ 0 & \lambda_2 & \ldots & 0 \\ \vdots & & \ddots & \vdots \\ 0 & \ldots & 0 & \lambda_m \end{bmatrix}. \tag{12}$$

(See, for example, Mitchell,[134] page 5.) Pre-multiplying both sides of (11) by H^{-1}, and defining $\mathbf{d}_n$ by

$$\tilde{\mathbf{e}}_n = H\mathbf{d}_n, \tag{13}$$

we obtain

$$\sum_{j=0}^{k} H^{-1}(\alpha_j I - h\beta_j J)H\mathbf{d}_{n+j} = H^{-1}\boldsymbol{\phi}.$$

Using (12), we have

$$\sum_{j=0}^{k} (\alpha_j I - h\beta_j \Lambda)\mathbf{d}_{n+j} = \boldsymbol{\psi}, \tag{14}$$

where $\boldsymbol{\psi} = H^{-1}\boldsymbol{\phi}$. Since I and Λ are diagonal matrices, the components of (14) are *uncoupled*; that is, we may write (14) in the form

$$\sum_{j=0}^{k} (\alpha_j - h\beta_j\lambda_i)\,{}^{i}d_{n+j} = {}^{i}\psi, \qquad i = 1, 2, \ldots, m, \tag{15}$$

where $\mathbf{d}_n$ and $\boldsymbol{\psi}$ have components ${}^{i}d_n$ and ${}^{i}\psi$ respectively, $i = 1, 2, \ldots, m$, and *each of the equations in* (15) *is independent of the others.* Each equation of (15) is now exactly of the form of equation (38) of chapter 3, the linearized error equation for a single equation, the only changes being that $\tilde{e}_n$ is replaced by ${}^{i}d_n$ and λ, the estimate for the scalar $\partial f/\partial y$, is replaced by λ_i, an eigenvalue of the (constant) estimate for the matrix $\partial \mathbf{f}/\partial \mathbf{y}$. Since, by (13), $\mathbf{d}_n$ will grow or decay with n if and only if $\tilde{\mathbf{e}}_n$ does, it follows that all of the weak stability theory developed in sections 3.6 and 3.11 holds for systems if we interpret λ as an eigenvalue of J. (Note that in (15) i runs from 1 to m; thus we must ensure that any criterion imposed on $\bar{h} = h\lambda$ is satisfied when λ is *any* eigenvalue of J.)

There is, however, one new feature. The Jacobian is not, in general, symmetric; hence its eigenvalues are not necessarily real. Thus, in the case of systems, the parameter $\bar{h}$ $(= h\lambda)$ which appears in the stability polynomial associated with a given method, may be complex. The definitions of absolute and relative stability are thus amended as follows.

Definition A linear multistep or predictor–corrector method is said to be **absolutely/relatively stable** *in a region $\mathscr{R}$ of the complex plane if, for all $\bar{h} \in \mathscr{R}$, all roots of the stability polynomial $\pi(r, \bar{h})$ associated with the method, satisfy*

$$|r_s| < 1, \qquad s = 1, 2, \ldots, k \quad / \quad |r_s| < |r_1|, \qquad s = 2, 3, \ldots, k.$$

For any given method, the associated stability polynomial is that previously derived in the case of a single equation. Thus it is given by equation (41) of chapter 3 in the case of a linear multistep method, and by equations (67) and (69) of chapter 3 for a predictor–corrector method in $P(EC)^mE$ and $P(EC)^m$ mode respectively.

Of the four methods, described in section 3.7, for finding intervals of stability, only the Schur criterion and the boundary locus method are applicable to a system. Indeed, since the latter already assumes that $\bar{h}$ is complex, it is immediately applicable, as described in section 3.7 for a linear multistep method and in section 3.11 for a predictor–corrector method. The root locus method, which calls for plots of the roots of the stability polynomial against $\bar{h}$, is clearly inapplicable to a system. The Routh–Hurwitz criteria, in the form described in section 3.7, is also inapplicable to a system. (A criterion, somewhat similar to the Wilf [182] criterion, but applicable to a system when the method is a predictor–corrector method in *PMECME* mode (see section 3.10), is derived by Abdel Karim.[1])

For all but the simplest of methods (see exercise 2 below), it is not possible to quantify the region of stability in a meaningful way other than by presenting diagrams showing the boundary $\partial\mathscr{R}$ of the region, plotted in the complex plane. (See, for example, the diagrams presented by Krogh.[90]) If such a diagram is not available (and we are not prepared to spend the computing effort involved in finding it), knowledge of the *interval* of stability, that is the intersection of $\partial\mathscr{R}$ on the real line, can still give some indication of a safe choice for h even when a system is to be solved.

Example 2 *Find the region of absolute stability of the Adams–Moulton method*

$$y_{n+3} - y_{n+2} = \frac{h}{24}(9f_{n+3} + 19f_{n+2} - 5f_{n+1} + f_n).$$

For this method (whose interval of absolute stability we know, by table 8 of chapter 3 (page 85), to be $(-3, 0)$), we have

$$\rho(r) = r^3 - r^2, \qquad \sigma(r) = \tfrac{1}{24}(9r^3 + 19r^2 - 5r + 1).$$

According to the boundary locus method, the locus of $\partial\mathscr{R}$ is given by

$$\bar{h}(\theta) = \frac{\rho(\exp(i\theta))}{\sigma(\exp(i\theta))} = \frac{24[\exp(3i\theta) - \exp(2i\theta)]}{9\exp(3i\theta) + 19\exp(2i\theta) - 5\exp(i\theta) + 1}$$
$$= x(\theta) + iy(\theta),$$

where we find, after some manipulation, that

$$x(\theta) = 24(-10 + 15\cos\theta - 6\cos 2\theta + \cos 3\theta)/D(\theta),$$
$$y(\theta) = 24(33\sin\theta - 6\sin 2\theta + \sin 3\theta)/D(\theta),$$
$$D(\theta) = 468 + 142\cos\theta - 52\cos 2\theta + 18\cos 3\theta.$$

Clearly $x(-\theta) = x(\theta)$ and $y(-\theta) = -y(\theta)$, so that the locus is symmetric about the x-axis. Evaluating x and y at intervals of θ of 30°, we obtain the following tabulation:

θ:	0°	30°	60°	90°	120°	150°	180°
x:	0	−0·000	−0·002	−0·185	−0·735	−1·955	−3·000
y:	0	0·523	1·026	1·477	1·838	1·707	0

These points are plotted in the complex $\bar{h}$-plane, and the curve $\partial\mathscr{R}$ joining them sketched in to give the region $\mathscr{R}$ of absolute stability indicated in figure 7.

Example 3 *The method of example 2 is to be applied to the problem*

$$y'' = -20y' - 200y, \quad y(0) = 1, \quad y'(0) = -10.$$

Determine, approximately, the largest value of the steplength which will ensure that the method is absolutely stable.

Writing the given differential equation as a first-order system, we set $y = u$ and $y' = v$ to obtain

$$u' = v$$
$$v' = -20v - 200u.$$

The Jacobian is

$$\begin{pmatrix} 0 & 1 \\ -200 & -20 \end{pmatrix},$$

and its eigenvalues are the roots of the quadratic $\lambda^2 + 20\lambda + 200$, that is, are $-10 \pm 10i$. The steplength h must be chosen so that both $\bar{h}_1 = (-10 + 10i)h$ and $\bar{h}_2 = (-10 - 10i)h$ lie within the region $\mathscr{R}$ given in figure 7. By approximate measurements on figure 7 we find that $\bar{h}_1$ and $\bar{h}_2$ fall just outside $\mathscr{R}$ if $h = 0{\cdot}2$, and just inside

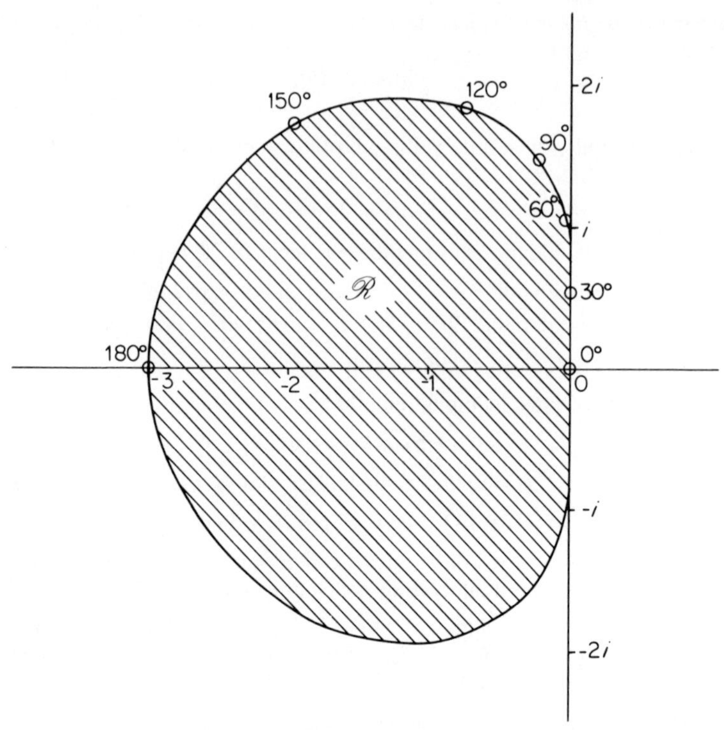

FIGURE 7

if $h = 0{\cdot}175$. We conclude that 0·175 is the approximate maximum allowable steplength. Note that the fact that the given problem is linear with constant coefficients has resulted in the Jacobian being constant. For a more general problem, the Jacobian would be a function of the solution, and we would have to re-evaluate its eigenvalues† from time to time as the solution progressed, to ascertain whether or not the choice of steplength was appropriate.

Exercises

1. Corroborate the findings of example 3 by computing solutions to the given problem with $h = 0{\cdot}1$ and $h = 0{\cdot}25$.

2. In examples 7, 8, 9, and 10 of section 3.7 we showed that the method

$$y_{n+2} - y_n = \tfrac{1}{2}h(f_{n+1} + 3f_n)$$

has an interval of absolute stability of $(-\frac{4}{3}, 0)$. Show that its region of stability is the circle on this interval as diameter.

† A heuristically based criterion for relative stability, applicable to a system, and not requiring computation of eigenvalues, is described by Krogh.[91]

3. What is the maximum allowable steplength if the method of exercise 2 is applied to the problem of example 3?

4. The parameter $\bar{h}$ is not necessarily complex for a system. Show that it is real for the system discussed in example 1 (page 219), and, indeed, for any first-order system arising from the equation $y'' = \phi(x, y, y')$ for which $\partial\phi/\partial y > 0$.

5. Find an expression for the locus of the boundary of the region $\mathscr{R}$ of absolute stability of the method

$$y_{n+2} - (1 + a)y_{n+1} + ay_n = \frac{h}{12}[(5 + a)f_{n+2} + 8(1 - a)f_{n+1} - (1 + 5a)f_n].$$

To what does this locus reduce when (i) $a = 1$, (ii) $a = -1$? Deduce $\mathscr{R}$ for these two cases.

6. In exercise 20 of chapter 3 (page 101) it was asserted that the *PECE* method which uses Simpson's rule as corrector and the method

$$y_{n+2} + 4y_{n+1} - 5y_n = h(4f_{n+1} + 2f_n)$$

as predictor, when applied to a single equation is, for small negative $\bar{h}$, relatively stable according to the criterion $|r_s| \leqslant |r_1|$, $s = 2, 3, \ldots, k$. This result is known (Krogh[92]) not to hold for a system. Demonstrate this by considering a system whose Jacobian has an eigenvalue which is purely imaginary.

8.3 Applicability of Runge–Kutta methods

The general explicit one-step method given by equation (1) of chapter 4 is applicable to a system, and may be written in the form

$$\mathbf{y}_{n+1} - \mathbf{y}_n = h\boldsymbol{\phi}(x_n, \mathbf{y}_n, h).$$

The method is of order p if p is the largest integer for which

$$\mathbf{y}(x + h) - \mathbf{y}(x) - h\boldsymbol{\phi}(x, \mathbf{y}(x), h) = \mathbf{O}(h^{p+1})$$

holds, where $\mathbf{y}(x)$ is the theoretical solution of (4); it is consistent if $\boldsymbol{\phi}(x, \mathbf{y}, 0) = \mathbf{f}(x, \mathbf{y})$. Theorem 4.1 of section 4.2 is also applicable to a system, if moduli of scalars are replaced by norms of corresponding vectors; a proof can be found in Henrici,[67] page 124.

In particular, the general Runge–Kutta method defined by equation (5) of chapter 4 is applicable to a system if the scalars y, f, ϕ, and k_r, $r = 1, 2, \ldots, R$, are replaced by vectors $\mathbf{y}, \mathbf{f}, \boldsymbol{\phi}$, and $\mathbf{k}_r$ respectively. However, the orders of certain Runge–Kutta methods do depend on whether they are applied to a single equation or a system—a fact first pointed out by Butcher.[14] In section 4.3 we saw that the order of a Runge–Kutta method for a single equation is determined by the extent to which we are able to match groups of derivatives in two expansions for y_{n+1}, one arising from a Taylor expansion and the other from the method. When a system is involved, the groups of derivatives become much more complicated. (It is easily checked that the derivation of section 4.3 is not, as it stands, applicable to a system in the sense defined in section 8.1.) In particular, when order greater than four is sought, we find that the number of groups of derivatives to be matched is greater in the case of a system than in the case of a single equation. Thus, for a Runge–Kutta method to be of order five for a system, its coefficients must satisfy one more condition than is necessary for it to be of order five for a single equation. However, we saw in section 4.3 that we generally have more coefficients than conditions, so that it is possible that a particular choice of coefficients which gives an order five method for a single equation will also satisfy the additional condition needed to ensure order five for a system. In general, *a Runge–Kutta method which has order p for a single equation also has order p for a system if* $p \leqslant 4$*; for* $p > 4$*, the order for a system may be less than p.* In fact, most of the commonly used Runge–Kutta methods of order greater than four have the same order for a system as for a single equation; in particular this is true of all such Runge–Kutta methods quoted in chapter 4.

The derivation of higher order Runge–Kutta methods by the elementary notation of section 4.3 is complicated enough for a single equation; for a system it becomes quite impracticable, and an improved notation is essential. An elegant notation, which involves ideas from graph theory, has been devised by Butcher.[13,22] Using this notation, Harris[66] discusses the use of a computer to give, automatically, the equations which define the coefficients of a Runge–Kutta method of prescribed order and number of stages.

The various error bounds quoted in section 4.5 are not, as they stand, applicable to a system, and, indeed, need considerable modification. Thus Lotkin[121] shows that, in the case of a 2×2 system, the bound given by equation (32) of chapter 4 for the popular fourth-order Runge–Kutta method (equation (20) of chapter 4) is replaced by

$$|{}^{i}T_{n+1}| < \tfrac{973}{720} h^5 P^4 Q, \qquad i = 1, 2,$$

where $\mathbf{T}_{n+1} = [{}^{1}T_{n+1}, {}^{2}T_{n+1}]^T$ is the (vector) local truncation error, and it is assumed that the components ${}^{1}f$ and ${}^{2}f$ of $\mathbf{f}$, and their partial derivatives with respect to x, ${}^{1}y$, and ${}^{2}y$ satisfy

$$|{}^{i}f| < Q \qquad \left| \frac{\partial^{l+m+n}({}^{i}f(x, y))}{\partial x^l \partial({}^{1}y^m)\partial({}^{2}y^n)} \right| < \frac{P^{l+m+n}}{Q^{m+n-1}}, \qquad l + m + n \leqslant 4, \qquad i = 1, 2.$$

An error bound for the same method, applicable to a general $m \times m$ system, is somewhat complicated; one can be found in Henrici,[67] page 130. However, of the error estimates quoted in section 4.6, all but that given by (46) are applicable to a system.

The weak stability theory for Runge–Kutta methods developed in section 4.7 is applicable to a system if we make the modification discussed in section 8.2, namely that the parameter $\bar{h} = h\lambda$, where λ is an eigenvalue of the Jacobian (assumed constant), and is allowed to take complex values. It follows from section 4.7 that for a given p, $p = 1, 2, 3, 4$, all p-stage Runge–Kutta methods of order p have the same region of absolute stability. These regions are sketched in figure 8.

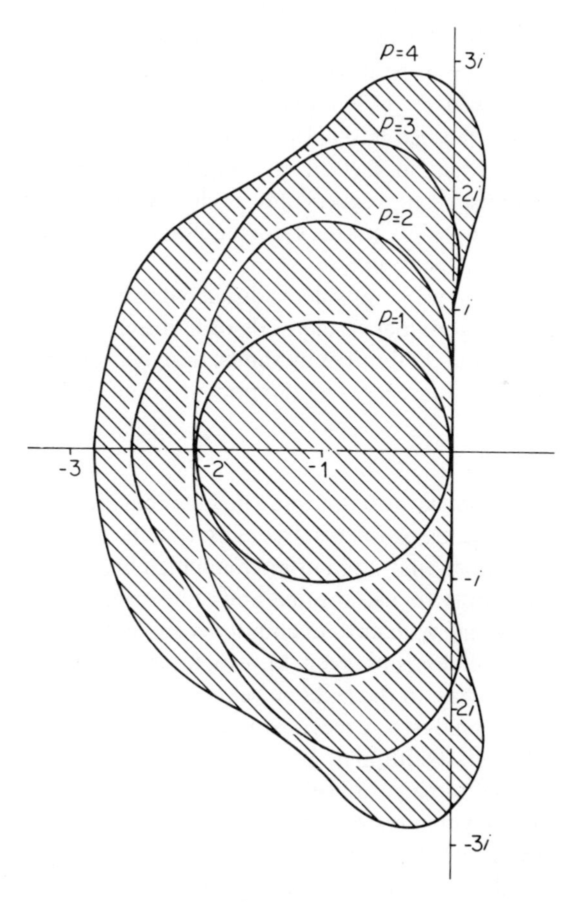

FIGURE 8

The general conclusions of section 4.9, which compares Runge–Kutta with predictor–corrector methods, remain valid when systems are involved. Finally, both the implicit Runge–Kutta methods of section 4.10 and the block methods of section 4.11 are applicable to a system. In connection with the former, convergence of the iteration (75)—with obvious notational modifications—is still guaranteed by (76).

8.4 Applicability of the methods of chapters 5, 6, and 7

The hybrid methods of chapter 5 are all applicable to a system. Section 5.4, on weak stability, is adapted in the way already described for linear multistep and Runge–Kutta methods.

Polynomial extrapolation, described in section 6.1, is clearly applicable if the scalars $A(h)$ and $a_s^{(m)}$ are replaced by vectors. Since Gragg's method (equation (14) of chapter 6) is applicable to a system, we conclude that this method, followed by polynomial extrapolation, is applicable to a system, in the sense defined in section 8.1. On the other hand, the rational extrapolation described in section 6.5 cannot be interpreted when the scalars $b_s^{(m)}$ are replaced by vectors; but it clearly can be applied to each component of a vector $\mathbf{b}_s^{(m)}$. Thus the GBS method is component-applicable, in the sense defined in section 8.1.

Of the methods for special problems discussed in chapter 7, Obrechkoff methods and the trigonometric methods of Gautschi are applicable to a system; the rational and adaptive methods, designed for problems whose solutions possess singularities, are component-applicable.

8.5 Stiff systems

Many fields of application, notably chemical engineering and control theory, yield initial value problems involving systems of ordinary differential equations which exhibit a phenomenon which has come to be known as 'stiffness'. Attempts to use the methods and techniques already discussed to solve such problems encounter very substantial difficulties. The problem of stiffness has been known for some time (Curtiss and Hirschfelder[34]), but has, in the last five years or so, attracted the attention of many numerical analysts. A recent survey of methods for stiff problems produced by Bjurel, Dahlquist, Lindberg, Linde, and Odén[7] (which, incidentally, claims already to be obsolete on publication) reports on over forty methods which have been proposed. Clearly we cannot present all of these here. Following a second survey by Sigurdsson,[168] we shall instead consider the properties that a successful method for stiff problems ought to possess, and then present only a small selection of specific methods. The reader who wishes to see a wider selection is referred to the survey by Bjurel *et al.*

Consider the following linear constant coefficient initial value problem:

$$\mathbf{y}' = A\mathbf{y}, \qquad \mathbf{y}(0) = [1, 0, -1]^T,$$

where (16)

$$A = \begin{bmatrix} -21 & 19 & -20 \\ 19 & -21 & 20 \\ 40 & -40 & -40 \end{bmatrix}.$$

Its theoretical solution is (see example 1 of chapter 1, page 6)

$$\begin{aligned} u(x) &= \tfrac{1}{2}e^{-2x} + \tfrac{1}{2}e^{-40x}(\cos 40x + \sin 40x), \\ v(x) &= \tfrac{1}{2}e^{-2x} - \tfrac{1}{2}e^{-40x}(\cos 40x + \sin 40x), \qquad (17) \\ w(x) &= -e^{-40x}(\cos 40x - \sin 40x), \end{aligned}$$

where

$$\mathbf{y}(x) = [u(x), v(x), w(x)]^T.$$

The graphs of $u(x)$, $v(x)$, and $w(x)$ are shown as continuous lines in figure 9. In the approximate range $0 \leqslant x < 0{\cdot}1$, all three components vary rapidly with x and, if we wish to follow these solutions by a numerical method, we could expect to have to use a rather small steplength. For $x \geqslant 0{\cdot}1$ however, two of the components, u and v, are practically identical and vary much more slowly with x, while the third component, w, is virtually zero. We would hope to be able to use a somewhat larger steplength to integrate the system numerically in this range. Let us attempt to solve the system numerically in the range $0{\cdot}1 \leqslant x \leqslant 1{\cdot}0$, using Euler's rule with a steplength of 0·04, and assuming, as initial condition, the theoretical solution $\mathbf{y}(0{\cdot}1)$ given by (17). The computed results for the component u for the first six steps are indicated by the circled points in figure 9, and are clearly unacceptable. Yet the problem

$$\mathbf{y}' = A\mathbf{y}, \qquad \mathbf{y}(0{\cdot}1) = [\tfrac{1}{2}e^{-0{\cdot}2}, \tfrac{1}{2}e^{-0{\cdot}2}, 0]^T,$$

where (18)

$$A = \begin{bmatrix} -2 & 0 & 0 \\ 0 & -2 & 0 \\ 0 & 0 & 0 \end{bmatrix},$$

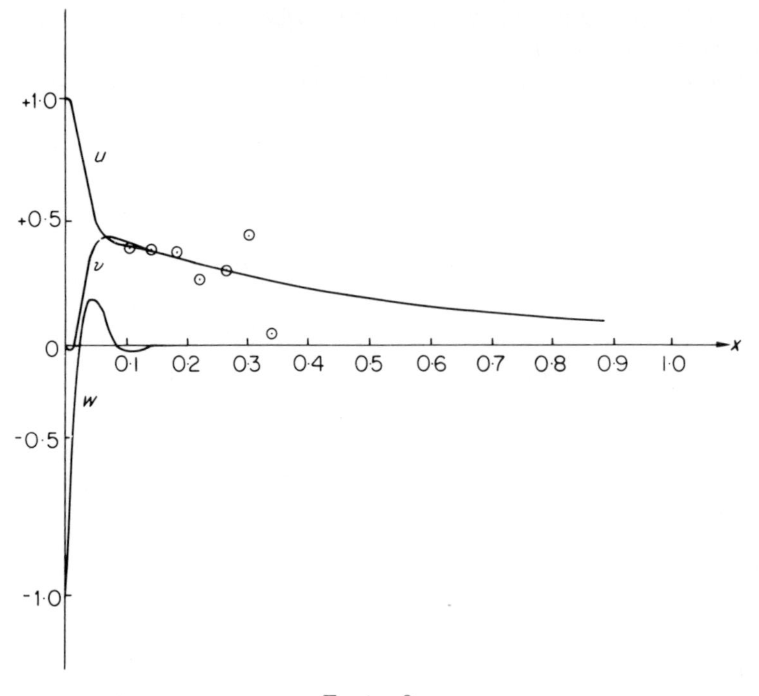

FIGURE 9

whose theoretical solution $\mathbf{y}(x) = [\frac{1}{2}e^{-2x}, \frac{1}{2}e^{-2x}, 0]^T$ is, in the range $0{\cdot}1 \leqslant x \leqslant 1{\cdot}0$, virtually indistinguishable from that of (16), is integrated perfectly satisfactorily by Euler's rule with steplength 0·04. The scatter of the circled points in figure 9 suggests (see, for example, table 7 of chapter 3) that for the problem (16) in the range $0{\cdot}1 \leqslant x \leqslant 1{\cdot}0$ the choice of $h = 0{\cdot}04$ causes $\bar{h}$ to lie outside the region of absolute stability, and indeed this is the case. For the problem (16), the Jacobian $\partial\mathbf{f}/\partial\mathbf{y}$ is the constant matrix A (given by (16)) whose eigenvalues λ are -2, $-40 \pm 40i$. The region $\mathscr{R}$ of absolute stability for Euler's rule (see figure 8 with $p = 1$) is the circle centre -1, radius 1, and it follows that for $\bar{h}\ (= h\lambda)$ to lie within $\mathscr{R}$ for all three values of λ, we must satisfy $h < 0{\cdot}025$. Note that the eigenvalues responsible for this severe restriction on h are $-40 \pm 40i$, that is, the very eigenvalues whose contributions to the theoretical solution (17) are negligible in the range $0{\cdot}1 \leqslant x \leqslant 1{\cdot}0$. On the other hand, the eigenvalues of the Jacobian of the problem (18) are -2, -2, 0, and for absolute stability we require only $h < 1{\cdot}0$. Note also that Euler's rule, although of very low order, does not have an excessively small region of absolute stability. It follows from figure 8 that, had we used instead a fourth-order four-stage Runge–Kutta method, then, in order to achieve absolute stability, the steplength would have had to be restricted to be less than approximately 0·05.

We saw in section 1.6 that the $m \times m$ linear system

$$\mathbf{y}' = A\mathbf{y} + \boldsymbol{\phi}(x), \tag{19}$$

where the matrix A has distinct eigenvalues λ_t and corresponding eigenvectors $\mathbf{c}_t$, $t = 1, 2, \ldots, m$, has a general solution of the form

$$\mathbf{y}(x) = \sum_{t=1}^{m} k_t e^{\lambda_t x} \mathbf{c}_t + \boldsymbol{\psi}(x). \tag{20}$$

Let us assume that $\operatorname{Re} \lambda_t < 0$, $t = 1, 2, \ldots, m$. Then the term

$$\sum_{t=1}^{m} k_t e^{\lambda_t x} \mathbf{c}_t \to 0 \quad \text{as} \quad x \to \infty;$$

we therefore call this term the *transient solution*, and call the remaining term $\boldsymbol{\psi}(x)$ the *steady-state solution*. Let λ_μ and λ_ν be two eigenvalues of A such that

$$|\operatorname{Re} \lambda_\mu| \geqslant |\operatorname{Re} \lambda_t| \geqslant |\operatorname{Re} \lambda_\nu|, \qquad t = 1, 2, \ldots, m.$$

If our aim is to find numerically the steady-state solution $\boldsymbol{\psi}(x)$, then we must pursue the numerical solution of (19) until the slowest decaying exponential in the transient solution, namely $e^{\lambda_\nu x}$, is negligible. Thus, the smaller $|\operatorname{Re} \lambda_\nu|$, the longer will be the range of integration. On the other hand, the presence of eigenvalues of A far out to the left in the complex plane will force us, as in the above example, to use excessively small steplengths in order that $\bar{h}$ will lie within the region of absolute stability of the method of our choice (unless, of course, the region is infinite and includes the left-hand half-plane). The further out such eigenvalues lie, the more severe the restriction on steplength. A rough measure of this difficulty is the magnitude of $|\operatorname{Re} \lambda_\mu|$. If $|\operatorname{Re} \lambda_\mu| \gg |\operatorname{Re} \lambda_\nu|$, we are forced into the highly undesirable computational situation of having to integrate numerically over a long range, using a steplength which is *everywhere* excessively small relative to the interval: this is the problem of stiffness. We can make the following somewhat heuristic definition:

Definition *The linear system* $\mathbf{y}' = A\mathbf{y} + \boldsymbol{\phi}(x)$ *is said to be* **stiff** *if* (i) $\operatorname{Re} \lambda_t < 0$, $t = 1, 2, \ldots, m$, *and* (ii) $\max_{t=1,2,\ldots,m} |\operatorname{Re} \lambda_t| \gg \min_{t=1,2,\ldots,m} |\operatorname{Re} \lambda_t|$, *where λ_t, $t = 1, 2, \ldots, m$, are the eigenvalues of A. The ratio*

$$\left[\max_{t=1,2,\ldots,m} |\text{Re}\,\lambda_t|\right] : \left[\min_{t=1,2,\ldots,m} |\text{Re}\,\lambda_t|\right]$$

is called the **stiffness ratio**.

Non-linear systems $\mathbf{y}' = \mathbf{f}(x, \mathbf{y})$ exhibit stiffness if the eigenvalues of the Jacobian $\partial\mathbf{f}/\partial\mathbf{y}$ behave in a similar fashion. The eigenvalues are no longer constant but depend on the solution, and therefore vary with x. Accordingly we say that the system $\mathbf{y}' = \mathbf{f}(x, \mathbf{y})$ is stiff in an interval I of x if, for $x \in I$, the eigenvalues $\lambda_t(x)$ of $\partial\mathbf{f}/\partial\mathbf{y}$ satisfy (i) and (ii) above.

Note that if the partial derivatives appearing in the Jacobian $\partial\mathbf{f}/\partial\mathbf{y}$ are continuous and bounded in an appropriate region (see the vector form of theorem 1.1), then the Lipschitz constant L of the system $\mathbf{y}' = \mathbf{f}(x, \mathbf{y})$ may be taken to be $L = \|\partial\mathbf{f}/\partial\mathbf{y}\|$. For any matrix A, $\|A\| \geqslant \rho(A)$, where $\rho(A)$, the spectral radius, is defined to be

$$\max_{t=1,2,\ldots,m} |\lambda_t|,$$

λ_t, $t = 1, 2, \ldots, m$, being the eigenvalues of A. (See, for example, Mitchell[134] page 13.) If $\max_{t=1,2,\ldots,m} |\text{Re}\,\lambda_t| \gg 0$ it follows that $L \gg 0$. Thus stiff systems are occasionally referred to as 'systems with large Lipschitz constants'.

The example (16) has a stiffness ratio of 20, which is considered very modest indeed. Stiffness ratios of the order of 10^6 frequently arise in practical problems.

Exercises

7. Find the stiffness ratio for the system

$$u' = -10u + 9v,$$

$$v' = 10u - 11v.$$

What is the largest steplength which can be used with a fourth-order four-stage Runge–Kutta method?

8. Consider the problem (19) in the case when all the eigenvalues are real. It is to be integrated from $x = 0$ to a value of x for which the terms $e^{\lambda_t x}$ in (20) satisfy $|e^{\lambda_t x}| < 0{\cdot}001$, $t = 1, 2, \ldots, m$. If the stiffness ratio is 10^6, show that Euler's rule would need to be applied approximately $3{\cdot}5 \times 10^6$ times.

9. (Dahlquist.) Find the theoretical solution of the problem $y' = -100y + \cos x$, $y(0) = 1$. In what sense could this *single* equation be said to exhibit stiffness? If Euler's rule were used to obtain a numerical solution, what would be the maximum allowable steplength?

8.6 The problem of stability for stiff systems

We have seen in section 8.5 that a basic difficulty (but not the only one) in the numerical solution of stiff systems is the satisfaction of the requirement of absolute stability. Thus several definitions, which call for the method to possess some 'adequate' region of absolute stability, have been proposed. These definitions, originally constructed for linear multistep methods, are applicable to any numerical method which involves a discretization with associated steplength h.

Definition (*Dahlquist*[37]) *A numerical method is said to be A-***stable** *if its region of absolute stability contains the whole of the left-hand half-plane* $\mathrm{Re}\, h\lambda < 0$.

If an A-stable method is applied to a stiff system (and it is therefore implied that $\mathrm{Re}\,\lambda_t < 0, t = 1, 2, \ldots, m$), then the difficulties described in section 8.5 disappear, since, no matter how large $\max_{t=1,2,\ldots,m} |\mathrm{Re}\,\lambda_t|$, no stability restriction on h can result. However, A-stability is a severe requirement to ask of a numerical method, as the following somewhat depressing theorem of Dahlquist[37] shows.

Theorem 8.1 (*i*) *An explicit linear multistep method cannot be A-stable.* (*ii*) *The order of an A-stable implicit linear multistep method cannot exceed two.* (*iii*) *The second-order A-stable implicit linear multistep method with smallest error constant is the Trapezoidal rule.*

The restriction on order implied by (ii) is a severe one. (Note that the principal local truncation error of the Trapezoidal rule is $-\frac{1}{12}h^3\mathbf{y}^{(3)}(x_n)$; for a stiff system, we can expect the components of $\mathbf{y}^{(3)}(x)$ to be very large, at least in an interval on which the transient solution is not negligible.) In view of this, several less demanding stability definitions have been proposed; we present two here.

Definition (*Widlund*[181]) *A numerical method is said to be* $A(\alpha)$-**stable**, $\alpha \in (0, \pi/2)$, *if its region of absolute stability contains the infinite wedge* $W_\alpha = \{h\lambda \mid -\alpha < \pi - \arg h\lambda < \alpha\}$; *it is said to be* $A(0)$-**stable** *if it is* $A(\alpha)$-*stable for some* (*sufficiently small*) $\alpha \in (0, \pi/2)$.

The region W_α of the complex $h\lambda$-plane is shown in figure 10. Note that for a given λ, with $\mathrm{Re}\,\lambda < 0$, the point $h\lambda$ either lies inside W_α *for all positive* h, or lies outside it *for all positive* h. Thus, if we can ascertain in advance that all the eigenvalues of a stiff system lie in a certain wedge W_{α^*}, then an $A(\alpha^*)$-stable method can be used without any stability restriction on the steplength. In particular, any $A(0)$-stable method can be so used if all the

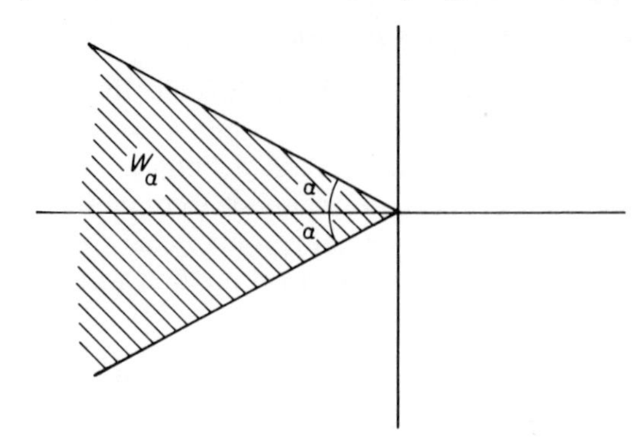

FIGURE 10

eigenvalues are known to be real (which will be the case, for example, if the Jacobian is symmetric). Corresponding to theorem 8.1, the following theorem has been established by Widlund.[181]

Theorem 8.2 (*i*) *An explicit linear multistep method cannot be $A(0)$-stable.* (*ii*) *There is only one $A(0)$-stable linear k-step method whose order exceeds $k + 1$, namely the Trapezoidal rule.* (*iii*) *For all $\alpha \in [0, \pi/2)$, there exist $A(\alpha)$-stable linear k-step methods of order p for which $k = p = 3$ and $k = p = 4$.*

An alternative slackening of the A-stability requirement is incorporated in the following definition.

Definition (*Gear*[52,53]) *A numerical method is said to be* **stiffly stable** *if* (*i*) *its region of absolute stability contains $\mathscr{R}_1$ and $\mathscr{R}_2$ and* (*ii*) *it is accurate for all $h \in \mathscr{R}_2$ when applied to the scalar test equation $y' = \lambda y$, λ a complex constant with* $\operatorname{Re} \lambda < 0$, *where*

$$\mathscr{R}_1 = \{h\lambda |\operatorname{Re} h\lambda < -a\},$$

$$\mathscr{R}_2 = \{h\lambda | -a \leqslant \operatorname{Re} h\lambda \leqslant b, -c \leqslant \operatorname{Im} h\lambda \leqslant c\},$$

and a, b, and c are positive constants.†

The regions $\mathscr{R}_1$ and $\mathscr{R}_2$ of the complex $h\lambda$-plane are shown in figure 11. The motivation for this definition is that those eigenvalues which represent rapidly decaying terms in the transient solution will correspond to values of $h\lambda$ in $\mathscr{R}_1$; we generally have no interest in representing such terms

† Note that, since $b > 0$, eigenvalues with small positive real parts are catered for; thus Gear's definition of stiffness is somewhat wider than ours.

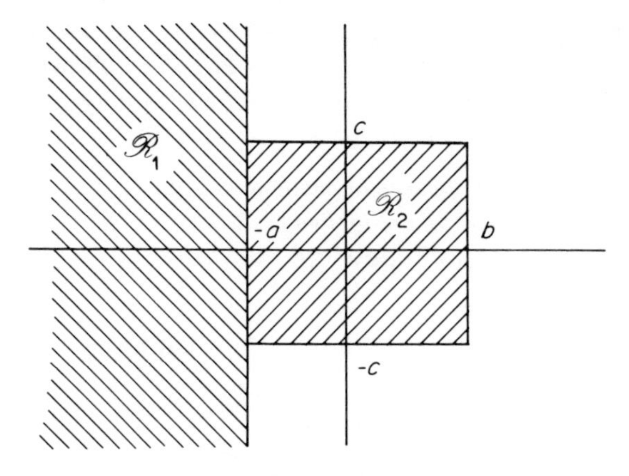

FIGURE 11

accurately but only stably—that is, we will be satisfied with any representation of these terms which decays sufficiently rapidly. The remaining eigenvalues of the system represent terms in the solution which we would like to represent accurately as well as stably; by suitable choice of h, values of $h\lambda$ corresponding to these eigenvalues can be made to lie in $\mathscr{R}_2$.

Severe as the requirement of A-stability is, in one sense it is not severe enough. Consider, for example, the application of the Trapezoidal rule to the scalar test equation $y' = \lambda y$, λ a complex constant with $\operatorname{Re}\lambda < 0$. We obtain

$$y_{n+1}/y_n = \left(1 + \frac{h\lambda}{2}\right) \bigg/ \left(1 - \frac{h\lambda}{2}\right). \tag{21}$$

Since the Trapezoidal rule is A-stable, $y_n \to 0$ as $n \to \infty$ for all fixed positive h. However,

$$\left|\frac{y_{n+1}}{y_n}\right| = \left|\frac{1 + \frac{1}{2}h\lambda}{1 - \frac{1}{2}h\lambda}\right| = \left[\frac{1 + h\operatorname{Re}\lambda + \frac{1}{4}h^2|\lambda|^2}{1 - h\operatorname{Re}\lambda + \frac{1}{4}h^2|\lambda|^2}\right]^{\frac{1}{2}}, \tag{22}$$

and, if $|\operatorname{Re}\lambda| \gg 0$ and h is not small—it is, after all, our aim to avoid such restrictions—then $|y_{n+1}/y_n|$ will be close to $+1$. (Indeed, in the limit as $\operatorname{Re}\lambda \to -\infty$, the right-hand side of (22) tends to $+1$.) Thus $|y_n|$ will decay to zero only very slowly, and it follows that an A-stable method may be unsatisfactory for an excessively stiff system. Contrast the behaviour of the method $y_{n+1} - y_n = hf_{n+1}$ (the backward Euler method) for which

$$\left|\frac{y_{n+1}}{y_n}\right| = \left|\frac{1}{1 - h\lambda}\right| = \left[\frac{1}{1 - 2h\operatorname{Re}\lambda + h^2|\lambda|^2}\right]^{\frac{1}{2}}. \tag{23}$$

The right-hand side of (23) tends to 0 as $\operatorname{Re}\lambda \to -\infty$, and we can expect a rapid decay of $|y_n|$ even for moderately large h. Following Ehle,[39] we make the following definition of L-stability (left-stability). (Axelsson[4] calls the same property 'stiff A-stability'.)

Definition *A one-step numerical method is said to be* **L-stable** *if it is A-stable and, in addition, when applied to the scalar test equation $y' = \lambda y$, λ a complex constant with $\operatorname{Re}\lambda < 0$, it yields $y_{n+1} = R(h\lambda)y_n$, where $|R(h\lambda)| \to 0$ as $\operatorname{Re} h\lambda \to -\infty$.*

Note that L-stability $\Rightarrow$ A-stability $\Rightarrow$ $A(0)$-stability.

Exercises

10. Show from first principles that the Trapezoidal rule is A-stable and that the backward Euler method is L-stable.

11. Show that a stiffly stable method is necessarily $A(\alpha)$-stable, but not conversely. What is the minimum value of α in terms of the parameters a, b, c appearing in the definition of stiff stability?

8.7 Some properties of rational approximations to the exponential

It is easy to verify that the rational expression appearing on the right-hand side of (21) is an approximation to $e^{h\lambda}$ in error by a term $O((h\lambda)^3)$ (cf. equation (43) of chapter 3). It is worth considering briefly some properties of such rational approximations, since known results concerning them can be useful in ascertaining whether certain numerical methods satisfy one or other of the stability requirements we have discussed. Let q be a complex scalar and let $R_T^S(q)$, where $S \geqslant 0$, $T \geqslant 0$, be given by

$$R_T^S(q) = \left(\sum_{i=0}^{S} a_i q^i\right) \bigg/ \left(\sum_{i=0}^{T} b_i q^i\right), \quad a_0 = b_0 = 1, \tag{24}$$

where all the a_i and b_i are real. We say that $R_T^S(q)$ is an (S, T) *rational approximation of order p to the exponential* e^q if $R_T^S(q) = e^q + O(q^{p+1})$. The maximum possible order is $S + T$, and is achieved by expressing e^q as a power series and equating coefficients of q^i, $i = 1, 2, \ldots, S + T$, in the equation

$$\sum_{i=0}^{S} a_i q^i = \left(\sum_{i=0}^{T} b_i q^i\right)\left(\sum_{i=0}^{\infty} \frac{1}{i!} q^i\right)$$

to define uniquely the coefficients a_i, b_i appearing in (24). The resulting order $S + T$ approximation, which we shall denote by $\hat{R}_T^S(q)$, is called the

(S, T) Padé approximation to e^q (see, for example, Ralston[150], page 278). Following Ehle,[39] we make the following definition:

Definition *A rational approximation* $R_T^S(q)$ *to* e^q *is said to be (i) A*-**acceptable**, *if* $|R_T^S(q)| < 1$ *whenever* $\operatorname{Re} q < 0$, *(ii) A(0)*-**acceptable**, *if* $|R_T^S(q)| < 1$ *whenever q is real and negative, and (iii) L*-**acceptable**, *if it is A-acceptable and, in addition, satisfies* $|R_T^S(q)| \to 0$ *as* $\operatorname{Re} q \to -\infty$.

It follows immediately that if a one-step method, applied to the usual scalar test equation $y' = \lambda y$, λ a complex constant, yields $y_{n+1} = R_T^S(h\lambda)y_n$, then the method is *A*-, *A*(0)-, or *L*-stable according as the approximation $R_T^S(h\lambda)$ to $e^{h\lambda}$ is *A*-, *A*(0)-, or *L*-acceptable. The following results concerning Padé approximations are known.

Theorem 8.3 *Let* $\hat{R}_T^S(q)$ *be the (S, T) Padé approximation to* e^q. *Then, (i) (Birkhoff and Varga*[6]*) if* $T = S$, $\hat{R}_T^S(q)$ *is A-acceptable, (ii) (Varga*[178]*) if* $T \geqslant S$, $\hat{R}_T^S(q)$ *is A(0)-acceptable, and (iii) (Ehle*[39]*) if* $T = S + 1$ *or* $T = S + 2$, $\hat{R}_T^S(q)$ *is L-acceptable.*

Further results, due to Liniger and Willoughby,[115] concern (1, 1) and (2, 2) rational approximations (not necessarily Padé approximations). Consider the (1, 1) rational approximation which contains a free parameter α,

$$R_1^1(q;\alpha) = \frac{1 + \frac{1}{2}(1 - \alpha)q}{1 - \frac{1}{2}(1 + \alpha)q}. \tag{25}$$

It has order one in general, and order two (Padé approximation) if $\alpha = 0$. Consider also the (2, 2) rational approximation containing two free parameters α and β,

$$R_2^2(q;\alpha,\beta) = \frac{1 + \frac{1}{2}(1 - \alpha)q + \frac{1}{4}(\beta - \alpha)q^2}{1 - \frac{1}{2}(1 + \alpha)q + \frac{1}{4}(\beta + \alpha)q^2}. \tag{26}$$

It has order two in general, order three if $\alpha \neq 0$, $\beta = \frac{1}{3}$, and order four (Padé approximation) if $\alpha = 0$, $\beta = \frac{1}{3}$.

Theorem 8.4 (Liniger and Willoughby[115]*)* *Let* $R_1^1(q;\alpha)$ *and* $R_2^2(q;\alpha,\beta)$ *be defined by (25) and (26) respectively. Then (i)* $R_1^1(q;\alpha)$ *is A-acceptable if and only if* $\alpha \geqslant 0$, *and L-acceptable if and only if* $\alpha = 1$, *and (ii)* $R_2^2(q;\alpha,\beta)$ *is A-acceptable if and only if* $\alpha \geqslant 0$, $\beta \geqslant 0$, *and L-acceptable if and only if* $\alpha = \beta > 0$.

Exercises

12. Find the (2, 2) Padé approximation to e^q.

13. Show that the implicit Runge–Kutta method (74) of chapter 4 applied to $y' = \lambda y$ equates y_{n+1}/y_n to the (2, 2) Padé approximation to $e^{h\lambda}$. Deduce that the method is A-stable.

14. Show that the method in exercise 18 of chapter 4 (page 155) similarly yields the (2, 1) Padé approximation. Deduce that the method is not $A(0)$-stable.

15. Show that the *explicit* rational method (23) of chapter 7 applied to $y' = \lambda y$ equates y_{n+1}/y_n to the $(s, 1)$ Padé approximation to $e^{h\lambda}$. Deduce that for $s = 0$ the method is L-stable, for $s = 1$ it is A-stable, and for $s > 1$ it is not even $A(0)$-stable.

8.8 The problem of implicitness for stiff systems

For many classes of numerical method (certainly for the class of linear multistep methods), A-stability, and even $A(0)$-stability, imply implicitness. Thus, for a linear multistep method for example, we must solve, at each integration step, a set of simultaneous non-linear equations of the form

$$\mathbf{y}_{n+k} = h\beta_k \mathbf{f}(x_{n+k}, \mathbf{y}_{n+k}) + \mathbf{g}, \tag{27}$$

where $\mathbf{g}$ is a known vector. Predictor–corrector techniques of the type hitherto considered prove to be inadequate when the system is stiff. If we attempt to use a $P(EC)^mE$ or $P(EC)^m$ mode with fixed m, then, as we have seen in section 3.11, the absolute stability region of the method is no longer that of the corrector alone—in general, the A- or $A(0)$-stability is lost. Nor is the mode of correcting to convergence (which would preserve A- or $A(0)$-stability of the corrector) feasible, since in order that the iteration should converge we would require, by the vector form of theorem 1.2, that

$$L|h\beta_k| < 1, \tag{28}$$

where L is the Lipschitz constant of $\mathbf{f}(x, \mathbf{y})$ with respect to $\mathbf{y}$. We saw in section 8.5 that when the system is stiff this Lipschitz constant is very large, and consequently (28) imposes a severe restriction on steplength; in practice, it is of the same order of severity as that imposed by stability requirements when a method with a finite region of absolute stability is employed.

An alternative method for handling (27) is the vector form of the well-known Newton method $y^{[s+1]} = y^{[s]} - F(y^{[s]})/F'(y^{[s]})$, $s = 0, 1, \ldots$ for the iterative solution of the scalar equation $F(y) = 0$. If a set of m simultaneous equations in m unknowns, ${}^iF({}^1y, {}^2y, \ldots, {}^my) = 0$, $i = 1, 2, \ldots, m$ is written in the vector form $\mathbf{F}(\mathbf{y}) = \mathbf{0}$, then the Newton method may be written as

$$\mathbf{y}^{[s+1]} = \mathbf{y}^{[s]} - J^{-1}(\mathbf{y}^{[s]})\mathbf{F}(\mathbf{y}^{[s]}), \qquad s = 0, 1, \ldots,$$

where $J(\mathbf{y})$ is the Jacobian matrix $\partial\mathbf{F}(\mathbf{y})/\partial\mathbf{y}$. If this method is applied to (27), we obtain

$$\mathbf{y}_{n+k}^{[s+1]} = \mathbf{y}_{n+k}^{[s]} - \left[I - h\beta_k \frac{\partial\mathbf{f}(x_{n+k}, \mathbf{y}_{n+k}^{[s]})}{\partial\mathbf{y}}\right]^{-1} [\mathbf{y}_{n+k}^{[s]} - h\beta_k \mathbf{f}(x_{n+k}, \mathbf{y}_{n+k}^{[s]}) - \mathbf{g}]$$

$$s = 0, 1, \ldots. \quad (29)$$

Sufficient conditions for the convergence of Newton's method for a system are rather complicated (Isaacson and Keller,[80] page 115). However, when (29) is applied to a stiff system, convergence is usually obtained without a restriction on h of comparable severity to that implied by (28), *provided that we can supply a sufficiently accurate initial estimate*, $\mathbf{y}_{n+k}^{[0]}$: a separate predictor can be used for this last purpose. Note, however, that (29) calls for the re-evaluation of the Jacobian $\partial\mathbf{f}/\partial\mathbf{y}$ and the consequent re-inversion of the matrix $I - h\beta_k\,\partial\mathbf{f}/\partial\mathbf{y}$ *at each step of the iteration*. This can be very expensive in terms of computing time, and a commonly used device is to hold the value of $\partial\mathbf{f}/\partial\mathbf{y}$ in (29) constant for a number of consecutive iteration steps. If the iteration converges when so modified, then it will converge to the theoretical solution of (27); if after a few steps (typically three) it appears not to be converging, then the Jacobian is re-evaluated and the corresponding matrix re-inverted. For problems for which $\partial\mathbf{f}/\partial\mathbf{y}$ varies only slowly, it may even prove possible to hold the same value for $\partial\mathbf{f}/\partial\mathbf{y}$ for a number of consecutive integration steps.

Implicit Runge–Kutta methods applied to stiff systems run into similar difficulties if we attempt to use direct iteration. (See, for example, the restriction implied by equation (76) of chapter 4.) Once again, Newton iteration is necessary.

Exercises

16. Consider a non-linear problem $\mathbf{y}' = \mathbf{f}(x, \mathbf{y})$, whose Jacobian $\partial\mathbf{f}/\partial\mathbf{y}$, at the initial point, coincides with the matrix A given by (16). Taking the Lipschitz constant L of the Jacobian to be $\|A\|_\infty = \max_i \sum_j |a_{ij}|$ (see Mitchell[134] page 12), show that the restriction imposed by (28) when the A-stable implicit Trapezoidal rule is applied with direct iteration, is actually more severe than that imposed by stability when the A-unstable explicit Euler's rule is used.

17. Newton's method for a scalar equation is frequently illustrated geometrically by observing that if $w = F(y)$ is plotted in the y–w plane, then the tangent to $w = F(y)$ at the point $(y^{[s]}, F(y^{[s]}))$ cuts the y-axis at $y^{[s+1]}$. Illustrate on such a diagram the effect of keeping $F'(y)$ constant for a number of consecutive steps; sketch a case which suggests that this might on occasion *improve* the rate of convergence.

8.9 Linear multistep methods for stiff systems

What A-stable linear multistep methods are available? By theorem 8.1, we know that such methods must be implicit, and have order not greater than two. The best known A-stable method is the Trapezoidal rule, which has the additional advantage of possessing an asymptotic expansion in even powers of the steplength (see equations (15) and (16) of chapter 6), thus permitting efficient use of the extrapolation processes described in chapter 6. Its disadvantage, discussed in section 8.6, is that, if a moderate steplength is used in the initial phase, then fast decaying components of the theoretical solution are represented numerically by slowly decaying components, resulting in a slowly decaying oscillatory error. (In applications where we are interested only in the steady-state solution, this may not be important.) The difficulty can be avoided either by choosing a very small steplength in the initial phase, or, as shown by Lindberg,[110,111] by using a moderate steplength and applying in the first few steps the same smoothing procedure as is used in Gragg's method (equation (14) of chapter 6); that is, $\mathbf{y}_n$ is replaced by $\hat{\mathbf{y}}_n$, where $\hat{\mathbf{y}}_n = \frac{1}{4}\mathbf{y}_{n+1} + \frac{1}{2}\mathbf{y}_n + \frac{1}{4}\mathbf{y}_{n-1}$. This procedure preserves the form of the asymptotic expansion.

The class of linear one-step methods of order one is given by

$$\mathbf{y}_{n+1} - \mathbf{y}_n = h[(1-\theta)\mathbf{f}_{n+1} + \theta\mathbf{f}_n], \tag{30}$$

often referred to as the 'θ-method'. It follows easily from theorem 8.4 that (30) is A-stable if and only if $\theta \leqslant \frac{1}{2}$. One way in which the free parameter θ may be used to effect is to achieve 'exponential fitting', a concept proposed by Liniger and Willoughby.[115]

Definition *A numerical method is said to be* **exponentially fitted** *at a (complex) value* λ_0 *if, when the method is applied to the scalar test problem* $y' = \lambda y$, $y(x_0) = y_0$, *with exact initial conditions, it yields the exact theoretical solution in the case when* $\lambda = \lambda_0$.

The method (30) applied to the above test equation yields

$$y_n = y_0\left[\frac{1 + \theta h\lambda}{1 - (1-\theta)h\lambda}\right]^n.$$

This coincides with the theoretical solution in the case $\lambda = \lambda_0$ if we choose θ such that $(1 + \theta h\lambda_0)/[1 - (1 - \theta)h\lambda_0] = e^{h\lambda_0}$, or

$$\theta = -\frac{1}{h\lambda_0} - \frac{e^{h\lambda_0}}{1 - e^{h\lambda_0}}.$$

Note that we can only exponentially fit the method (30) to one value of λ, whereas for a general stiff system, the Jacobian will have m eigenvalues. Strategies for choosing the value at which (30) should be exponentially fitted, when we have some *a priori* knowledge of the distribution of the eigenvalues of the Jacobian, are discussed by Liniger and Willoughby.[115] If we have no such knowledge, Liniger[112] proposes that θ be chosen to minimize

$$\max_{-\infty \leqslant h\lambda \leqslant 0} \left| e^{h\lambda} - \frac{1 + \theta h\lambda}{1 - (1 - \theta)h\lambda} \right|.$$

The value of θ which achieves this minimization is $\theta = 0{\cdot}122$.

There is no point in looking for A-stable methods in the one-parameter family of implicit linear two-step methods quoted in section 2.10, since they all have order at least three, and so cannot be A-stable. However, if we retain a second parameter, we may write the family in the form

$$\begin{aligned}\mathbf{y}_{n+2} - (1 + a)\mathbf{y}_{n+1} + a\mathbf{y}_n = h\{[\tfrac{1}{2}(1 + a) + \theta]\mathbf{f}_{n+2} \\ + [\tfrac{1}{2}(1 - 3a) - 2\theta]\mathbf{f}_{n+1} + \theta\mathbf{f}_n\}, \quad (31)\end{aligned}$$

which now has order two in general. It is shown by Liniger[113] that methods of this family are A-stable if and only if $-1 < a < 1$ and $a + 2\theta > 0$. (Note that the first of these conditions implies zero-stability.) Strategies for exponentially fitting methods of class (31) may be found in Liniger.[114]

Let us now consider linear multistep methods which are not necessarily A-stable, but are $A(\alpha)$- or stiffly stable. Since stiff stability implies $A(\alpha)$-stability for some α, we need, in view of theorem 8.2, look only at implicit linear multistep methods. With the usual notation for the characteristic polynomials of a linear multistep method, the associated stability polynomial is (see section 3.6) $\pi(r, \bar{h}) = \rho(r) - \bar{h}\sigma(r)$. Both $A(\alpha)$- and stiff stability require that the roots of $\pi(r, \bar{h})$ be inside the unit circle when $\bar{h}$ is real and $\bar{h} \to -\infty$. In this limit, the roots of $\pi(r, \bar{h})$ approach those of $\sigma(r)$, and it is thus natural to choose $\sigma(r)$ so that its roots lie within the unit circle. In particular, the choice $\sigma(r) = \beta_k r^k$, which has all its roots at the origin, is appropriate. The resulting class of methods

$$\sum_{j=0}^{k} \alpha_j \mathbf{y}_{n+j} = h\beta_k \mathbf{f}_{n+k}, \tag{32}$$

is known as the class of *backward differentiation methods*. Their use for stiff systems goes back to Curtis and Hirschfelder.[34] The coefficients of kth order k-step methods of class (32) are given in table 27 for $k = 1, 2, \ldots, 6$.

Table 27

k	β_k	α_6	α_5	α_4	α_3	α_2	α_1	α_0	$a_{\min}$	$\alpha_{\max}$
1	1						1	-1	0	90°
2	$\frac{2}{3}$					1	$-\frac{4}{3}$	$\frac{1}{3}$	0	90°
3	$\frac{6}{11}$				1	$-\frac{18}{11}$	$\frac{9}{11}$	$-\frac{2}{11}$	0·1	88°
4	$\frac{12}{25}$			1	$-\frac{48}{25}$	$\frac{36}{25}$	$-\frac{16}{25}$	$\frac{3}{25}$	0·7	73°
5	$\frac{60}{137}$		1	$-\frac{300}{137}$	$\frac{300}{137}$	$-\frac{200}{137}$	$\frac{75}{137}$	$-\frac{12}{137}$	2·4	51°
6	$\frac{60}{147}$	1	$-\frac{360}{147}$	$\frac{450}{147}$	$-\frac{400}{147}$	$\frac{225}{147}$	$-\frac{72}{147}$	$\frac{10}{147}$	6·1	18°

Regions of absolute stability for these methods may be found in Gear;[52] for $k = 1, 2, \ldots, 6$, all regions are infinite, and the corresponding methods are stiffly stable and $A(\alpha)$-stable. The approximate minimum values for the parameter a, appearing in the definition of stiff stability (see figure 11) are also quoted in table 27. Norsett[138] has derived a theoretical criterion for $A(\alpha)$-stability from which it is possible to deduce those maximum values of α for which members of the class (32) are $A(\alpha)$-stable (see figure 10); these values are also quoted in table 27.

Gear's method,[54,55] a package for the automatic solution of initial value problems which we discussed in section 3.14, contains an option for use when the system of equations is stiff. In this option, Adams–Moulton correctors are replaced by methods of class (32), with $k = 1, 2, \ldots, 6$, and direct iteration of the corrector is replaced by a form of Newton iteration suitable for incorporation into the one-step formulation described in section 3.14: both steplength and order are variable. At the time of writing, the stiff option in Gear's method represents the most highly developed available package for the treatment of stiff systems; but this area of numerical analysis is a very active field of research, and one can confidently predict the emergence of new methods and further automatic packages.†

† An extension of Gear's method to deal with simultaneous algebraic-differential systems is discussed in Gear.[56]

Finally, if we settle for something less than $A(\alpha)$-stability, the methods proposed by Robertson[152] are of interest. These comprise a one-parameter family obtained by taking the following linear combination of Simpson's rule and the two-step Adams–Moulton method:

$$(1-\alpha)\left[\mathbf{y}_{n+2}-\mathbf{y}_n-\frac{h}{3}(\mathbf{f}_{n+2}+4\mathbf{f}_{n+1}+\mathbf{f}_n)\right]$$

$$+\alpha\left[\mathbf{y}_{n+2}-\mathbf{y}_{n+1}-\frac{h}{12}(5\mathbf{f}_{n+2}+8\mathbf{f}_{n+1}-\mathbf{f}_n)\right],\qquad 0\leqslant\alpha<2. \quad (33)$$

These methods have order three if $\alpha \neq 0$, and the regions of absolute stability are large, almost circular, regions in the half-plane $\mathrm{Re}\,\bar{h} \leqslant 0$, the intervals of absolute stability being $[6\alpha/(\alpha-2), 0]$. (Note that as $\alpha \to 2$, zero-instability threatens.) Such methods are appropriate for moderately stiff systems.

Exercises

18. Show that the method (30) with $\theta = \frac{1}{2}$ (Trapezoidal rule) is exponentially fitted at 0, and with $\theta = 0$ (backward Euler method) it is exponentially fitted at $-\infty$.

19. Show that any one-step L-stable method is exponentially fitted at $-\infty$.

20. Show that if $a = 1$ and $\theta = -\frac{1}{2}$, (31) reduces to a differenced form of the Trapezoidal rule, which is zero-unstable.

21. Use the forms (30) and (31) to prove that the methods of class (32) with coefficients given by table 27 when $k = 1, 2$ are A-stable.

8.10 Runge–Kutta methods for stiff systems

Explicit Runge–Kutta methods have rather small regions of absolute stability (see figure 8); even those designed to have extended regions of stability (equation (63) of chapter 4) are inadequate unless the system is only very mildly stiff. On the other hand it is rather easier to find A-stable implicit Runge–Kutta methods than to find A-stable implicit linear multistep methods. For example, Ehle[39] has shown that Butcher's R-stage implicit Runge–Kutta methods of order $2R$, considered in section 4.10, are all A-stable; thus there exist A-stable methods of this type of arbitrarily high order. L-stable implicit Runge–Kutta methods are also possible (see exercise 22 below). Several authors have considered A- or L-stable implicit Runge–Kutta methods associated with various types of quadrature; the reader is referred to the survey by Bjurel *et al.*[7] for appropriate references. However all such methods suffer a serious practical disadvantage in that the solution of the implicit non-linear equations at each step is considerably harder to achieve in the case of implicit Runge–Kutta methods than in the case of implicit linear multistep methods. If we consider the R-stage fully implicit Runge–Kutta given by equation (67) of chapter 4, applied to an m-dimensional stiff system, then it is clear that the $\mathbf{k}_r$, $r = 1, 2, \ldots, R$ are also m-vectors. It follows that at each step we have to solve a system of mR simultaneous non-linear equations by some form of Newton iteration—and this will converge only if we can find a suitably accurate initial iterate. This constitutes a formidable computational task, and at the present time implicit Runge–Kutta methods do not appear to be competitive with methods, such as Gear's, based on linear multistep methods. In special situations they may be competitive; much will depend on the structure of the implicit equations arising from a particular problem.

If the Runge–Kutta method is semi-explicit in the sense of section 4.10, then the mR simultaneous equations split into R distinct sets of equations, each set containing m equations—a less daunting prospect. The class of semi-explicit methods developed by Butcher (see equation (77) of chapter 4) are not, however, A-stable. An example of an A-stable semi-explicit method is given in exercise 24 below; it would not appear to have any advantage over the Trapezoidal rule.

It is clear from the above discussion that our troubles with stiff systems are not over when we find an A- or L-stable method; the real test is the efficiency with which we can handle the resultant implicitness.

Exercises

22. Use the results of section 8.7 to show that the following third-order implicit Runge–Kutta method (Ehle[39]) is L-stable.

$$\mathbf{y}_{n+1} - \mathbf{y}_n = \frac{h}{4}(\mathbf{k}_1 + 3\mathbf{k}_2), \qquad \mathbf{k}_1 = \mathbf{f}(x_n, \mathbf{y}_n + \tfrac{1}{4}h\mathbf{k}_1 - \tfrac{1}{4}h\mathbf{k}_2),$$

$$\mathbf{k}_2 = \mathbf{f}(x_n + \tfrac{2}{3}h, \mathbf{y}_n + \tfrac{1}{4}h\mathbf{k}_1 + \tfrac{5}{12}h\mathbf{k}_2).$$

(Similar 3-stage fifth-order L-stable methods can be found in Ehle.[39])

23. Show that the semi-explicit method (77) of chapter 4 is not $A(0)$-stable.

24. (Ehle.[39]) Show that the following semi-explicit method of order two is A-stable.

$$\mathbf{y}_{n+1} - \mathbf{y}_n = \tfrac{1}{2}h(\mathbf{k}_1 + \mathbf{k}_2), \qquad \mathbf{k}_1 = \mathbf{f}(x_n, \mathbf{y}_n), \qquad \mathbf{k}_2 = \mathbf{f}(x_n + h, \mathbf{y}_n + \tfrac{1}{2}h\mathbf{k}_1 + \tfrac{1}{2}h\mathbf{k}_2).$$

8.11 Other methods for stiff systems

In the two preceding sections we have considered the application to stiff systems of the two best known classes of numerical methods. Many special techniques for stiff systems, which do not fall into these classes, have been suggested, and in this section we shall briefly consider a small selection of these.

The methods for stiff systems so far discussed are all implicit, and demand the use of some form of Newton iteration; thus the Jacobian $\partial \mathbf{f}/\partial \mathbf{y}$ has to be evaluated, usually at each step. It is thus natural to consider the possibility of methods which incorporate the Jacobian directly. Liniger and Willoughby[115] advocate the use of the following family of one-step Obrechkoff methods (see section 7.2).

$$\mathbf{y}_{n+1} - \mathbf{y}_n = \frac{h}{2}[(1 + \alpha)\mathbf{y}_{n+1}^{(1)} + (1 - \alpha)\mathbf{y}_n^{(1)}] - \frac{h^2}{4}[(\beta + \alpha)\mathbf{y}_{n+1}^{(2)} - (\beta - \alpha)\mathbf{y}_n^{(2)}], \quad (34)$$

which has order two in general, order three if $\beta = \frac{1}{3}$, and order four if, in addition, $\alpha = 0$. (Note that this last case corresponds to the method (4) of chapter 7.) The evaluation of $\mathbf{y}^{(2)}$ involves the Jacobian $\partial \mathbf{f}/\partial \mathbf{y}$ since $\mathbf{y}^{(2)} = \partial \mathbf{f}/\partial x + (\partial \mathbf{f}/\partial \mathbf{y})\mathbf{f}$. It follows immediately from theorem 8.4 that (34) is A-stable if and only if $\alpha \geqslant 0, \beta \geqslant 0$. Liniger and Willoughby develop a modification of Newton iteration for the solution of (34) which involves no derivatives higher than those appearing in the Jacobian. Strategies, based on exponential fitting, for the choice of the parameters α and β are considered.

Obrechkoff methods of the type

$$\mathbf{y}_{n+1} - \mathbf{y}_n = \sum_{i=1}^{l} h^i[a_i\mathbf{y}_{n+1}^{(i)} + b_i\mathbf{y}_n^{(i)}] \quad (35)$$

are considered by Makinson[126] for use with the special class of linear stiff systems

$$\mathbf{y}' = A\mathbf{y} + \boldsymbol{\phi}(x), \quad (36)$$

where A is a constant matrix which has some special structure—is tridiagonal, for example. Clearly, (35) applied to (36) yields an equation of the form

$$\left[I - \sum_{i=1}^{l} h^i a_i A^i\right]\mathbf{y}_{n+1} = \left[I + \sum_{i=1}^{l} h^i b_i A^i\right]\mathbf{y}_n + \boldsymbol{\psi}, \quad (37)$$

where ψ is a known function. The coefficients a_i and b_i, $i = 1, 2, \ldots, l$, are chosen to give A-stability, and also to cause the matrix coefficient of $\mathbf{y}_{n+1}$ in (37) to factorize into linear factors. If A is, for example, tridiagonal, the solution of (37) for $\mathbf{y}_{n+1}$ can then be performed very economically. A third-order method with these properties is obtained by choosing

$$l = 2, \quad a_1 = \theta, \quad b_1 = 1 - \theta, \quad a_2 = \tfrac{1}{6} - \tfrac{1}{2}\theta,$$
$$b_2 = \tfrac{1}{3} - \tfrac{1}{2}\theta, \quad \theta = 1 + \sqrt{3}/3. \tag{38}$$

The coefficient of $\mathbf{y}_{n+1}$ is now $[I - h(\tfrac{1}{2} + \sqrt{3}/6)A]^2$, and $\mathbf{y}_{n+1}$ can be found by twice solving a linear system with the same coefficient matrix.

Linear multistep methods whose coefficients depend directly on approximations to the Jacobian are proposed by Lambert and Sigurdsson.[106] The general class of methods, which have variable matrix coefficients, is

$$\sum_{j=0}^{k}\left[a_j^{(0)}I + \sum_{s=1}^{S} a_j^{(s)}h^sQ_n^s\right]\mathbf{y}_{n+j} = h\sum_{j=0}^{k}\left[b_j^{(0)}I + \sum_{s=1}^{S-1} b_j^{(s)}h^sQ_n^s\right]\mathbf{f}_{n+j}, \tag{39}$$

where Q_n is an $m \times m$ matrix such that, for all n, $\|Q_n\|$ is bounded and $a_k^{(0)}I + \sum_{s=1}^{S} a_k^{(s)}h^sQ_n^s$ is non-singular. The order of the method is independent of the choice made for Q_n, but, in practice, Q_n is chosen to be an approximation to the negative Jacobian $-\partial\mathbf{f}/\partial\mathbf{y}|_{x=x_n}$ of the given system; the methods are thus adaptive. The approximation is not necessarily re-evaluated at each step. There exist methods of class (39), of order up to $2S$, which, when applied with $Q_n = -A$ to the test equation $\mathbf{y}' = A\mathbf{y}$, A a constant matrix all of whose eigenvalues lie in the left half-plane, yield solutions which all tend to zero as $n \to \infty$ for all positive h; such methods are thus A-stable. It turns out that A-stability can be obtained even when $b_k^{(s)} = 0$, $s = 0, 1, \ldots, S$, that is, when the right-hand side of (39) does not depend on $\mathbf{y}_{n+k}$. It is misleading to call such methods explicit, since the coefficient of $\mathbf{y}_{n+k}$ on the left-hand side is a matrix, and it is still necessary to invert this matrix to obtain $\mathbf{y}_{n+k}$. It is more appropriate to call such methods *linearly implicit*, since they require the solution of a set of linear equations at each step, whereas a fully implicit method requires the solution of a set of non-linear equations. The following is an example of a third-order linearly implicit method of class (39).

$$(I + \tfrac{2}{3}hQ_n + \tfrac{1}{6}h^2Q_n^2)\mathbf{y}_{n+3} - (I + \tfrac{19}{12}hQ_n + \tfrac{1}{2}h^2Q_n^2)\mathbf{y}_{n+2}$$
$$+ (\tfrac{4}{3}hQ_n + \tfrac{1}{2}h^2Q_n^2)\mathbf{y}_{n+1} - (\tfrac{5}{12}hQ_n + \tfrac{1}{6}h^2Q_n^2)\mathbf{y}_n$$
$$= h\,\{(\tfrac{23}{12}I + \tfrac{1}{2}hQ_n)\mathbf{f}_{n+2} - (\tfrac{4}{3}I + \tfrac{1}{2}hQ_n)\mathbf{f}_{n+1} + (\tfrac{5}{12}I + \tfrac{1}{6}hQ_n)\mathbf{f}_n\}. \tag{40}$$

It is A-stable if Q is set equal to the negative of the Jacobian.

The Jacobian may also be introduced directly into the coefficients of Runge–Kutta methods, a device first proposed by Rosenbrock.[154] Let the given system be written in the form $\mathbf{y}' = \mathbf{f}(\mathbf{y})$. (The general system (2) can be so written at the cost of increasing the dimension from m to $m + 1$ by adding the new equation ${}^0y' = 1$, ${}^0y(x_0) = x_0$, which identifies ${}^0y$ with x.) The class of R-stage methods proposed by Rosenbrock may then be written in the form

$$\begin{aligned} \mathbf{y}_{n+1} - \mathbf{y}_n &= h \sum_{r=1}^{R} c_r \mathbf{k}_r, \\ \mathbf{k}_1 &= \mathbf{f}(\mathbf{y}_n) + \alpha_1 h \frac{\partial \mathbf{f}(\mathbf{y}_n)}{\partial \mathbf{y}} \mathbf{k}_1, \\ \mathbf{k}_r &= \mathbf{f}\left(\mathbf{y}_n + h \sum_{s=1}^{r-1} b_{rs} \mathbf{k}_s\right) \\ &\quad + \alpha_r h \frac{\partial \mathbf{f}(\mathbf{y}_n + h \sum_{s=1}^{r-1} \beta_{rs} \mathbf{k}_s)}{\partial \mathbf{y}} \mathbf{k}_r, \qquad r = 2, 3, \ldots, R. \end{aligned} \tag{41}$$

They can be regarded either as modifications of explicit Runge–Kutta methods or as linearizations of semi-explicit Runge–Kutta methods. Whereas semi-explicit Runge–Kutta methods would call for the solution of R sets of non-linear equations, the methods (41) call for the solution of R sets of linear equations; they are thus linearly implicit. A two-stage third-order A-stable method of class (41) is given by

$$\begin{aligned} c_1 &= -0{\cdot}413{,}154{,}32, & c_2 &= 1{\cdot}413{,}154{,}32, \\ \alpha_1 &= 1{\cdot}408{,}248{,}29, & \alpha_2 &= 0{\cdot}591{,}751{,}71, \\ b_{21} &= \beta_{21} = 0{\cdot}173{,}786{,}67. \end{aligned} \tag{42}$$

Haines[62] derives methods similar to those of Rosenbrock, but containing a built-in error estimate, of the type considered in section 4.6.

Finally, we mention two techniques of a quite different sort which have been suggested for dealing with stiff systems. One, which has been proposed by several authors, consists of pre-transforming the given system by a transformation of the form $\mathbf{z} = \exp(-xA)\mathbf{y}$, where A is an $m \times m$ matrix, and $\exp(-xA) \equiv I - xA + \frac{1}{2}(xA)^2 - \ldots$ is a matrix exponential. By choosing A to be an approximation to the Jacobian of the original system, a new problem, which is no longer stiff, may be obtained. In order to solve the new problem, however, the matrix exponential must be approximated, usually by a rational function in A. The computations can be somewhat unwieldy, but Lawson[109] obtains a neat simplification in the case where an explicit Runge–Kutta method is used to solve the transformed problem.

The second technique, known as the SAPS method, has been developed by Dahlquist[38] and Odén.[140] The idea is to partition the vector $\mathbf{y}$ into two vectors $\mathbf{u}$ and $\mathbf{v}$, the sum of whose dimensions is m and to model the stiff system by the system

$$\mathbf{u}' = A\mathbf{u} + \boldsymbol{\phi}(x, \mathbf{u}, \mathbf{v}), \tag{43i}$$

$$\mathbf{v}' = \boldsymbol{\psi}(x, \mathbf{u}, \mathbf{v}), \tag{43ii}$$

where A is a (piecewise) constant matrix with all its eigenvalues in the left half-plane, and $\boldsymbol{\phi}$ and $\boldsymbol{\psi}$ have Lipschitz constants which are very small compared with $\|A\|$. In other words, all of the stiffness of the original system is concentrated in (43*i*). Smooth **a**pproximate **p**articular **s**olutions (hence SAPS) of (43*i*) are obtained by a semi-analytical approach, while a conventional numerical method is used to integrate the non-stiff system (43*ii*).

Example 4 Compare the third-order methods given by (a) equation (32) with $k = 3$ (see table 27), (b) equation (34) with $\alpha = \beta = \frac{1}{3}$ and (c) equation (40) with $Q_n = -\partial\mathbf{f}/\partial\mathbf{y}|_{x=x_n}$, applied to the following non-linear stiff system, which arose from a problem of reaction kinetics (Liniger and Willoughby[115]):

$$u' = 0{\cdot}01 - (0{\cdot}01 + u + v)[1 + (u + 1000)(u + 1)],$$

$$v' = 0{\cdot}01 - (0{\cdot}01 + u + v)(1 + v^2),$$

$$u(0) = v(0) = 0, \qquad 0 \leqslant x \leqslant 100.$$

The eigenvalues of the Jacobian are -1012 and $-0{\cdot}01$ at $x = 0$, and $-21{\cdot}7$ and $-0{\cdot}089$ at $x = 100$. Thus the system is initially very stiff, but much less so for large x. No theoretical solution is available, and in order to be able to compare errors, we first compute an 'exact' solution using a fourth-order explicit Runge–Kutta method. It follows from figure 8 that the steplength must be less than $2{\cdot}8 \times 10^{-3}$. Let us take it to be 5×10^{-4}. (Note that to achieve such a solution for $0 \leqslant x \leqslant 100$, the right-hand sides of the given equations have to be evaluated no less than 800,000 times!) The system is then solved by (a), (b), and (c), each with steplength 0·1. The first two methods are implicit and are solved by Newton iteration applied five times at each step, the value of the Jacobian being held constant for the step; the method (c) is linearly implicit, and calls for one linear inversion at each step. Errors at $x = 100$ in the components u and v respectively turn out to be as follows:

Method (a)	-20×10^{-8},	23×10^{-8};
Method (b)	22×10^{-7},	-25×10^{-7};
Method (c)	34×10^{-7},	-35×10^{-7}.

For all three methods, the Jacobian was re-evaluated at each integration step. Methods (a) and (b) required five matrix inversions per step while method (c) needed only one but required the square of the Jacobian to be calculated at each step. The total derivative $\mathbf{y}^{(2)}$ had to be calculated at each step of method (b). Note that we have applied these methods in an unsophisticated way: an automatic package, such as Gear's, would be much more efficient.

Exercises

25. Show that (40) may be written in the form $\mathbf{L}_3 + hQ_n\mathbf{L}_2 + h^2Q_n^2\mathbf{L}_1 = 0$, where $\mathbf{L}_i = \sum_{j=0}^{3}(\alpha_{ij}\mathbf{y}_{n+j} - h\beta_{ij}\mathbf{f}_{n+j})$, $i = 1, 2, 3$, and the order of the linear multistep method $\mathbf{L}_i = 0$ is at least i. Deduce that the order of (40) is independent of Q_n.

26. Show that the method defined by (41) and (42), applied to the test equation $y' = \lambda y$, yields $y_{n+1}/y_n = (1 - \bar{h} - \frac{2}{3}\bar{h}^2)/(1 - 2\bar{h} + \frac{5}{6}\bar{h}^2)$, $\bar{h} = h\lambda$. (Allow for the fact that the coefficients in (42) are rounded.) Use theorem 8.4 to deduce that the method is A-stable.

8.12 Applications to partial differential equations

The theories of numerical methods for partial differential equations and of those for ordinary differential equations have developed along different paths, although they have many basic ideas in common. Similar properties do not, however, always have the same name, and it is the purpose of this section to point out the relationships that exist between commonly used stability definitions in the two separate subjects. Our notation throughout will be as close as possible to that of Mitchell,[134] to whom the reader who is unfamiliar with finite difference methods for partial differential equations is referred.

It is enough to consider the simple parabolic problem

$$\frac{\partial u}{\partial t} = \frac{\partial^2 u}{\partial x^2}$$

$$u(x, 0) = \phi(x),\ \ 0 \leqslant x \leqslant 1; \qquad u(0, t) = \psi_0(t),\ \ u(1, t) = \psi_1(t),\ \ t \geqslant 0. \qquad (44)$$

Consider a rectangular grid with sides parallel to the x- and t-axes, and with grid spacings Δx and Δt, where $M\Delta x = 1$. The *mesh ratio* r is defined by $r = \Delta t/(\Delta x)^2$. Let U_m^n be the approximation to the theoretical solution $u(m\Delta x, n\Delta t)$ at the point $x = m\Delta x$, $t = n\Delta t$, given by a finite difference method. The simplest such explicit method (Mitchell[134] page 20) is

$$U_m^{n+1} = (1 - 2r)U_m^n + r(U_{m+1}^n + U_{m-1}^n), \qquad (45)$$

while the simplest such implicit method (Mitchell[134] page 26) is the well-known Crank–Nicolson method

$$(1 + r)U_m^{n+1} - \tfrac{1}{2}r(U_{m+1}^{n+1} + U_{m-1}^{n+1}) = (1 - r)U_m^n + \tfrac{1}{2}r(U_{m+1}^n + U_{m-1}^n). \quad (46)$$

Let us derive these methods by reducing (44) to an initial value problem involving a system of ordinary differential equations. Denote $u(i\Delta x, t)$ by ${}^iu(t)$, $i = 1, 2, \ldots, M - 1$. Then

$$\frac{\partial^2 u(i\Delta x, t)}{\partial x^2} = [{}^{i+1}u(t) - 2\,{}^iu(t) + {}^{i-1}u(t)]/(\Delta x)^2 + O((\Delta x)^2),$$

and the problem (44) can be approximated by

$$\left.\begin{aligned} \frac{\mathrm{d}\,{}^iu(t)}{\mathrm{d}t} &= [{}^{i+1}u(t) - 2\,{}^iu(t) + {}^{i-1}u(t)]/(\Delta x)^2 \\ {}^iu(0) &= \phi(i\Delta x) \end{aligned}\right\} \quad i = 1, 2, \ldots, M - 1. \quad (47)$$

Since ${}^0u(t) = \psi_0(t)$ and ${}^Mu(t) = \psi_1(t)$ are known functions, (47) is a system of $M - 1$ first-order ordinary differential equations in the $M - 1$ unknowns ${}^iu(t)$, $i = 1, 2, \ldots, M - 1$. The process of deriving (47) from (44) is known as *semi-discretization* or the *method of lines*. Let us now apply Euler's rule with steplength Δt to (47) to obtain

$${}^iu_{n+1} - {}^iu_n = \frac{\Delta t}{(\Delta x)^2}[{}^{i+1}u_n - 2\,{}^iu_n + {}^{i-1}u_n]$$

or

$${}^iu_{n+1} = (1 - 2r){}^iu_n + r({}^{i+1}u_n + {}^{i-1}u_n), \qquad i = 1, 2, \ldots, M - 1.$$

On identifying ${}^mu_n$ with U_m^n we see that the finite difference method (45) applied to the original problem (44) is equivalent to applying Euler's rule to the semi-discretized problem (47). Similarly, the Crank–Nicolson method (46) applied to (44) is equivalent to the Trapezoidal rule applied to (47). (Note that this correspondence between methods is not one-to-one; for example, the semi-explicit Runge–Kutta method of exercise 24 (page 244) applied to (47) is also equivalent to the Crank-Nicolson method applied to (44).)

Let us now examine the stability of Euler's rule and the Trapezoidal rule applied to the system (47), which can be written in the form

$$\mathbf{u}' = \frac{1}{(\Delta x)^2}[A\mathbf{u} + \boldsymbol{\psi}(t)], \quad (48)$$

where $\mathbf{u} = [{}^1u, {}^2u, \ldots, {}^{M-1}u]^T$, $\boldsymbol{\psi}(t) = [\psi_0(t), 0, \ldots, 0, \psi_1(t)]^T$, and A is the $(M - 1) \times (M - 1)$ tridiagonal matrix

$$\begin{bmatrix} -2 & 1 & 0 & 0 & \cdots & 0 \\ 1 & -2 & 1 & 0 & & 0 \\ \vdots & & & & & \vdots \\ 0 & & 0 & 1 & -2 & 1 \\ 0 & \cdots & 0 & 0 & 1 & -2 \end{bmatrix}.$$

(The prime now indicates differentiation with respect to t.) The eigenvalues of A are known (Mitchell[134] page 10) to be $\lambda_t = -2 + 2\cos t\pi/M$, $t = 1, 2, \ldots, M - 1$. Hence the eigenvalues $\lambda_t/(\Delta x)^2$ of the Jacobian of (48) are real, and lie within the interval $(-4/(\Delta x)^2, 0)$ of the real line. The region of absolute stability of Euler's rule (see figure 8 with $p = 1$) intersects the real line in the interval $[-2, 0]$. It follows that for absolute stability, our choice for the (positive) steplength must satisfy $-2 \leqslant -4\Delta t/(\Delta x)^2 < 0$; that is, the mesh ratio r must satisfy

$$0 < r \leqslant \tfrac{1}{2}. \tag{49}$$

Correspondingly, it is shown in Mitchell[134] (page 36) that (45) is *conditionally stable*, the condition being precisely (49). Similarly, the Trapezoidal rule, being A-stable, gives absolute stability for the problem (48) with any positive steplength. Correspondingly (Mitchell[134] page 37), the Crank–Nicolson method is said to be *unconditionally stable.* Note that since the eigenvalues of the system (48) are always real and negative, A-stability of the method for (48) is not essential for unconditional stability of the corresponding method for (44); $A(0)$-stability is enough.

Thus, 'stability' of finite difference methods for parabolic partial differential equations corresponds to 'absolute stability' of the associated methods for the semi-discretized problem; 'unconditional stability' similarly corresponds to '$A(0)$-stability'. A similar interpretation of 'stability' for certain finite difference methods for hyperbolic systems (Mitchell[134] page 159) can be made. Of course, by no means all finite difference methods for partial differential equations can be so interpreted.

9

Linear multistep methods for a special class of second-order differential equations

9.1 Introduction

Using the approach described in section 1.5, it is clearly possible to express the second-order differential equation

$$y'' = f(x, y, y') \tag{1}$$

in the form of a first-order system $u' = v$, $v' = f(x, u, v)$, where $u = y$, $v = y'$. If, however, (1) has the special form

$$y'' = f(x, y), \tag{2}$$

it is natural to ask whether there exist direct methods which do not require us to introduce the first derivative explicitly into an equation in which it does not already appear. We might ask the same sort of question about special higher-order equations of the form $y^{(m)} = f(x, y)$: we shall consider, however, only special equations of the form (2), since these arise in a number of important applications, particularly in mechanics. For the purposes of this chapter, we take as the standard initial value problem

$$y'' = f(x, y), \qquad y(a) = \eta, \qquad y'(a) = \hat{\eta}. \tag{3}$$

Although there exist methods of Runge–Kutta type which tackle this problem directly (Collatz,[28] page 61, de Vogelaere,[179] Scraton[160]), we shall consider only linear k-step methods of the form

$$\sum_{j=0}^{k} \alpha_j y_{n+j} = h^2 \sum_{j=0}^{k} \beta_j f_{n+j}, \tag{4}$$

where $\alpha_k = +1$, and α_0 and β_0 do not both vanish. Since we cannot approximate y'' with less than three discrete values of y, we intuitively expect that k must be at least two, and indeed this turns out to be the case.

The direct application of methods of class (4) to problem (3), rather than the application of a conventional linear multistep method to an equivalent first-order system, is usually recommended. Ash[3] studies asymptotic errors by both approaches, for a subclass of methods, and finds theoretical backing for this recommendation.

Following previous practice in this book, we shall consider only the scalar equation of the form (2), although all of the results we obtain, except where otherwise indicated, extend, with obvious notational modifications, to a system of such equations.

9.2 Order, consistency, zero-stability, and convergence

With the linear multistep method (4) we associate the linear difference operator

$$\mathscr{L}[y(x);h] = \sum_{j=0}^{k} [\alpha_j y(x+jh) - h^2\beta_j y''(x+jh)],$$

where $y(x)$ is an arbitrary function, continuously differentiable on an interval $[a, b]$. If we assume that $y(x)$ has as many higher derivatives as we require, then, on Taylor expanding about the point x, we obtain

$$\mathscr{L}[y(x);h] = C_0 y(x) + C_1 h y^{(1)}(x) + \ldots + C_q h^q y^{(q)}(x) + \ldots,$$

where

$$\begin{aligned}
C_0 &= \alpha_0 + \alpha_1 + \alpha_2 + \ldots + \alpha_k,\\
C_1 &= \alpha_1 + 2\alpha_2 + \ldots + k\alpha_k,\\
C_2 &= \frac{1}{2!}(\alpha_1 + 2^2\alpha_2 + \ldots + k^2\alpha_k) - (\beta_0 + \beta_1 + \beta_2 + \ldots + \beta_k),\\
C_q &= \frac{1}{q!}(\alpha_1 + 2^q\alpha_2 + \ldots + k^q\alpha_k)\\
&\quad - \frac{1}{(q-2)!}(\beta_1 + 2^{q-2}\beta_2 + \ldots + k^{q-2}\beta_k), \qquad q = 3, 4, \ldots.
\end{aligned} \tag{5}$$

Following Henrici,[67] we say that the method has *order p* if

$$C_0 = C_1 = \ldots = C_p = C_{p+1} = 0, \qquad C_{p+2} \neq 0;$$

C_{p+2} is then the *error constant*, and $C_{p+2}h^{p+2}y^{(p+2)}(x_n)$ the *principal local truncation error* at the point x_n. As in section 2.6, the Taylor expansion may be made about a general point $x + th$, yielding an alternative set of coefficients to (5), which, by suitable choice of t, can simplify the derivation of particular methods. Once again, order and error constant are independent of the choice made for t.

The method is said to be *consistent* if it has order at least one. If we define the first and second characteristic polynomials

$$\rho(\zeta) = \sum_{j=0}^{k} \alpha_j \zeta^j, \qquad \sigma(\zeta) = \sum_{j=0}^{k} \beta_j \zeta^j,$$

it is easily verified that method (4) is consistent if and only if

$$\rho(1) = \rho'(1) = 0, \qquad \rho''(1) = 2\sigma(1).$$

(Note that if $k = 1$, $\rho(1) = \rho'(1) = 0$ implies $\alpha_1 = \alpha_0 = 0$; it follows that a consistent method is meaningful only if $k \geqslant 2$.) The polynomial $\rho(\zeta)$ associated with a consistent method thus has a *double* root at $+1$; this is the *principal root*, the other roots being *spurious*. Zero-stability is then defined as follows:

Definition The linear multistep method (4) is said to be **zero-stable** *if no root of the first characteristic polynomial $\rho(\zeta)$ has modulus greater than one, and if every root of modulus one has multiplicity not greater than two.*

The definition of convergence is precisely that given in section 2.5, with the addition of the hypothesis that $\lim_{h\to 0} [\eta_\mu(h) - \eta_0(h)]/h = \hat{\eta}$, where the starting values satisfy $y_\mu = \eta_\mu(h)$, $\mu = 0, 1, \ldots, k - 1$. With these definitions, theorem 2.1, which states that consistency and zero-stability are together necessary and sufficient for convergence, holds for methods of class (4). Moreover, theorem 2.2, which limits the order of a zero-stable linear k-step method to be not more than $k + 1$ when k is odd and not more than $k + 2$ when k is even, also holds. Proofs of these results can be found in Henrici.[67] Zero-stable k-step methods of order $k + 2$ are again called *optimal*. A necessary condition for optimality is that k be even and that all the roots of $\rho(\zeta)$ have modulus one.

Exercises

1. Derive the optimal 2-step method of class (4), and find its error constant.

2. Derive the method $y_{n+2} - 2y_{n+1} + y_n = h^2 f_{n+1}$ by analogues of each of the derivations discussed in sections 2.2, 2.3, and 2.4.

9.3 Specification of methods

As in the case of linear multistep methods for first-order equations, certain families of methods of class (4) were originally derived in terms of power series involving finite difference operators. Two families, both of which equate $y_{n+2} - 2y_{n+1} + y_n$ to such power series, are associated with the names of Störmer and Cowell. Thus the family of methods (4) for which $\rho(\zeta) = \zeta^k - 2\zeta^{k-1} + \zeta^{k-2}$ are frequently referred to as *Störmer–Cowell* methods. The best known such method is the optimal two-step method

$$y_{n+2} - 2y_{n+1} + y_n = \frac{h^2}{12}(f_{n+2} + 10f_{n+1} + f_n), \tag{6}$$

known as *Numerov's method*, and also, somewhat extravagantly, as the *royal road formula*.

We quote below explicit and implicit methods of class (4) with stepnumbers 2, 3, and 4. As in section 2.10, sufficient free parameters are retained to enable us to control the values taken by all of the spurious roots of $\rho(\zeta)$. The methods have the highest orders that can be attained whilst retaining the given number of free parameters. The order p and error constant C_{p+2} are quoted in each case.

Explicit methods

$k = 2$:

$$\begin{aligned} \alpha_2 &= 1, & & \\ \alpha_1 &= -2, & \beta_1 &= 1, \\ \alpha_0 &= 1, & \beta_0 &= 0, \\ p &= 2; & C_4 &= \tfrac{1}{12}. \end{aligned}$$

$k = 3$:

$$\begin{aligned} \alpha_3 &= 1, & & \\ \alpha_2 &= -2 - a, & \beta_2 &= (13 - a)/12, \\ \alpha_1 &= 1 + 2a, & \beta_1 &= (-1 - 5a)/6, \\ \alpha_0 &= -a, & \beta_0 &= (1 - a)/12, \\ p &= 3; & C_5 &= \tfrac{1}{12}, \quad \text{for all } a. \end{aligned}$$

$k = 4$:

$$\begin{aligned} \alpha_4 &= 1, & & \\ \alpha_3 &= -2 - a, & \beta_3 &= (14 - a)/12, \\ \alpha_2 &= 1 + 2a + b, & \beta_2 &= (-5 - 10a + b)/12, \\ \alpha_1 &= -a - 2b, & \beta_1 &= (4 - a + 10b)/12, \\ \alpha_0 &= b, & \beta_0 &= (-1 + b)/12, \\ p &= 4; & C_6 &= (19 + a - b)/240. \end{aligned}$$

There exist no values for a and b which cause the order to exceed 4 and the method to be zero-stable.

Implicit methods

$k = 2$:

$$\alpha_2 = 1, \qquad \beta_2 = \tfrac{1}{12},$$
$$\alpha_1 = -2, \qquad \beta_1 = \tfrac{5}{6},$$
$$\alpha_0 = 1, \qquad \beta_0 = \tfrac{1}{12},$$
$$p = 4; \qquad C_6 = -\tfrac{1}{240}.$$

$k = 3$:

$$\alpha_3 = 1, \qquad \beta_3 = \tfrac{1}{12},$$
$$\alpha_2 = -2 - a, \qquad \beta_2 = (10 - a)/12,$$
$$\alpha_1 = 1 + 2a, \qquad \beta_1 = (1 - 10a)/12,$$
$$\alpha_0 = -a, \qquad \beta_0 = -a/12,$$

$p = 4; C_6 = (a - 1)/240$. (Note that $a = 1$ gives a zero-unstable method.)

$k = 4$:

$$\alpha_4 = 1, \qquad \beta_4 = (19 + a - b)/240,$$
$$\alpha_3 = -2 - a, \qquad \beta_3 = (51 - 6a + b)/60,$$
$$\alpha_2 = 1 + 2a + b, \qquad \beta_2 = (7 - 97a + 7b)/120,$$
$$\alpha_1 = -a - 2b, \qquad \beta_1 = (1 - 6a + 51b)/60,$$
$$\alpha_0 = b, \qquad \beta_0 = (-1 + a + 19b)/240.$$

If $b \neq 1$, then $p = 5; C_7 = (-1 + b)/240$.

If $b = 1$, then $p = 6; C_8 = (-190 - 31a)/60{,}480$.

There exist no values for a and b which cause the order to exceed 6 and the method to be zero-stable.

Exercise

3. Show that there exists an infinite family of optimal 4-step methods of class (4). Derive that member of the family whose first characteristic polynomial has two pairs of double roots on the unit circle.

9.4 Application to initial value problems

If a linear multistep method of class (4) is applied to the problem (3), it is necessary to supply additional starting values y_μ, $\mu = 1, 2, \ldots, k - 1$.

(Note that this is necessary even when k takes the smallest possible value, namely 2.) Such values can be obtained by Taylor expansion, since

$$y(a+h) = y(a) + hy^{(1)}(a) + \frac{h^2}{2!}y^{(2)}(a) + \frac{h^3}{3!}y^{(3)}(a) + \dots$$

$$= \eta + h\hat{\eta} + \frac{h^2}{2!}f(a,\eta) + \frac{h^3}{3!}\left[\frac{\partial f(a,\eta)}{\partial x} + \hat{\eta}\frac{\partial f(a,\eta)}{\partial y}\right] + \dots, \tag{7}$$

$$y^{(1)}(a+h) = y^{(1)}(a) + hy^{(2)}(a) + \frac{h^2}{2!}y^{(3)}(a) + \dots$$

$$= \hat{\eta} + hf(a,\eta) + \frac{h^2}{2!}\left[\frac{\partial f(a,\eta)}{\partial x} + \hat{\eta}\frac{\partial f(a,\eta)}{\partial y}\right] + \dots. \tag{8}$$

$$\vdots \qquad\qquad \vdots$$

It is now possible to compute $y(x_0 + 2h)$ from the expansion

$$y(a+2h) = y(a+h) + hy^{(1)}(a+h) + \frac{h^2}{2!}y^{(2)}(a+h) + \dots,$$

substituting the approximations already obtained for $y(a+h)$ and $y'(a+h)$. Continuing in this way, and truncating the expansions at appropriate points, additional starting values of any desired order of accuracy can be obtained. Alternatively, one can apply an explicit Runge–Kutta method to a first-order system equivalent to (3).

A bound for the global error when (4) is applied to (3), similar to the bound obtained in section 3.4, may be found in Henrici[67] page 314. The bound possesses salient features analogous to those discussed in section 3.5, and reflects the separate influences of starting error, local truncation error, and local round-off error. Now, however, if the local truncation error is bounded by GYh^{p+2}, the corresponding bound for the global truncation error is proportional to GYh^p.

A theory of weak stability for methods of class (4) can be developed along lines parallel to those of section 3.6. In place of equation (38) of chapter 3, we obtain, under the usual assumptions that $\partial f/\partial y = \lambda$, constant, and the local error $= \phi$, constant, the linearized error equation

$$\sum_{j=0}^{k} (\alpha_j - h^2\lambda\beta_j)\tilde{e}_{n+j} = \phi,$$

where $\tilde{e}_n = y(x_n) - \tilde{y}_n$, $\{\tilde{y}_n\}$ being the solution of (4) when round-off error has been committed. The general solution of this equation is

$$\tilde{e}_n = \sum_{s=1}^{k} d_s r_s^n - \phi/h^2\lambda \sum_{j=0}^{k} \beta_j,$$

where the d_s are arbitrary constants and the r_s are the roots, assumed constant, of the stability polynomial

$$\pi(r, \bar{h}) = \rho(r) - \bar{h}\sigma(r), \tag{9}$$

where

$$\bar{h} = h^2\lambda.$$

Now, however, *two* roots of (9)—let us call them r_1 and r_2—tend to the *double* principal root $\zeta_1 = +1$. Whereas for the first-order equation we found that $r_1 = \exp(h\lambda) + O(h^{p+1})$, where p is the order of the method, we now find that

$$r_1 = \exp(h\lambda^{\frac{1}{2}}) + O(h^{p+2}) \quad \text{and} \quad r_2 = \exp(-h\lambda^{\frac{1}{2}}) + O(h^{p+2}).$$

Since for many methods r_1 and r_2 lie *on* the unit circle when $\bar{h}$ is small and negative, we modify slightly our definitions of absolute and relative stability as follows.

Definition *The linear multistep method (4) is said to be* **absolutely stable** *for a given $\bar{h}$ if, for that $\bar{h}$, all the roots r_s of (9) satisfy $|r_s| \leqslant 1$, $s = 1, 2, \ldots, k$; it is said to be* **relatively stable** *if, for that $\bar{h}$, $|r_s| \leqslant \min(|r_1|, |r_2|)$, $s = 3, 4, \ldots, k$. An interval $[\alpha, \beta]$ of the real line is said to be an* **interval of absolute (relative) stability** *if the method is absolutely (relatively) stable for all $\bar{h} \in [\alpha, \beta]$.*

Once again, every zero-stable consistent linear multistep method of class (4) is absolutely unstable for small positive $\bar{h}$.

If a system of equations $\mathbf{y}'' = \mathbf{f}(x, \mathbf{y})$ is to be solved, then λ is taken to be an eigenvalue, assumed constant, of the Jacobian $\partial\mathbf{f}/\partial\mathbf{y}$, and may, of course, take complex values; intervals of absolute (relative) stability are then replaced by regions of absolute (relative) stability.

Example 1 *Show that Numerov's method (6) has a region of absolute stability consisting of the segment $[-6, 0]$ of the real line.*

Using the boundary locus method, we have that the boundary of the region of absolute stability is given by

$$\bar{h}(\theta) = \frac{\rho(e^{i\theta})}{\sigma(e^{i\theta})} = 12\frac{e^{2i\theta} - 2e^{i\theta} + 1}{e^{2i\theta} + 10e^{i\theta} + 1} = 12\frac{-18 + 16\cos\theta + 2\cos 2\theta}{102 + 40\cos\theta + 2\cos 2\theta}.$$

Thus the boundary of the region is a segment of the real axis. An elementary calculation shows that the derivative of the above expression vanishes when $\theta = 0$ or $\theta = \pi$, the former giving a maximum and the latter a minimum. These turn out to be 0 and -6 respectively, and the required result follows.

Predictor–corrector pairs based on methods of class (4) can be employed. The modes $P(EC)^m E$ and $P(EC)^m$ are defined by equations (58) and (59) of chapter 3, with h replaced by h^2. Once again, if predictor and corrector both have the same order p, Milne's device is applicable, and the principal local truncation error $C_{p+2}h^{p+2}y^{(p+2)}(x_n)$ of the pair in $P(EC)^m E$ or $P(EC)^m$ mode is estimated by

$$\frac{C_{p+2}}{C^*_{p+2} - C_{p+2}}[y^{[m]}_{n+k} - y^{[0]}_{n+k}].$$

Suitable predictors for use with Numerov's method (6) are given by the two-parameter family of four-step explicit methods quoted in section 9.3; popular choices for the parameters are $a = b = 0$, or $a = -2, b = 1$.

Exercises

4. Construct the (quadratic) stability polynomial for Numerov's method (6). Sketch the position of the roots r_1, r_2 in the complex plane, for real $\bar{h}$ in the ranges $\bar{h} < -6$, $-6 \leqslant \bar{h} \leqslant 0, \bar{h} > 0$. Thus corroborate that the interval of absolute stability is indeed $[-6, 0]$.

5. Investigate the weak stability of the method

$$y_{n+2} - 2y_{n+1} + y_n = h^2 f_{n+1}.$$

6. Show that the optimal 4-step method derived in exercise 3 (page 256) has no interval of absolute stability. (Compare with Simpson's rule.)

7. Let a predictor P and a corrector C of class (4) have orders p^* and p respectively. Derive results, analogous to those of section 3.10, concerning the local truncation error of the pair P and C in $P(EC)^m E$ or $P(EC)^m$ mode. In what main respect do these results differ from those of section 3.10?

9.5 Application to boundary value problems

If the special second-order equation (2) satisfies conditions at two separate points $x_0 = a$ and $x_N = a + Nh$, then the initial value problem (3) is replaced by a *two-point boundary value problem*. We shall consider the example

$$y'' = f(x, y), \qquad y(x_0) = A, \qquad y(x_N) = B. \tag{10}$$

If f is continuous and satisfies a Lipschitz condition, and if $\partial f/\partial y$ is continuous and non-negative, for all $x \in [x_0, x_N]$ and for all y, then there exists a unique solution to the problem (10) (Henrici[67] page 347). Let us consider how a method of class (4) can be used to solve this problem. Applying such a method in the interval $[x_0, x_N]$ gives

$$\sum_{j=0}^{k} \alpha_j y_{n+j} = h^2 \sum_{j=0}^{k} \beta_j f_{n+j}, \qquad n = 0, 1, 2, \ldots, N-k. \tag{11}$$

Since $y_0 = y(x_0)$, $f_0 = f(x_0, y_0)$, $y_N = y(x_N)$, and $f_N = f(x_N, y_N)$ are known, (11) affords $N - k + 1$ equations in the $N - 1$ unknowns $y_1, y_2, \ldots, y_{N-1}$. If k takes the smallest possible value, 2, then we have as many equations as unknowns, but if $k > 2$, we have more unknowns than equations. We cannot now use Taylor expansions to obtain additional 'starting' values as we did for the initial value problem, since the expansions (7) and (8) require knowledge of $y'(x_0)$ which we do not possess. Nor can we use a method, such as Runge–Kutta, applied to an equivalent first-order system, since, due to our ignorance of $y'(x_0)$, we no longer have an initial vector for this system. Thus we are forced to choose $k = 2$: there is no reason for not choosing the most accurate 2-step method available to us, namely Numerov's method, (6). Applying (6) to (10), we obtain the following system of $N - 1$ equations in $N - 1$ unknowns:

$$\begin{bmatrix} -2 & 1 & 0 & 0 & \ldots & 0 \\ 1 & -2 & 1 & 0 & \ldots & 0 \\ \vdots & & & & & \vdots \\ 0 & 0 & \ldots & & 1 & -2 \end{bmatrix} \begin{bmatrix} y_1 \\ y_2 \\ \vdots \\ y_{N-1} \end{bmatrix}$$

$$= \frac{h^2}{12} \begin{bmatrix} 10 & 1 & 0 & 0 & \ldots & 0 \\ 1 & 10 & 1 & 0 & \ldots & 0 \\ \vdots & & & & & \vdots \\ 0 & 0 & \ldots & & 1 & 10 \end{bmatrix} \begin{bmatrix} f_1 \\ f_2 \\ \vdots \\ f_{N-1} \end{bmatrix} - \begin{bmatrix} y_0 - \frac{h^2}{12} f_0 \\ 0 \\ \vdots \\ 0 \\ y_N - \frac{h^2}{12} f_N \end{bmatrix}.$$

Recalling that $f_j = f(x_j, y_j)$, we see that this is a system of simultaneous *non-linear* equations. One way of solving it would be Newton's method for a system, described in section 8.8; there are other methods. The study of numerical methods for systems of non-linear equations is a large and well-developed area of numerical analysis, and for this reason, we do not intend in this book to study boundary value problems in detail. The reader is referred to the books by Fox[43] and Keller[84] which are devoted to this topic. We make one exception to this rule, and describe the *shooting method* which essentially converts boundary value problems into initial value problems, and can thus be regarded as an application of previous work in this book. However, since it is applicable to a more general equation than (2), it is appropriate to present it in an appendix.

Appendix

The shooting method for two-point boundary value problems

A detailed study of numerical methods for boundary value problems involving ordinary differential equations is outside the scope of this book; the reader is referred to the specialized books by Fox[43] and Keller.[84] One technique for boundary value problems, the *shooting method*, is, however, briefly considered here, since it can be regarded as an application of the methods for initial value problems considered previously. We illustrate the method by considering the two-point boundary value problem

$$y'' = f(x, y, y'), \qquad y(a) = A, \qquad y(b) = B, \qquad a < b, \tag{1}$$

which we shall assume possesses a unique solution. We make an initial guess s for $y'(a)$, and denote by $y(x; s)$ the solution of the initial value problem

$$y'' = f(x, y, y'), \qquad y(a) = A, \qquad y'(a) = s. \tag{2}$$

Writing $u(x; s)$ for $y(x; s)$ and $v(x; s)$ for $\partial y(x; s)/\partial x$, (2) may be written as the first-order system

$$\begin{aligned} \frac{\partial u(x; s)}{\partial x} &= v(x; s), \qquad u(a; s) = A, \\ \frac{\partial v(x; s)}{\partial x} &= f(x, u(x; s), v(x; s)), \qquad v(a; s) = s. \end{aligned} \tag{3}$$

The solution $u(x; s)$ of (3) will coincide with the solution of (1) if we can find a value of s such that

$$\phi(s) \equiv u(b; s) - B = 0. \tag{4}$$

This equation can be solved numerically by the Newton–Raphson method; that is, we compute the sequence $\{s_\nu\}$ defined by

$$s_{\nu+1} = s_\nu - \phi(s_\nu)/\phi'(s_\nu), \qquad s_0 \text{ arbitrary}. \tag{5}$$

If s_0 is a sufficiently good first approximation, the sequence $\{s_\nu\}$ will converge to the required root of (4); but how do we compute $\phi'(s_\nu)$? Let us define new variables

$$\xi(x; s) = \frac{\partial u(x; s)}{\partial s}, \qquad \eta(x; s) = \frac{\partial v(x; s)}{\partial s}$$

and differentiate the initial value problem (3) partially with respect to s to obtain a second initial value problem

$$\begin{aligned} \frac{\partial \xi(x; s)}{\partial x} &= \eta(x; s), \qquad \xi(a, s) = 0, \\ \frac{\partial \eta(x; s)}{\partial x} &= p(x; s)\xi(x; s) + q(x; s)\eta(x; s), \qquad \eta(a, s) = 1, \end{aligned} \tag{6}$$

where

$$p(x; s) = \frac{\partial f(x, u(x; s), v(x; s))}{\partial u}, \qquad q(x; s) = \frac{\partial f(x, u(x; s), v(x; s))}{\partial v}. \tag{7}$$

If we assign a value s_0 to s, the two initial value problems (3) and (6) can be solved by any suitable numerical method such as predictor–corrector or Runge–Kutta. From the solution of (3) we obtain $u(b; s_0)$ and hence, from (4), $\phi(s_0)$, while from the solution of (6) we obtain $\xi(b; s_0) = \phi'(s_0)$. Thus, from (5), we obtain a new estimate s_1 and the process may be repeated. Note that, by (7), the two initial value problems (3) and (6) are *coupled* and therefore must be solved simultaneously; in effect, (3) and (6) constitute an initial value problem for a system of four coupled simultaneous differential equations. Note also that each iteration of the Newton–Raphson method calls for a new numerical solution of this initial value problem.

Shooting methods are applicable to more general problems than (1) (see Keller[84]). Two particular difficulties can arise in practice. Firstly, it can be difficult to find an initial estimate sufficiently accurate for the Newton–Raphson iteration to converge. Secondly, the initial value problem generated by the shooting method is frequently unstable, in the sense of being very sensitive to perturbations in the initial conditions. (This should not be confused with instability of the numerical method used to solve the initial value problem.) Both of these difficulties are to some extent alleviated by a recent modification, known as the *multiple shooting method* described by Osborne.[142]

Exercise

Show that the initial value problems (3) and (6) are uncoupled if and only if (1) is linear in the sense that $f(x, y, y') = P(x)y + Q(x)y' + R(x)$.

References

1. Abdel Karim, A. I., 'A theorem for the stability of general predictor–corrector methods for the solution of systems of differential equations', *J. Assoc. Comput. Mach.*, **15**, 706–711 (1968).
2. Aitken, A. C., 'On interpolation by iteration of proportional parts', *Proc. Edinburgh Math. Soc.*, **2**, 56–76 (1932).
3. Ash, J. H., 'Analysis of multistep methods for special second-order ordinary differential equations', *Ph.D. thesis, University of Toronto* (1969).
4. Axelsson, O., 'A class of A-stable methods', *BIT*, **9**, 185–199 (1969).
5. Bieberbach, L., *Theorie der Differentialgleichungen*, Springer, 1930.
6. Birkhoff, G., and R. S. Varga, 'Discretization errors for well-set Cauchy problems: I', *J. Math. and Phys.*, **44**, 1–23 (1965).
7. Bjurel, G., G. Dahlquist, B. Lindberg, S. Linde, and L. Odén, 'Survey of stiff ordinary differential equations', *Royal Institute of Technology, Stockholm, Department of Information Processing, Computer Science Report NA 70.11* (1970).
8. Blum, E. K., 'A modification of the Runge–Kutta fourth-order method', *Math. Comp.*, **16**, 176–187 (1962).
9. Brown, R. R., J. D. Riley, and M. M. Bennett, 'Stability properties of Adams–Moulton type methods', *Math. Comp.*, **19**, 90–96 (1965).
10. Brush, D. G., J. J. Kohfeld, and G. T. Thompson, 'Solution of ordinary differential equations using two "off-step" points', *J. Assoc. Comput. Mach.*, **14**, 769–784 (1967).
11. Bulirsch, R., and J. Stoer, 'Fehlerabschätzungen und Extrapolation mit rationalen Funktionen bei Verfahren vom Richardson-Typus', *Numer. Math.*, **6**, 413–427 (1964).
12. Bulirsch, R., and J. Stoer, 'Numerical treatment of ordinary differential equations by extrapolation methods', *Numer. Math.*, **8**, 1–13 (1966).
13. Butcher, J. C., 'Coefficients for the study of Runge–Kutta integration processes', *J. Austral. Math. Soc.*, **3**, 185–201 (1963).
14. Butcher, J. C., 'On the integration processes of A. Huťa', *J. Austral. Math. Soc.*, **3**, 202–206 (1963).
15. Butcher, J. C., 'Implicit Runge–Kutta processes', *Math. Comp.*, **18**, 50–64 (1964).
16. Butcher, J. C., 'Integration processes based on Radau quadrature formulas', *Math. Comp.*, **18**, 233–243 (1964).

17. Butcher, J. C., 'On Runge–Kutta processes of high order', *J. Austral. Math. Soc.*, **4**, 179–194 (1964).
18. Butcher, J. C., 'On the attainable order of Runge–Kutta methods', *Math. Comp.*, **19**, 408–417 (1965).
19. Butcher, J. C., 'A modified multistep method for the numerical integration of ordinary differential equations', *J. Assoc. Comput. Mach.*, **12**, 124–135 (1965).
20. Butcher, J. C., 'On the convergence of numerical solutions to ordinary differential equations', *Math. Comp.*, **20**, 1–10 (1966).
21. Butcher, J. C., 'A multistep generalization of Runge–Kutta methods with four or five stages', *J. Assoc. Comput. Mach.*, **14**, 84–99 (1967).
22. Butcher, J. C., 'An algebraic theory of integration methods', *Math. Comp.*, **26** (1972). (To appear.)
23. Carr, J. W. III., 'Error bounds for Runge–Kutta single-step integration processes', *J. Assoc. Comput. Mach.*, **5**, 39–44 (1958).
24. Ceschino, F., 'Une méthode de mise en oeuvre des formules d'Obrechkoff pour l'intégration des équations différentielles', *Chiffres*, **2**, 49–54 (1961).
25. Ceschino, F., 'Modification de la longueur du pas dans l'intégration numérique par les méthodes à pas liés', *Chiffres*, **2**, 101–106 (1961).
26. Chase, P. E., 'Stability properties of predictor–corrector methods for ordinary differential equations', *J. Assoc. Comput. Mach.*, **9**, 457–468 (1962).
27. Clark, N., 'A study of some numerical methods for the integration of first-order ordinary differential equations', *Argonne National Lab. Report No. 7428* (1968).
28. Collatz, L., *The Numerical Treatment of Differential Equations*, Springer, 1960.
29. Conte, S. D., and R. F. Reeves, 'A Kutta third-order procedure for solving differential equations requiring minimum storage', *J. Assoc. Comput. Mach.*, **3**, 22–25 (1956).
30. Copson, E. T., 'On a generalisation of monotonic sequences', *Proc. Edinburgh Math. Soc.*, **17**, 159–164 (1970).
31. Crane, P. C., and P. A. Fox, 'A comparative study of computer programs for integrating differential equations', *Numerical Mathematics Program Library Project, Bell Telephone Lab.* (1969).
32. Crane, R. L., 'Stability and local accuracy of numerical methods for ordinary differential equations', *Dissertation, Iowa State University of Science and Technology* (1962).
33. Crane, R. L., and R. W. Klopfenstein, 'A predictor–corrector algorithm with an increased range of absolute stability', *J. Assoc. Comput. Mach.*, **12**, 227–241 (1965).
34. Curtiss, C. F., and J. O. Hirschfelder, 'Integration of stiff equations', *Proc. Nat. Acad. Sci. U.S.A.*, **38**, 235–243 (1952).
35. Dahlquist, G., 'Convergence and stability in the numerical integration of ordinary differential equations', *Math. Scand.*, **4**, 33–53 (1956).
36. Dahlquist, G., 'Stability and error bounds in the numerical integration of ordinary differential equations', *Kungl. Tekniska Hogskolans Handlingar No. 130* (1959).
37. Dahlquist, G., 'A special stability problem for linear multistep methods', *BIT*, **3**, 27–43 (1963).

38. Dahlquist, G., 'A numerical method for some ordinary differential equations with large Lipschitz constants', *Information Processing 68*, ed. A. J. H. Morrell, North Holland Publishing Co., 183–186, 1969.
39. Ehle, B. L., 'On Padé approximations to the exponential function and A-stable methods for the numerical solution of initial value problems', *University of Waterloo Dept. Applied Analysis and Computer Science, Research Rep. No. CSRR 2010* (1969).
40. Emanuel, G., 'The Wilf stability criterion for numerical integration', *J. Assoc. Comput. Mach.*, **10**, 557–561 (1963).
41. England, R., 'Error estimates for Runge–Kutta type solutions to systems of ordinary differential equations', *Comput. J.*, **12**, 166–170 (1969).
42. Fehlberg, E., 'New high-order Runge–Kutta formulas with step size control for systems of first- and second-order differential equations', *Z. Angew. Math. Mech.*, **44**, 17–29 (1964).
43. Fox, L., *The Numerical Solution of Two-Point Boundary Value Problems in Ordinary Differential Equations*, Oxford University Press, 1957.
44. Fox, L., *Numerical Solution of Ordinary and Partial Differential Equations*, Pergamon, 1962.
45. Fox, L., and E. T. Goodwin, 'Some new methods for the numerical integration of ordinary differential equations', *Proc. Cambridge Philos. Soc.*, **45**, 373–388 (1949).
46. Fraboul, F., 'Un critère de stabilité pour l'intégration numérique des équations différentielles', *Chiffres*, **1**, 55–63 (1962).
47. Fyfe, D. J., 'Economical evaluation of Runge–Kutta formulae', *Math. Comp.*, **20**, 392–398 (1966).
48. Galler, B. A., and D. P. Rozenberg, 'A generalization of a theorem of Carr on error bounds for Runge–Kutta procedures', *J. Assoc. Comput. Mach.*, **7**, 57–60 (1960).
49. Gautschi, W., 'Numerical integration of ordinary differential equations based on trigonometric polynomials', *Numer. Math.*, **3**, 381–397 (1961).
50. Gear, C. W., 'Hybrid methods for initial value problems in ordinary differential equations', *SIAM J. Numer. Anal.*, **2**, 69–86 (1965).
51. Gear, C. W., 'The numerical integration of ordinary differential equations', *Math. Comp.*, **21**, 146–156 (1967).
52. Gear, C. W., 'Numerical integration of stiff ordinary differential equations', *University of Illinois Dept. of Computer Science Report No. 221* (1967).
53. Gear, C. W., 'The automatic integration of stiff ordinary differential equations', *Information Processing 68*, ed. A. J. H. Morrell, North Holland Publishing Co., 187–193, 1969.
54. Gear, C. W., 'The automatic integration of ordinary differential equations', *Comm. ACM.*, **14**, 176–179 (1971).
55. Gear, C. W., 'Algorithm 407, DIFSUB for solution of ordinary differential equations', *Comm. ACM.*, **14**, 185–190 (1971).
56. Gear, C. W., 'Simultaneous numerical solution of differential–algebraic equations', *IEEE Transactions on Circuit Theory*, **18**, 89–95 (1971).
57. Gill, S., 'A process for the step-by-step integration of differential equations in an automatic digital computing machine', *Proc. Cambridge Philos. Soc.*, **47**, 95–108 (1951).
58. Gillespie, R. P., *Partial Differentiation*, Oliver and Boyd, 1951.

59. Gragg, W. B., 'Repeated extrapolation to the limit in the numerical solution of ordinary differential equations', *Ph.D. thesis, University of California, Los Angeles* (1964).
60. Gragg, W. B., 'On extrapolation algorithms for ordinary initial value problems', *SIAM J. Numer. Anal.*, **2**, 384–403 (1965).
61. Gragg, W. B., and H. J. Stetter, 'Generalized multistep predictor–corrector methods', *J. Assoc. Comput. Mach.*, **11**, 188–209 (1964).
62. Haines, C. F., 'Implicit integration processes with error estimates for the numerical solution of differential equations', *Comput. J.*, **12**, 183–187 (1969).
63. Hall, G., 'The stability of predictor–corrector methods', *Comput. J.*, **9**, 410–412 (1967).
64. Hammer, P. C., and J. W. Hollingsworth, 'Trapezoidal methods of approximating solutions of differential equations', *M.T.A.C.*, **9**, 92–96 (1955).
65. Hamming, R. W., 'Stable predictor–corrector methods for ordinary differential equations', *J. Assoc. Comput. Mach.*, **6**, 37–47 (1959).
66. Harris, R. P., 'Runge–Kutta processes', *Proc. Fourth Australian Computer Conference, Adelaide*, 429–433 (1969).
67. Henrici, P., *Discrete Variable Methods in Ordinary Differential Equations*, John Wiley and Sons, 1962.
68. Henrici, P., *Error Propagation for Difference Methods*, John Wiley and Sons, 1963.
69. Hermite, C., *Oeuvres*, *Tome III*, Gauthier-Villars, 1912.
70. Heun, K., 'Neue Methode zur approximativen Integration der Differentialgleichungen einer unabhängigen Veränderlichen', *Z. Math. Physik*, **45**, 23–38 (1900).
71. Hull, T. E., 'A search for optimum methods for the numerical integration of ordinary differential equations', *SIAM Rev.*, **9**, 647–654 (1967).
72. Hull, T. E., and A. L. Creemer, 'Efficiency of predictor–corrector procedures', *J. Assoc. Comput. Mach.*, **10**, 291–301 (1963).
73. Hull, T. E., W. H. Enright, B. M. Fellen, and A. E. Sedgwick, 'Comparing numerical methods for ordinary differential equations', *University of Toronto Dept. of Computer Science Tech. Rep. No. 29* (1971).
74. Hull, T. E., and R. L. Johnston, 'Optimum Runge–Kutta methods', *Math. Comp.*, **18**, 306–310 (1964).
75. Hull, T. E., and A. C. R. Newbery, 'Error bounds for three-point integration procedures', *J. Soc. Indust. Appl. Math.*, **7**, 402–412 (1959).
76. Hull, T. E., and A. C. R. Newbery, 'Integration procedures which minimize propagated errors', *J. Soc. Indust. Appl. Math.*, **9**, 31–47 (1961).
77. Hull, T. E., and A. C. R. Newbery, 'Corrector formulas for multistep integration methods', *J. Soc. Indust. Appl. Math.*, **10**, 351–369 (1962).
78. Hurewicz, W., *Lectures on Ordinary Differential Equations*, M.I.T. Press, 1958.
79. Huťa, A., 'Contribution à la formule de sixième ordre dans la méthode de Runge–Kutta–Nyström', *Acta Fac. Rerum Natur. Univ. Comenian. Math.*, **2**, 21–24 (1957).
80. Isaacson, E., and H. B. Keller, *Analysis of Numerical Methods*, John Wiley and Sons, 1966.
81. Jankovič, V., 'Vzorce pre numerické riešenie diferenciálnej rovnice $y' = f(t, y)$, ktoré obsahujú vyššie derivácie', *Apl. Mat.*, **10**, 469–480 (1965).

82. Jones, D. S., and D. W. Jordan, *Introductory Analysis, Vols. I and II*, John Wiley and Sons, 1969.
83. Joyce, D. C., 'A survey of extrapolation processes in numerical analysis', *Tech. Rep. Series No. 8, University of Newcastle-upon-Tyne Computing Laboratory* (1970).
84. Keller, H. B., *Numerical Methods for Two-Point Boundary Value Problems*, Blaisdell, 1968.
85. King, R., 'Runge–Kutta methods with constrained minimum error bounds', *Math. Comp.*, **20**, 386–391 (1966).
86. Klopfenstein, R. W., and R. S. Millman, 'Numerical stability of a one evaluation predictor–corrector algorithm for numerical solution of ordinary differential equations', *Math. Comp.*, **22**, 557–564 (1968).
87. Kohfeld, J. J., and G. T. Thompson, 'Multistep methods with modified predictors and correctors', *J. Assoc. Comput. Mach.*, **14**, 155–166 (1967).
88. Kohfeld, J. J., and G. T. Thompson, 'A modification of Nordsieck's method using an "off-step" point', *J. Assoc. Comput. Mach.*, **15**, 390–401 (1968).
89. Kopal, Z., *Numerical Analysis*, Chapman and Hall, 1955.
90. Krogh, F. T., 'Predictor–corrector methods of high order with improved stability characteristics', *J. Assoc. Comput. Mach.*, **13**, 374–385 (1966).
91. Krogh, F. T., 'A test for instability in the numerical solution of ordinary differential equations', *J. Assoc. Comput. Mach.*, **14**, 351–354 (1967).
92. Krogh, F. T., 'A note on conditionally stable correctors', *Math. Comp.*, **21**, 717–719 (1967).
93. Krogh, F. T., 'Variable order integrators for the numerical solution of ordinary differential equations', *Jet Propulsion Lab. Tech. Memo., California Institute of Technology* (1969).
94. Krogh, F. T., 'On testing a subroutine for the numerical integration of ordinary differential equations', *Jet Propulsion Lab. Tech. Memo. No. 217, California Institute of Technology* (1970).
95. Krogh, F. T., 'Suggestions on conversion (with listings) of the variable order integrators VODQ, SVDQ and DVDQ', *Jet Propulsion Lab. Tech. Memo. No. 278, California Institute of Technology* (1971).
96. Krückeberg, F., 'Zur numerischen Integration und Fehlenerfassung bei Anfangswertaufgaben gewönlicher Differentialgleichungen', *Schriften des Rheinisch-Westfälischen Institutes für instrumentalle Mathematik an der Universität Bonn, No. 1* (1961).
97. Kuntzmann, J., 'Evaluation de l'erreur sur un pas dans les méthodes a pas séparés', *Chiffres*, **2**, 97–102 (1959).
98. Kunz, K. S., *Numerical Analysis*, 1st edition, McGraw-Hill, 1957.
99. Kutta, W., 'Beitrag zur näherungsweisen Integration totaler Differentialgleichungen', *Z. Math. Phys.*, **46**, 435–453 (1901).
100. Lambert, J. D., 'Linear multistep methods with mildly varying coefficients', *Math. Comp.*, **24**, 81–94 (1970).
101. Lambert, J. D., 'Predictor–corrector algorithms with identical regions of stability', *SIAM J. Numer. Anal.*, **8**, 337–344 (1971).
102. Lambert, J. D., and A. R. Mitchell, 'On the solution of $y' = f(x, y)$ by a class of high accuracy difference formulae of low order', *Z. Angew. Math. Phys.*, **13**, 223–232 (1962).

103. Lambert, J. D., and B. Shaw, 'On the numerical solution of $y' = f(x, y)$ by a class of formulae based on rational approximation', *Math. Comp.*, **19**, 456–462 (1965).
104. Lambert, J. D., and B. Shaw, 'A method for the numerical solution of $y' = f(x, y)$ based on a self-adjusting non-polynomial interpolant', *Math. Comp.*, **20**, 11–20 (1966).
105. Lambert, J. D., and B. Shaw, 'A generalisation of multistep methods for ordinary differential equations', *Numer. Math.*, **8**, 250–263 (1966).
106. Lambert, J. D., and S. T. Sigurdsson, 'Multistep methods with variable matrix coefficients', *SIAM J. Numer. Anal.*, **9**, (1972). (To appear.)
107. Lawson, J. D., 'An order five Runge–Kutta process with extended region of stability', *SIAM J. Numer. Anal.*, **3**, 593–597 (1966).
108. Lawson, J. D., 'An order six Runge–Kutta process with extended region of stability', *SIAM J. Numer. Anal.*, **4**, 620–625 (1967).
109. Lawson, J. D., 'Generalized Runge–Kutta processes for stable systems with large Lipschitz constants', *SIAM J. Numer. Anal.*, **4**, 372–380 (1967).
110. Lindberg, B., 'On smoothing and extrapolation for the Trapezoidal rule', *BIT*, **11**, 29–52 (1971).
111. Lindberg, B., 'On smoothing for the Trapezoidal rule', *Royal Institute of Technology, Stockholm, Department of Information Processing, Computer Science Report NA 71.31* (1971).
112. Liniger, W., 'Optimization of a numerical integration method for stiff systems of ordinary differential equations', *IBM Research Report RC2198* (1968).
113. Liniger, W., 'A criterion for *A*-stability of linear multistep integration formulae', *Computing*, **3**, 280–285 (1968).
114. Liniger, W., 'Global accuracy and *A*-stability of one- and two-step integration formulae for stiff ordinary differential equations', *Conference on the Numerical Solution of Differential Equations, Dundee, 1969*, ed. J. Ll. Morris, 188–193, Springer, 1969.
115. Liniger, W., and R. A. Willoughby, 'Efficient numerical integration methods for stiff systems of differential equations', *IBM Research Report RC-1970* (1967).
116. Loscalzo, F. R., 'Numerical solution of ordinary differential equations by spline functions (SPLINDIF)', *Tech. Report 842, Math. Res. Center, Madison* (1968).
117. Loscalzo, F. R., 'On the use of spline functions for the numerical solution of ordinary differential equations', *Tech. Report 869, Math. Res. Center, Madison* (1968).
118. Loscalzo, F. R., and I. J. Schoenberg, 'On the use of spline functions for the approximation of solutions of ordinary differential equations', *Tech. Report 723, Math. Res. Center, Madison* (1967).
119. Loscalzo, F. R., and T. D. Talbot, 'Spline function approximations for solutions of ordinary differential equations', *Bull. Amer. Math. Soc.*, **73**, 438–442 (1967).
120. Loscalzo, F. R., and T. D. Talbot, 'Spline function approximations for solutions of ordinary differential equations', *SIAM J. Numer. Anal.*, **4**, 433–445 (1967).
121. Lotkin, M., 'On the accuracy of Runge–Kutta's method', *M.T.A.C.*, **5**, 128–132 (1951).

122. Luther, H. A., 'Further explicit fifth-order Runge–Kutta formulas', *SIAM Rev.*, **8**, 374–380 (1966).
123. Luther, H. A., 'An explicit sixth-order Runge–Kutta formula', *Math. Comp.*, **22**, 434–436 (1968).
124. Luther, H. A., and H. P. Konen, 'Some fifth-order classical Runge–Kutta formulas', *SIAM Rev.*, **7**, 551–558 (1965).
125. Lyche, T., 'Optimal order methods with an arbitrary number of nonstep points', *Conference on the Numerical Solution of Differential Equations, Dundee, 1969*, ed. J. Ll. Morris, Springer, 1969.
126. Makinson, G. J., 'Stable high order implicit methods for the numerical solution of systems of differential equations', *Comput. J.*, **11**, 305–310 (1968).
127. Merson, R. H., 'An operational method for the study of integration processes', *Proc. Symp. Data Processing, Weapons Research Establishment, Salisbury, S. Australia* (1957).
128. Miller, J. J. H., 'On the location of zeros of certain classes of polynomials with applications to numerical analysis', *J. Inst. Math. Appl.*, **8**, 397–406 (1971).
129. Milne, W. E., 'Numerical integration of ordinary differential equations', *Amer. Math. Monthly*, **33**, 455–460 (1926).
130. Milne, W. E., 'A note on the numerical integration of differential equations', *J. Res. Nat. Bur. Standards*, **43**, 537–542 (1949).
131. Milne, W. E., *Numerical Solution of Differential Equations*, John Wiley and Sons, 1953.
132. Milne, W. E., and R. R. Reynolds, 'Stability of a numerical solution of differential equations', *J. Assoc. Comput. Mach.*, **6**, 193–203 (1959); Part II, *J. Assoc. Comput. Mach.*, **7**, 46–56 (1960).
133. Milne, W. E., and R. R. Reynolds, 'Fifth-order methods for the numerical solution of ordinary differential equations', *J. Assoc. Comput. Mach.*, **9**, 64–70 (1962).
134. Mitchell, A. R., *Computational Methods in Partial Differential Equations*, John Wiley and Sons, 1969.
135. Morris, J. Ll., *Computational Methods in Elementary Numerical Analysis*, John Wiley and Sons. (To appear.)
136. Neville, E. H., 'Iterative interpolation', *J. Ind. Math. Soc.*, **20**, 87–120 (1934).
137. Nordsieck, A., 'On numerical integration of ordinary differential equations', *Math. Comp.*, **16**, 22–49 (1962).
138. Norsett, S. P., A criterion for $A(\alpha)$-stability of linear multistep methods', *BIT*, **9**, 259–263 (1969).
139. Obrechkoff, N., 'Neue Quadraturformeln', *Abh. Preuss. Akad. Wiss. Math. Nat. Kl.*, **4**, (1940).
140. Odén, L., 'An experimental and theoretical analysis of the SAPS method for stiff ordinary differential equations', *Royal Institute of Technology, Stockholm, Department of Information Processing, Computer Science Report NA 71.28* (1971).
141. Osborne, M. R., 'On Nordsieck's method for the numerical solution of ordinary differential equations', *BIT*, **6**, 51–57 (1966).
142. Osborne, M. R., 'On shooting methods for boundary value problems', *J. Math. Anal. Appl.*, **27**, 417–433 (1969).
143. Pereyra, V., 'On improving an approximate solution of a functional equation by deferred corrections', *Numer. Math.*, **8**, 376–391 (1966).

144. Pereyra, V., 'Iterated deferred corrections for nonlinear operator equations', *Numer. Math.*, **10**, 316–323 (1967).
145. Pereyra, V., 'Iterated deferred corrections for nonlinear boundary value problems', *Numer. Math.*, **11**, 111–125 (1968).
146. Rahme, H. S., 'A new look at the numerical integration of ordinary differential equations', *J. Assoc. Comput. Mach.*, **16**, 496–506 (1969).
147. Ralston, A., 'Some theoretical and computational matters relating to predictor–corrector methods of numerical integration', *Comput. J.*, **4**, 1–4 (1961).
148. Ralston, A., 'Runge–Kutta methods with minimum error bounds', *Math. Comp.*, **16**, 431–437 (1962).
149. Ralston, A., 'Relative stability in the numerical solution of ordinary differential equations', *SIAM Rev.*, **7**, 114–125 (1965).
150. Ralston, A., *A First Course in Numerical Analysis*, McGraw-Hill, 1965.
151. Richardson, L. F., 'The deferred approach to the limit, I—single lattice', *Trans. Roy. Soc. London*, **226**, 299–349 (1927).
152. Robertson, H. H., 'Some new formulae for the numerical integration of differential equations', *Information Processing UNESCO, Paris*, 106–108 (1959).
153. Rodabaugh, D. J., 'On stable correctors', *Comput. J.*, **13**, 98–100 (1968).
154. Rosenbrock, H. H., 'Some general implicit processes for the numerical solution of differential equations', *Comput. J.*, **5**, 329–330 (1963).
155. Rosser, J. B., 'A Runge–Kutta for all seasons', *SIAM Rev.*, **9**, 417–452 (1967).
156. Runge, C., 'Über die numerische Auflösung von Differentialgleichungen', *Math. Ann.*, **46**, 167–178 (1895).
157. Sanchez, D. A., *Ordinary Differential Equations and Stability Theory*, W. H. Freeman and Co., 1968.
158. Sarafyan, D., 'Multistep methods for the numerical solution of ordinary differential equations made self-starting', *Tech. Report 495*, *Math. Res. Center*, *Madison* (1965).
159. Schur, J., 'Über Potenzreihen die im Innern des Einheitskreises beschränkt sind', *J. Reine Angew. Math.*, **147**, 205–232 (1916).
160. Scraton, R. E., 'The numerical solution of second-order differential equations not containing the first derivative explicitly', *Comput. J.*, **6**, 368–370 (1964).
161. Scraton, R. E., 'Estimation of the truncation error in Runge–Kutta and allied processes', *Comput. J.*, **7**, 246–248 (1964).
162. Shampine, L. F., and H. A. Watts, 'Block implicit one-step methods', *Math. Comp.*, **23**, 731–740 (1969).
163. Shaw, B., 'The use of non-polynomial interpolants in the numerical solution of ordinary differential equations', *Ph.D. thesis*, *University of St. Andrews* (1965).
164. Shaw, B., 'Modified multistep methods based on a non-polynomial interpolant', *J. Assoc. Comput. Mach.*, **14**, 143–154 (1967).
165. Shintani, H., 'Approximate computation of errors in numerical integration of ordinary differential equations by one-step methods', *J. Sci. Hiroshima Univ. Ser. A-1 Math.*, **29**, 97–120 (1965).
166. Shintani, H., 'On a one-step method of order 4', *J. Sci. Hiroshima Univ. Ser. A-1 Math.*, **30**, 91–107 (1966).
167. Shintani, H., 'Two-step processes by one-step methods of order 3 and of order 4', *J. Sci. Hiroshima Univ. Ser. A-1 Math.*, **30**, 183–195 (1966).

168. Sigurdsson, S. T., 'Study of numerical methods for integration of stiff systems of ordinary differential equations', *M.Sc. dissertation, University of Dundee* (1970).
169. Spijker, M. N., 'Convergence and stability of step-by-step methods for the numerical solution of initial value problems', *Numer. Math.*, **8**, 161–177 (1966).
170. Stetter, H. J., 'Stabilizing predictors for weakly unstable correctors', *Math. Comp.*, **9**, 84–89 (1965).
171. Stetter, H. J., 'A study of strong and weak stability in discretization algorithms', *SIAM J. Numer. Anal.*, **2**, 265–280 (1965).
172. Stetter, H. J., 'Improved absolute stability of predictor–corrector schemes', *Computing*, **3**, 286–296 (1968).
173. Stetter, H. J., 'Stability properties of the extrapolation methods', *Conference on the Numerical Solution of Differential Equations, Dundee, 1969*, ed. J. Ll. Morris, Springer, 1969.
174. Stetter, H. J., 'Symmetric two-step algorithms for ordinary differential equations', *Computing*, **5**, 267–280 (1970).
175. Stewart, N. F., 'The comparison of numerical methods for ordinary differential equations', *University of Toronto Dept. of Computer Science Tech. Rep. No. 3* (1968).
176. Stoer, J., 'Über zwei Algorithm zur Interpolation mit rationalen Funktionen', *Numer. Math.*, **3**, 285–304 (1961).
177. Turnbull, H., *Theory of Equations, 4th ed.*, Oliver and Boyd, 1947.
178. Varga, R. S., 'On higher order stable implicit methods for solving parabolic partial differential equations', *J. Math. and Phys.*, **40**, 220–231 (1961).
179. Vogelaere, R. de, 'A method for the numerical integration of differential equations of second order without explicit first derivatives', *J. Res. Nat. Bur. Standards*, **54**, 119–125 (1955).
180. Watt, J. M., 'The asymptotic discretization error of a class of methods for solving ordinary differential equations', *Proc. Cambridge Philos. Soc.*, **63**, 461–472 (1967).
181. Widlund, O. B., 'A note on unconditionally stable linear multistep methods', *BIT*, **7**, 65–70 (1967).
182. Wilf, H. S., 'A stability criterion for numerical integration', *J. Assoc. Comput. Mach.*, **6**, 363–366 (1959).

Index

The abbreviations LM, PC, and RK are used to denote linear multistep, predictor–corrector, and Runge–Kutta respectively. Major text references are in **bold type**.